R. GARETH WYN JONES

ENERGY AND POWER
Our perilous obsessions

YNNI A PHŴER
Ein chwantau peryglus

Acknowledgements

With many thanks to Ali Anwar from the H'mm Foundation,

i/to Euros, Aled, Eirig ac Owain.

First published by The H'mm Foundation in 2024

© R Gareth Wyn Jones

ISBN: 978-1-0685053-0-0

All rights reserved. No part of this publication may be reproduced, stored in a retrieval system or transmitted in any form, or by any means, electronic, mechanical. photocopying, recording or otherwise without the explicit permission of the publishers.

The rights of the contributors to be identified as authors of their contributions has been asserted in accordance with the Copyright, Design and Patents Act, 1988.

Cover photo: ©NASA
The Earth seen from Apollo 17
Typeset and designed by Andy Dark
Printed in Wales by Gwasg Gomer,
Llandysul Enterprise Park, Llandysul, SA44 4JL

Published by the H'mm Foundation,
c/o Bevan Buckland, Langdon House, Langdon Road,
Swansea SA1 8QY.

Contents

Foreword/Rhagair: *Mark Mansfield* — 2

Preface/Rhagarweiniad: *Robin Grove-White* — 4

1 An Introduction — 10

2 Energy and Power: Essential, Addictive, Toxic — 14

3 Ynni a Phŵer: Hanfodol, Caethiwus a Gwenwynig — 38

4 Nation.Cymru: The History of Energy — 64

5 Energy and Power: Main Scientific Essay — 120

6 Overshooting the Limits — 190

7 Croesi Ffiniau — 246

8 Ynni, Gwaith a Chymhlethdod — 306

Profiles: R Gareth Wyn Jones and Robin Grove-White — 360

Acknowledgements/Diolchiadau — 364

Foreword

Mark Mansfield, CEO
Nation.Cymru

It is entirely fitting and urgently timely that the inaugural Nation.Cymru lecture considers the history of energy within the context of climate change, a change which has become so rapid that it has escalated into an emergency and now posits an existential challenge to life as we know it.

We are delighted that a series of articles by Emeritus Professor Gareth Wyn Jones from Bangor University forms the basis for this lecture, delivered at the Senedd in September 2024. It is now presented here in book form, alongside the original essays and other material which add both context and depth to his engaging analysis.

We are especially grateful to Llywydd, Rt. Hon. Elin Jones MS for sponsoring the event in the Neuadd.

We are also hugely grateful to Ali Anwar at the H'mm Foundation for sponsoring and arranging its publication, with copies being given to all Senedd members as part of the lecture's dissemination.

Nation.Cymru was established seven years ago to be a new and independent voice within the ecology of the Welsh media and we are very proud of our journalism. We now have over a million visits every month and we continue to expand our news and cultural offering.

This lecture by one of our most distinguished scientists is part of our ongoing conversation with readers in Wales and, increasingly, elsewhere. In telling the history of energy it also tells the story of human profligacy whilst also suggesting ways we can remedy the situation while there is still time.

I encourage you to read it.

Rhagair

Mark Mansfield, Prif Weithredwr
Nation.Cymru

Mae'n gwbl synhwyrol a hynod amserol bod darlith gyntaf Nation.Cymru yn ystyried hanes egni yng nghyd-destun newid hinsawdd, newid sydd wedi bod mor sydyn fel ei bod bellach yn argyfwng sydd yn herio bywyd fel yr ydym yn ei hadnabod.

Rydym yn hynod falch bod cyfres o erthyglau gan Yr Athro Emeritws Gareth Wyn Jones o Brifysgol Bangor yn gosod seiliau'r ddarlith hon, wedi ei chyflwyno yn y Senedd mis Medi 2024. Cyflwynir y ddarlith yma, yn ffurf llyfr, ynghyd â'r traethodau gwreiddiol ac ysgrifau eraill, sydd yn ychwanegu cyd-destun i'w dadansoddiad deniadol.

Rydym yn hynod ddiolchgar i'r Llywydd, Y Gw. Anrh. Elin Jones A.S. am noddi'r digwyddiad yn y Neuadd.

Rydym hefyd yn eithriadol o ddiolchgar i Ali Anwar o'r elusen H'mm Foundation am noddi a threfnu cyhoeddi'r gyfrol, gyda chopïau wedi eu rhoi i holl Aelodau o'r Senedd, fel rhan o ledaeniad y ddarlith.

Sefydlwyd Nation.Cymru saith mlynedd yn ôl i weithredu fel llais newydd ac annibynnol o fewn ecoleg gyfryngol Cymru, ac rydym yn hynod falch o'n newyddiaduraeth. Mae gan Nation bellach dros filiwn o ymwelwyr yn fisol ac rydym yn parhau i ehangu ein cynigion newyddiadurol a ddiwylliannol.

Mae'r ddarlith hon gan un o'n gwyddonwyr mwyaf enwog yn rhan o'n sgwrs barhaus gyda darllenwyr yng Nghymru ac, yn gynyddol, tu hwnt i'n ffiniau. Wrth adrodd hanes egni, mae'n adrodd hanes gormodedd dynol hefyd gan awgrymu yn ogystal ffyrdd medrwn adfer y sefyllfa, tra mae yna amser o hyd i'w wneud.

Rwyf yn eich annog i'w ddarllen.

Preface

Prof Robin Grove-White

*'The more laws and restrictions there are,
The poorer people become.
The sharper men's weapons,
The more trouble in the land.
The more ingenious and clever men are,
The more strange things happen.
The more rules and regulations,
The more robbers and thieves.'*
(Lao Tzu ~600 BC)[1]

There is an urgent need to think about the future role of energy in society in a sharply different way.

A significant new scientifically-grounded hypothesis by Professor Gareth Wyn Jones suggests that sustaining worthwhile human life on planet Earth will require *less* energy consumption by society in future, not more.

This challenges the current mainstream consensus that moving to non-carbon-based sources of energy will allow world-wide economic and so

Professor Wyn Jones's book – titled *'Energy the Great Driver: Seven Revolutions and the Challenge of Climate Change'*[2] – argues that the multiplying societal complexities and incoherencies consequent upon escalating production and use of energy (whether conventional or *renewable*) now make it imperative that societies learn to adapt to using less energy overall, if civilised human life is to continue. This argument rests on historical analysis of the biosphere's experience of successive energy 'revolutions' up to and including the present day, using concepts derived from physics and biology.

The arguments, and some questions arising, are summarised in both the transcript of his lecture in the Senedd and the more comprehensive paper which forms the body of this volume

These arguments demand cross-disciplinary examination and discussion, involving scholars from the humanities and social sciences, as much as natural and physical scientists from different specialisms. The conclusions appear to go with the grain of emergent work on the margins of a number of disciplines, as well as broader indications of unease within society about aspects of dominant economic and techn

The overall thesis needs to be shared and tested widely, before broader claims can be substantiated.

Ostensibly however, it could provide state-of-the-art scientifically-based grounding for government pursuit of more ambitious and imaginative sustainability-targeted policies, affecting multiple spheres of human activity.

Against this background, that the scientific merits, and second, the potentially wider social implications of the arguments must be examined and explored.

---

1. Lao Tzo. https://www.goodreads.com/quotes/7360224-the-more-laws-and-restrictions-there-are-the-poorer-people
2. R. Gareth Wyn Jones (2019) *Energy The Great Driver: Seven Revolutions and the Challenges of Climate Change.* UoWP

Rhagairweiniad

Yr Athro Robin Grove-White

'Po fwyaf niferus y deddfau a chyfyngiadau,
Tlotaf daw'r bobl.
Po finiocaf arfau dynion,
Mwy o drafferth yn y tir a geir.
Po glyfrwch a dawnus y mae dynion,
Mwy o bethau diarth digwyddith.
Po fwyaf niferus y rheolau a'r mesurau,
Mwyaf niferus daw'r gwylliaid a'r lladron.'
(Lao Tzu ~600 CC)

Mae yna angen brys i ystyried rôl egni yn ein cymdeithas mewn modd gwrthgyferbyniol.

Mae damcaniaeth newydd, arwyddocaol, sydd wedi'i seilio ar wyddoniaeth gan Yr Athro Gareth Wyn Jones yn awgrymu y bydd angen i gymdeithas defnyddio lai o egni yn y dyfodol, nid mwy, er mwyn cynnal bywyd dynol safonol ar Y Ddaear.

Mae hyn yn herio'r consenws prif ffrwd; wneith symud tuag at ffynonellau egni di-garbon caniatáu twf cymdeithasol ac economaidd byd-eang ac fel o'r blaen.

Mae llyfr Yr Athro Wyn Jones *'Energy the Great Driver: Seven Revolutions and the Challenge of Climate Change'* – yn dadlau bod y cymhlethdodau ac anghydlyniadau cymdeithasol cynyddol o ganlyniad i gynhyrchiant a defnydd egni cynyddol (boed yn danwydd ffosil neu'n ynni adnewyddol) nawr yn golygu ei fod yn hollbwysig i gymdeithas dysgu sut i addasu i ddefnyddio llai o egni ar y cyfan, os yw bywyd dynol ar y cyd â chymdeithas wâr am barhau. Mae'r ddadl hon yn seiliedig ar ddadansoddiad hanesyddol o brofiad y biosffer o'r chwyldroadau egni olynol, gan gynnwys y presennol, wrth ddefnyddio cysyniadau yn seiliedig ar ffiseg a bywydeg.

Mae'r dadleuon, a rhai o'r cwestiynau a godir, yn cael eu crynhoi yn ei adysgrif o'i ddarlith yn y Senedd a'r papur mwy cynhwysfawr sydd yn ffurfio corff y gyfrol hon.

Mae'r dadleuon yma'n mynnu trafodaeth ac archwiliad trawsddisgyblaethol, gan gynnwys ysgolheigion o'r dyniaethau a'r gwyddorau cymdeithasol, cymaint â'r gwyddorau naturiol a ffisegol o arbenigaethau gwahanol. Mae'r casgliadau i'w weld yng nghyd-fynd â gwaith allddodol sydd ar gyrion disgyblaethau niferus, yn ogystal ag arwyddion o annifyrdod ehangach o fewn cymdeithas ynghylch agweddau o daflwybrau economaidd a thechnolegol dominyddol.

Mae angen lledaenu'r ddadl hon a'i phrofi yn eang, cyn bod modd cadarnhau honiadau ehangach.

Eto i gyd, mae yna posibilrwydd yma o ddarparu seiliau cadarn, blaengar, yn seiliedig ar wyddoniaeth, wneith caniatau polisiau cynaliadwy mwy ddychmygol ac uchelgeisiol gan lywodraeth, gan ddylanwadau ar sawl sffêr o weithgarwch dynol.

Yn erbyn y gefndir hwn, sef haeddiant y wyddonol, ac yn ail, mae rhaid craffu ac archwilio'r goblygiadau cymdeithasol ehangach bosib.

1
An Introduction

1
An Introduction

This book is something of a hybrid. Linguistically some articles are in both Welsh and English while some are presented only in one language. Some are original, specifically the text from which the Nation.Cymru lecture in the Senedd on Tuesday 24th of September 2024 was distilled. Others are reprinted from previously published material and I am especially grateful to Sioned Rowlands and O'r Pedwar Gwynt for her ready cooperation.

One chapter is composed of the short Nation.Cymru articles (Chapter 4). It tries to explain the ideas and their implications in terms that, hopefully, can be widely understood. In contrast the 'Main Scientific Essay' in Chapter 5 is more demanding and is rooted in the science. It touches on some basic thermodynamics as well as challenging issues such as the relationships between energy, entropy and information and aspects of the social and behavioural sciences. As the individual Chapters were originally written with different audiences in mind, there is, unavoidably, some overlap between them. Hopefully this will not detract from the overall message.

The central theme of the book is the danger, as I perceive it, implicit in humanity's long obsession with ever increasing energy and the power which results from the control that energy. This, when set against both the environmental consequences of and our material, social and cultural dependence on the exploitation of that energy and power, highlights a core human paradox. This book seeks to explore this fundamental and potentially fatal paradox, while, I must stress, recognising the limitations of the author and the enormity of the task.

My interest in this dilemma has grown gradually over the years ago largely out of my experiences working with communities and colleagues from range of disciplines in southern Africa and in Syria and my own research and reading.

Initially I tested some ideas with two of Wales' most renowned, scientists, Sir John Meurig Thomas and Sir John Houghton, both sadly now deceased. They encouraged me to continue my search.

A little later I was invited by Y Coleg Cymraeg Cenedlaethol to give the annual Edward Lhuyd Lecture. This allowed me to outline my emerging hypothesis although then largely from a biological perspective https://www.porth.ac.uk/en/collection/darlith-edward-llwyd-2017-ynni-gwaith-a-chymhlethdod. This, in turn, led to the two articles in O'r Pedawr Gwynt in 2018 and the interview with Cynog Dafis which are reprinted in this book (Chapter 7).

In the autumn of 2019 my book, Energy The Great Driver: Seven Revolution and the Challenges of Climate Change, was published https://www.uwp.co.uk/book/energy-the-great-driver/. The response was muted. Partly as plans for a major launch in the spring of 2020 were thwarted by the arrival of Covid.

However during the early part of this decade I have been encouraged by colleagues, especially Robin Grove-White and his colleagues from the Lancaster University Centre for the Study of Environmental Change, to persevere. I also found a number of authors, particularly Eric Chaisson, the astrophysics and Bob Ayres, the physicist turned economist whose papers, embarrassingly, I was previously unaware of, but which added substantial weight to my conjectures. I've sought to summarise the additional evidence which substantiates and expands the original ideas in the extended essay which forms the fifth chapter of this book.

Crucially Jon Gower and Nation.Cymru added their support. First by requesting a series of short, more popular pieces for the web site and, then, by inviting me to deliver the first Nation.Cymru lecture in Y Senedd in Cardiff. It was through this association that Ali Anwar and the H'mm Foundation became involved and, with amazing generosity, have funded the production of this book at very short notice. I am deeply grateful to Ali Anwar.

Chapters 6 and 7 have arisen from a conference jointly sponsored by The Institute of Welsh Affairs and the University of Wales Trinity St David's in 2012 on the theme "Wales' Central Organising Principle – Legislating for Sustainable Development". My contribution was entitled, 'Overshooting Limits – Seeking a new Paradigm' and predates the work on energy transformations *per se*. Nevertheless it is complementary to that work. See chapter 10 in https://www.iwa.wales/wp-content/media/2012/01/centorgsd.pdf. A closely related paper which appeared in *Y Traethodydd* under title 'Croesi Ffiniau' is also reprinted here.

This material, as indeed does much of the "green" agenda, highlights both the acute dangers of humanity's current trajectory and the profound problems associated with any substantial change of direction. I am suggesting that key to any resolution is the assertion and hope that we can live well on less energy. As this will be vigorously contested, some will despair.

Let me therefore reassert two quotations from 'Overshooting Limits'.

'Hope is not the conviction that something will turn up but the certainty that something makes sense regardless of how things turn out.'
Vaclav Havel

'– to be truly radical is to make hope possible rather than despair convincing.'
Raymond Williams.

2
Energy and Power
Essential, Addictive and Toxic

This chapter contains the English text on which the lecture given in Y Senedd on the 24th of September 2024 was based. The event was arranged by Jon Gower on behalf by Nation.Cymru who intend to support a series of annual invited lectures on contemporary issues. I was profoundly honoured and grateful to have been invited to present the inaugural lecture and wish to record my personal thanks to Jon, specifically, and to Nation.Cymru for their continuing interest in these hypotheses and their implications.

The event was formally sponsored by the Llywydd, the Rt. Hon Elin Jones, for whose support we are very grateful. It was a great honour to be permitted to use the facilities of the Senedd and we received the invaluable advice and assistance of the Senedd staff. The occasion, to which all Senedd members were invited, was made possible by the enthusiastic support of Ali Anwar and the H'mm Foundation, whose generosity have also led to the publication of this volume.

The lecture and this chapter attempt to convey, without burdening the reader or the listener with too many technicalities, the fundament 'energy' dilemma facing humanity. As noted in the Induction (Chapter 1), greater detail to justify both these profound concerns and the range of evidence to validate them, can be found in other Chapters especially Chapter 5.

While the prognosis is undoubtedly highly challenging, the lecture and this chapter seek to emphasise that there is an alternative future for both Wales and wider global society. This indeed requires major changes in our political-economy and a reassessment of some of our aspirations as humans; but dreams can turn into reality.

Energy and Power:
Essential, Addictive and Toxic

It is a great honour to be invited to give this first Nation.Cymru lecture. Even more so as it is to be given in the Senedd.

I am old enough to remember when "Senedd i Gymru" was but a distant dream and the dreamers thought of as cranks. I was one of those cranks.

But **Dreams** can become **Reality** – please, please remember that as I speak. The dreams of a few can turn into new realities AND I will argue we, as humanity, are in desperate need of both a new dream and a new reality.

This lecture is to celebrate the success of the Nation.Cymru web site and to honour those who have had the enterprise and drive to initiate and make a success of it.

My congratulations and thanks to them and to Ali Anwar and the H'mm Foundation for allowing me a platform to turn my dream into a physical reality.

My topic this evening is Energy and Power. They are essential to all life, but, I will argue, dangerously addictive and potentially toxic.

They are fundamental to all human enterprise; energy to our travel, to heating our houses, to our health, to making steel, plastic and cement and indeed everything else.

We need energy to run our private and public businesses and computers. Increasingly, I am told, in Cardiff Bay to run the air-con in overheated flats.

The energy in our food to allows us to grow and reproduce and indeed to prepare this event: for me to talk to you and for you to listen.

Mental energy is a physical reality as well a turn of phrase.

Energy is one of science's most fundamental and pervasive concepts.

A branch of science called thermodynamics tells us that everything –

all events – depend on energy transactions.

But I am not here to celebrate the importance energy but to suggest that, while energy transactions are essential, in human society energy and power, in excess, can be, indeed are, toxic.

If I am correct in this then we, as humanity, face some very hard decisions and must rethink some of our most cherished assumptions, globally and locally.

I realise of course that I am joining a long parade of old men crying woe; perhaps as an old horseman trying to stop a runaway car would shout – wow! But I can also can foresee a positive outcome.

It will be for you the audience here in Y Senedd and on the web to decide whether I am an old crank, or making a persuasive case.

Some of what I have to say will seem bleak. But I also hope to inspire you with a dream: to advocate a fresh approach to humankind growing problems.

Carwyn James famously said "Get your retaliation in first". So In case any of you have to leave early or get bored or fall fast asleep, let me start, not so much with my retaliation, but with my take-home message.

I am asserting that our fundamental problems run even deeper than the potential impacts of catastrophic climate change, and the other profound environmental and social issues with which we are grappling. Sadly many in our society harbour deepening resentments. They are disillusioned with democracy, careless of our freedoms and open to exploitation. Many feel that the world is not delivering for them personally.

I am arguing that underlying many of our problems is our fateful relationship with energy, **irrespective of the source of that energy**.

I am arguing that, even we succeed in making huge cuts in our greenhouse gas emissions even reaching net zero by mid-century, sadly an ambition that is in serious doubt, we won't solve some of our fundament issues if we continue to the path of seeking to exploit more and more energy.

I am actually arguing that, if we find the nirvana of limitless pollution-free cheap energy, humanity and this planet and humanity will still be in serious trouble.

I am also arguing that there is another way which will respect our humanity and that of others and secure a healthier planet.

The failing health of our planet is illustrated dramatically in Figure 1. It shows that in six of nine areas of major environmental concern which include biosphere integrity and pollution as well as climate change, safe limits are being exceeded.

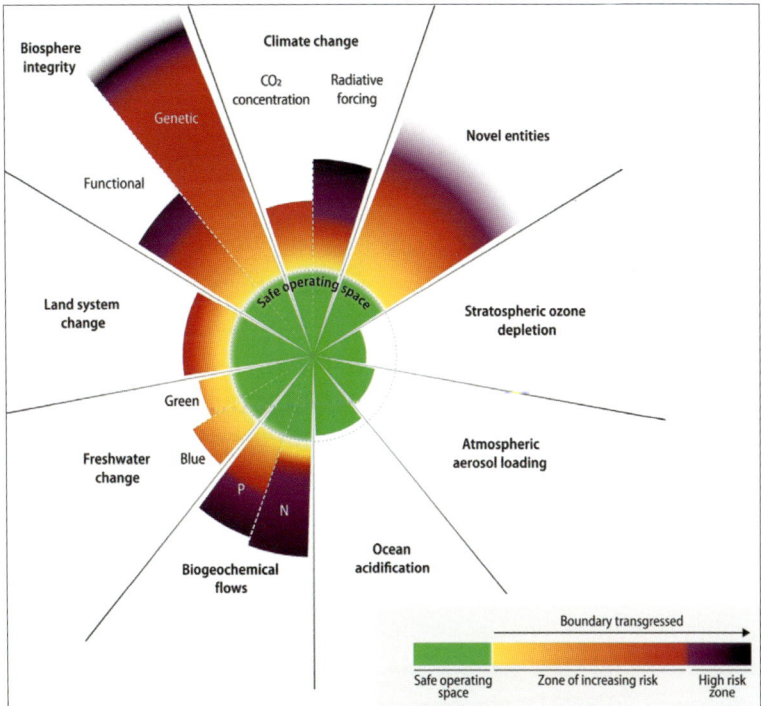

Figure 1. Estimates of the extent by which human activity has exceeded safe limits in 9 areas.
Safe operating space shown as green and zones of increasing risk showing as yellow to dark purple. Note climate change is but one of 6 activities which exceed safe limits. *https://www.science.org/doi/10.1126/sciadv.adh2458*

Let me first try and explain – very succinctly – some of the back ground science that has influenced my thinking.

Energy is a physical concept and the science tells us that the more energy, properly termed free energy transformations, that can be exploited or used with much greater efficiency, the more work that can be accomplished in a given time and the more power generated.

Power is defined as work per unit time and work is what makes thing happen.

Free in this context has nothing to do with money but means that the energy is free to do work and bring about transformations.

Universally energy, or rather energy and mass after Einstein's famous equation $e=mc^2$, is conserved but gradually when transformation occur, some energy is lost into forms unable to do work. Think, for example, of the waste frictional heat released by an engine or the warmth produced by your computer and indeed the metabolic waste heat lost by our bodies- a lot from our heads.

This "lost" component of energy is termed entropy. It cannot do useful work. Some interpret it as an increase in disorder, others in terms of a lack of information to define a given situation.

Nevertheless it's been shown, both experimentally and theoretically, that in systems far from equilibrium, energy flows can spontaneously give rise to ordered, complex structures and systems.

Such systems are intrinsically unstable and will collapse unless fed a constant energy supply – think about hurricanes dissipating over land.

A number of famous people have postulated that these concepts are applicable both to biology and human society. In my book and the booklet, I go into this in much more detail.

For now – simply let me assert – the more energy that can be tapped – the more work that can be done and power generated. This can then lead to structures and societies of greater dynamic but ordered complexity and potential.

An astrophysicist, Eric Chaisson uses the phrase "the arrow of complexity" to describe this observation.

To illustrate this consider how life on Earth has evolved over billions of years from a single, relatively simple cell to us, sophisticated, sentient, morally-aware humans living in enormously complex cities. But, let me stress, we remain entirely depend on simple unicellular organisms to keep the whole interdependent life-support system running.

The arrow of complexity is complemented by deep ocean of interdependence.

In biology and in our human world, stabilising this complexity in our chaotic world depends on both sources of energy and a parallel hierarchy of regulatory mechanisms – often called homeostatic mechanisms in biology – to promote the stability of each new emergent form.

In our societies this stability depends both our behavioural norms, some in all likelihood inherited from very ancient ancestors – aspects of what we often call our human nature.

And, in modern cultures and organised societies, on the laws and regulations we have devised, often in Seneddau such as this one, and as well as a plethora of customs, influences and beliefs. Without such regulations, laws and behavioural standards, no human society can operate.

Many of these "regulations" have developed in the last few hundred years post the Industrial Revolution, although, of course our religious beliefs and philosophies have deeper roots.

Let me stress some vital points.

First, as more energy is exploited, more work can be done in a given time.

So events will follow each other more and more quickly – that is everything will continuously speed up.

Indeed given the huge increase in energy exploitation by humans in last 200 years or so after the Industrial Revolution, it's not surprising people talk of "the great acceleration" – it's our living experience (see

Figures 2,3 and 4.)

Secondly, information and energy are interrelated. Increasing entropy implies an increase in disorder or lack of defining information.

Conversely, Erwin Schrödinger suggested that cells – the basic unit of all complex dynamic living systems can be characterised as accumulating negative entropy. That is both more energy and more information are embedded in them and in growing complexity.

Thirdly, looking at the long 4 billion year history of life on Earth, a pattern emerges of a sequence of revolutionary step changes in energy capture and use.

It is not feasible to even outline these this evening, but I must note that the last three have a very direct impact on our emergence; the evolution of *Homo sapiens*.

Initially our ancestor, *Homo erectus*, about one and a half million years ago invested in energy in increasing brain power. This investment in 'brain not brawn' started hominins on a new path. One of more sophisticated social interactions, in all likelihood language, as well as new technologies. Then about 8 to 10 thousand years ago, the agricultural revolutions made more food calories available, enabling denser, more complex societies to develop with larger, more stable populations. This led to the need for record keeping, writing and numeracy. And in last 250 years the Industrial Revolution, driven by burning fossil fuels, has been transformational as we discuss.

As a result of the leaps, although we, like our ancestors perhaps a million years ago, only need about 2,000 to 2,500 food calories per day of energy to maintain our basic metabolic health, our life styles, on average, now depend on the equivalent of a hundred times this much energy per day!

Finally these energy step changes have created new potentials. Investing in brain power, speech, writing, numeracy etc has transformed what an ape-like animal can achieve.

It also appears that each new potential allows a new emergent elite to flourish best able to exploit the new opportunities.

In my judgment the underlying physics – the thermodynamics – and the evidence from evolutionary biology, anthropology and our social and economic history are all consistent with these propositions.

This is the basis hypothesis of my book, 'Energy the Great Driver. https://www.uwp.co.uk/book/energy-the-great-driver/ and in the booklet.

Just to note, these simplified propositions are drawing on the work of some of greatest scientists of that last century Wilhelm Ostwald, Erwin Schrödinger, Ilya Prigogine, Daniel Kahneman and Lord Rayleigh – all Nobel Laureates as well as the neurophysiologist Antonio Damazio.

I see these conjectures as establishing a framework of understanding.

My main aim this evening is consider what this framework might mean to us and for our planet now and in the immediate future.

Just to repeat my basic proposition.

The more energy we can use or use much more efficiently – for example because of our brain power and tools and now likely AI, the more work we can do and the more power can be harnessed so the more dynamic ordered but unstable complexity can be created with its attendant new potentials and problems.

In a series of planetary major energy step-changes and many more smaller jumps, this growing complexity has allowed new opportunities to emerge.

In human terms these are exemplified by our brains, our social, material and latterly economic constructs, our growing technological and scientific prowess and our great cities and art. But above our consciousness and our philosophies and beliefs.

With this growing complexity, there been a massive acceleration in the rate of change.

It has also created a need for parallel increase in regulatory mechanisms to stabilise the complexity. All is now changing at a break-neck speed. Although in our biological history the evolution of biological homeostatic regulation is so efficient is almost invisible on a day to day basis, there are now good reasons to think effective regulatory stabilization is not now keeping up with the rate of change.

How do these hypotheses play out in our modern world experiencing global warming as well as other major environmental (see Figure 1.) as well as social challenges?

Undoubtedly our society, our affluence, our standard of living is entirely dependent on abundant cheap, convenient energy: that is burning fossil fuels – coal, oil and gas and eating cheap food.

A massive increase in global material wealth as measured by GNP (Figure 2.) and in the human population has occurred (Figure 3.) The latter from about 800 million in the 1770s to 8 billion now.

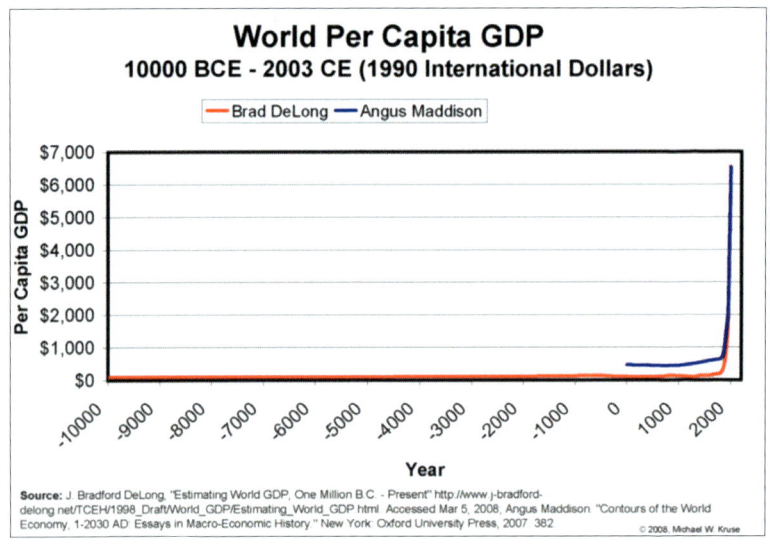

Figure 2. Estimates of Average Human Wealth measured as GDP per head since the last Ice Age.
https://www.google.co.uk/search?q=world+gdp+history+chart&tbm=isch&tbo=u&source=univ&sa=X&ved=2ahUKEwjTwfqxkL7eAhUDaVAKHdyiCsgQsAR6BAgFEAE&biw=1920&bih=887#imgrc=nDa6wP58RoifAM:

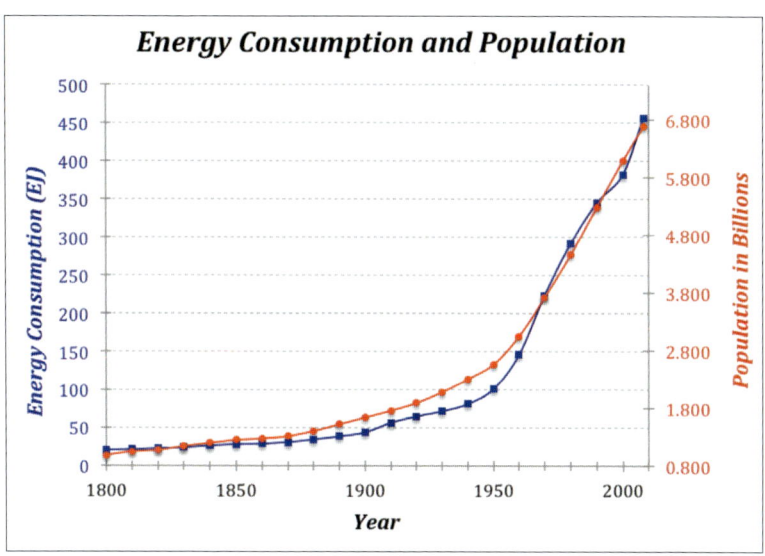

Figure 3. Growth in Global Population and Energy Use since 1800
https://www.e-education.psu.edu/earth104/node/1347

Energy use both in total and per head soared. Many, many, commodities, such as steel, plastics, cement including food follow this trend We all we depend on food – on cheap calorific energy as well as nutrients

More energy; more and more stuff, more and more quickly!!

Because prosperity and wealth are correlated strongly with energy use, wars have been fought over and through energy supply as in the Ukraine now. Resources have been stolen or commandeered. Controlling energy has made countries and individuals enormously rich.

The controllers of Saudi or Iraqi oil, Qatari or Russian gas are hugely powerful. Once it was Welsh coal – for 50 years coal was king – Cardiff and the Coal Exchange flourished (see Figure 4.)

Energy is the basic reason we are here today in Cardiff Bay, not in Machynlleth as Jan Morris once dreamt.

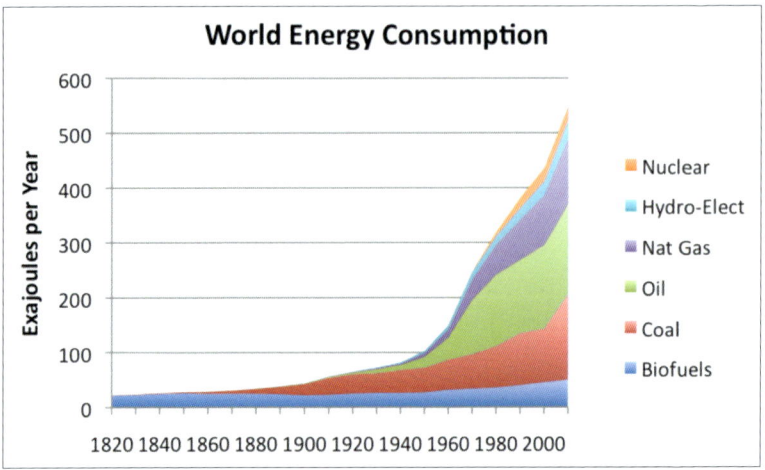

Figure 4. Changes in total world energy consumption and in energy sources since the early Industrial Revolution.
https://ourfiniteworld.com/2012/03/12/world-energy-consumption-since-1820-in-charts/

Fossil fuels have literally created our world with its massive but distorted wealth and its corruption.

We now also know beyond reasonable doubt that the release of Green House Gases, mainly carbon dioxide [CO_2], methane and nitrous oxide from burning fuels and from agriculture and land use change is causing global warming and climate change. This is happening at an extent and at a rate which threatens the global economic order and the very social fabric on our planet. A fabric, ironically, woven and underpinned by fossil fuel wealth.

It's important to emphasise, despite all the talk of combating climate change, 80% of our energy, globally and in the UK, still comes from burning fossil fuels. We are churning out greenhouse gasses at a record rate so causing rapid climate change and global warming.

We humans are using more and more of this planet's other resources including from photosynthesis at the expense of other life-forms and their habitats. So causing biodiversity loss as well as chemical pollution and plastic fouling.

We all we depend on food, on cheap calorific energy as well as nutrients. And maybe should shudder to realise that the agri-food chain accounts for 30% of all global emissions.

Our food production not only depends on the sun's energy and photosynthesis but on fossil fuels to make fertilizers and agrochemicals and to carry, refrigerate and process the food we eat from around the world as well as producing lots of methane and nitrous oxide, both highly potent GHGs.

Famously big agriculture has been called a system for turning oil into food.

The increasing global demand for energy means the installation of new renewable energy infrastructure is barely keeping up with increasing demand and the retiring of old nuclear plants.

Things are hopefully changing but almost certainly too slowly to avoid major chaos.

Undoubtedly combating Anthropogenic Global Warming is one of humanity's greatest failures.

Very broadly there are three types of responses out there.

Firstly. Deny, disparage and sow doubt. It's all a hoax or greater exaggerated! We see this especially in the Trumpian USA. In the UK this is exemplified by the Global Warming Policy Foundation, the anti-net zero lobby, Farage and his Reform colleagues and large parts of right wing press in London.

Secondly. The political mainstream, including the current UK government, agree that we need major cuts in GHG emissions, large increases in renewables and likely in nuclear electricity. In the near future, electricity from fission and later hopefully fusion. Increasingly this mainstream recognises that a dependence on carbon capture and storage and planetary geoengineering may become necessary as things get really bad.

The overall and over-riding objective of the mainstream is to ensure the continuation of the current economic model based on an annual, Indefinite growth in GDP, whether in the UK and globally. This because consumer-led growth is seen as a prerequisite to social stability and poverty reduction.

So yes! Let's cut emissions but, but, but!

The recent Tory administration did as little as possible despite a rhetorical allegiance to net zero. How rapidly Starmer's Labour will act remains to be seen. Welsh Government has only limited powers and often talked more than it's achieved. In China there is a massive investment in 'renewables' but fossil fuels remain vital. In India coal is still king. While in Russia they entertain the hope that maybe Siberia will benefit from climate change.

Overall significant change has been far too slow. But worse in the US elections Trump triumphs all bets are off. The USA, as the greatest world power, will then join the deniers and dissemblers.

Crucially, everywhere the beneficiaries of the current geo-political deal and the pervasive economic model, in one way or another, all wish it to continue regardless. They wish to maintain their status and we must count ourselves as part of this group.

Naturally, the less developed and the left-behind are bent on catching up.

Thirdly. Perhaps I can characterise the third group as the despairing, some the militant despairing.

They calculate that we are on a path to massive mean global temperature rises, to passing major tipping points and to global chaos. They have good cause to think this. Some are militant; other just resigned.

Given the political and economic power and influence of main stream and their tacit assumptions about using more and more energy, I would like to give more thought to their position and how it relates to my hypothesis.

As I've said the widely-shared assumption is that growth, as measured by GDP, can continue and will make us richer, more prosperous, more content. It will also allow the poor to gradually become less poor as growth, dependent on cheap reliable energy, raises the overall level of prosperity. Global and national trickle-down development.

If you want to find an unrestrained expression of this growth scenario, I suggest you read the techno-optimists manifesto by Marc Andreessen, one of the leading lights of Silicon Valley and now major Trump backer. https://a16z.com/the-techno-optimist-manifesto/

To quote.

"*We believe energy should be in an upward spiral. Energy is the foundational engine of our civilization. The more energy we have, the more people we can have, and the better everyone's lives can be. We should raise everyone to the energy consumption level we have, then increase our energy 1,000x, then raise everyone else's energy 1,000x as well.*" (Techno-Optimist Manifesto, Marc Andreessen).

Andreessen's manifesto is "conventional wisdom" on steroids.

While Andreessen and I agree that energy is 'the foundational engine of our civilisation', we agree on little else.

It is vital to consider how do both in the rampant version of Andreessen or the more restrained mainstream version espoused by virtually all politicians and most economists, relate to my conjectures.

To do this let us, for a moment, assume that climate change problem is miraculously "solved". But, of course, we will return to that issue in a moment.

What then will Andreessen's 'upward spiral' of energy mean even if there is no contribution to climate change?

I suggest:

A] More societal and material dynamic complexity.

B] A continuing increase in human demands on the resources of this planet, likely including photosynthesis. This will be at the expense of other organisms and their habitats. Andreessen even envisages a growing human population. Every one of whom will have their physical and social demands.

C] A further acceleration of the speed of change as well as the increase in dynamic complexity will be inevitable. This will likely have profound and unsettling effects on our social as well as our economic lives and on our psychological well-being and our mental health. There is already evidence of mounting metal health issues especially among the younger generation.

D] As Joseph Schumpeter realised, at the heart of capitalism and the continuous growth model lies 'creative destruction'. The cycle of destruction and creation will, indeed must, turn faster and faster as energy spirals upward and new technologies, devices and fashions replace their predecessors.

E] The new complexity will create new possibilities which will be taken up by the minority best able to take advantage of them. In all probability a powerful new elite will emerge. In all likelihood this will be based on the world of AI, digitised information processing and robots and maybe cyborgs and hybrids. Chips implanted in our brains?

F] As complexity increases at an accelerating rate, new regulatory systems must emerge, both locally and globally, to ensure and, at least seek to ensure, a degree of stability. There are good reasons to suspect that our regulatory systems are already unable to keep up with current rate of change e.g. regulating the social media and AI. It must be doubted whether our current democratic systems for generating new laws and regulation will be keep up with change which will be even faster than we are experiencing now.

G] There are real dangers that increasing regulation will be enacted at the expense of our rights and freedoms and possibly our democracies.

H] The emergent new elites, that arise to take advantage of the new opportunities related to the more complex and rapidly changing systems, will likely have at their disposal and use IT and AI to cement their status.

This may be what Andreessen wants and what the mainstream, perhaps inadvertently, will permit but the crucial questions remain:

Is this what people in general actually want and is it in their best interests?Indeed is this their dream for themselves and their children?

As I've indicated, my hypothesis suggests this high energy-world will have numerous and, in my judgement, negative impacts on our humanity, our freedoms, as well as on biodiversity and general planetary health.

Therefore I contend that in the real world we must contend with two interrelated but distinct issues.

The first being, of course, fossil fuel-driven climate change. It is a recognised, real and growing threat, although one we seem unable to respond to. However I am now highlighting a second, less appreciated threat. This more subtle threat that arises from humanity's ability to access and couple energy **from any source** to do work, create more power and accelerate complexity with corresponding need for more and more regulation. This threat will remain if human ingenuity succeeds in exploiting energy sources that do not emit greenhouse gases. Paradoxically it will increase if humanity finds the key to limitless cheap energy from say nuclear fusion, often touted as key to unfetter growth.

I wonder how many people now really believe in this dream of inevitable, continuous material progress.

I think the limitations of the *economicus* dream are getting more obvious as is the toxicity of our fossil fuel derived emissions of greenhouse gases as well as the more subtle and subversive impacts of accelerating energy exploitation.

Let me emphasise the human nature is more complex and indeed richer and more caring than *Homo economicus* would allow.

Our economy is an energy-dependent construct but set in a finite Earth as envisaged by Bob Ayres, Kate Rawoth and others (Figure 5.)

Is it therefore all just doom and gloom?

Well curiously I think not. I recommend TIAA, not TINA. '**There is an alternative' (TIAA) not Mrs. Thatcher's TINA 'there is no alterative'!!**

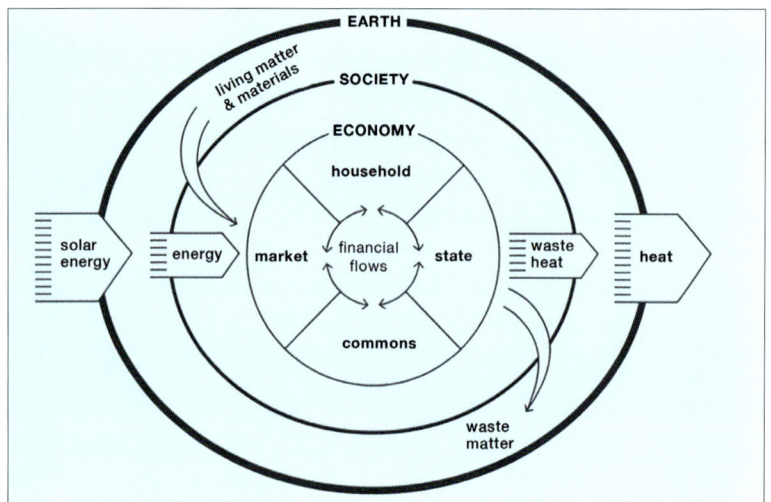

Figure 5. Model of the Economy driven by Energy but bounded by the Earth's resources
Kate Raworth. (2018) *Doughnut Economics: Seven ways to think like a 21st Century Economist.* Radom.

All is not lost although, of course, I recognise that the challenges are enormous.

My first hope is that the framework of understanding I've been advocating will stimulate a vigorous and informed debate. This may lead us, as the Greek philosophers advocated thousands of years ago, to our knowing ourselves better.

I hope this event in the Senedd will help kick start this debate.

Many voices are now raised to criticise the energy-driven growth model from various perspectives but so far these are minority. None have convincingly shown how we can get off or slow our current ever-accelerating economic tread-mill without crashing.

The old model partly because of its real historic successes, partly because of the reality of power and partly many of us buying into TINA, remains dominant. Nevertheless it appears to me to be tottering.

Paradoxically, aspects of the climate change crisis may be pointing a way forward. More low emissions energy is clearly essential. Specifically, in the richer countries more electricity is required to power our heating, transport and industry.

Unfortunately every renewable source has some disadvantages and limitations. These negative impacts catalyse nimbyism and slow progress.

Nuclear fission is horrendously expensive, has very many drawbacks, including a heavy demand for water cooling as well as the resolved long-term storage of highly radioactive waste.

Given the time-table of global warming, there is no prospect of it being will be rolled out in time in the UK still less worldwide.

Indeed, having worked in Aleppo and other places such a Boco Haram territory in north Nigeria, any proposal for tens of thousands on nuclear reactors worldwide, Small Modular Reactors or not, in a time of increasing social tensions, is too scary to contemplate.

Cheap fusion remains a distant dream. Green or geo-hydrogen has potential but will also require a vast new infrastructure.

Given these fundamental constraints is not the economic, social and environmental logic to work to build societies in which people can live well using less energy?

Could this not be both an effective way of this combatting climate change quickly, avoiding the worst excesses of modern society and the negative impacts of Andreessens' 'energy spiral'.

Why waste billions on new generation capacity and transmission if it's better spent on decreasing demand?

Should we not prioritise improving and insulating our housing stock and our offices. Should not ALL new build be carbon neutral and with a low energy demand?

Should we not prioritise providing better public transport and decreasing car use. We have an EV car but its construction is still relatively energy and materiel intensive. Encouraging walking and bikes, improving our public spaces and services, enjoying local holidays, our wild life and wonderful seaside and countryside and in investing in our cultural life.

We must ask ourselves, can we not live well on less energy?

It's a question to put to our sport teams? Can you really justify flying to South Africa so often?

If we reduce demand, we reduce, but do not eliminate, the problem of additional electricity generation. Maybe more local generation will reduce, but again not eliminate, the need for new transmission lines.

These are not new ideas. But this evening I am not arguing the specifics but for a new overall mind set.

A dream of not how can I acquire more and more but how can I live better on less energy?

This vision leads logically to an emphasis on the local community to counteract the centripetal force of big business and large conurbations.

The vision emphasises personal responsibility and freedom as well cooperation at the community as well as global level.

It is community-based renewable energy projects such as Ynni Ogwen and Anafon not Amazon.

However such a change in mindset is not straight forward and implementation will cost money; major public and cooperative investment.

It will certainly mean more taxes on the rich, reducing tax dodges and eliminating tax havens. And cooperation with our European neighbours.

Nevertheless I contend there is an alternative TIAA vision to that of human wellbeing and thriving being dependent on indefinite consumerist growth as measured by GDP per head. Critically it offers hope of rapid cuts in GHG emissions as well a pathway avoiding the worst pitfalls of the 'energy spiral'.

Let me summarise.

Without doubt energy is the *foundational engine of civilisation* to quote Andreessen; indeed of all life. In my book and in the booklet published today I suggest that our human flourishing is the result of a series of planetary energy revolutions over last 4 billion years.

Nevertheless Andreessen's projections and assumptions, and indeed those of the moderate mainstream, are deeply flawed and indeed toxic.

The dangers arise from two broad sources. Our society's dangerous addiction to cheap, convenient fossil fuel energy giving enormous and pernicious power to the providers. And also the growing energy-dependent complexity, accelerating change and regulatory challenges, which are our everyday experience, are doing serious damage to the subtle interactions and cycles that are essential to our health as humans and to our planet (see Figure 1).

My contention is that even if we succeeded in quickly replacing fossil fuel energy with other sources that create much less pollution, an improbability I must admit, we have still to address our relationship with energy, power, complexity and speed of change. I am simply suggesting that, as in many things, too much of a good thing can be toxic.

Therefore, seeking to build a low energy but flourishing society is, I suggest to you, the only way forward both in relation to global warming and the energy spiral.

This will require new economic priorities which I've barely sketched such as an alternative to GPD.

Undoubtedly, we face major difficulties in trying to working to turning this dream into a reality rectify.

I will mention only three.

This dream will be bitterly opposed by the power of great wealth, exemplified by Andreessen and his IT colleagues. They finance Trump and can spread misinformation and may well see themselves as the elite beneficiaries of an energy spiral.

Secondly the dream can only be realised in a fairer, more equitable world. There are very poor people, even in Wales, with very low per head energy use and Greenhouse Gas emissions. They have a legitimately claim on more energy, to more GHG-free energy.

Without greater fairness and local and global equity, we will flounder.

Finally I recognise that new technologies will be vital to solve many of humanity's problems but we must guard against hubris and be aware of the constant dangers of known and unknown unknowns.

The logical outcome of the patterns I've sketched might well be the emergence of a techno-elite or even cyborgs, able to exploit an emerging information-energy dense world and dominate the rest.

This to me is an Orwellian dystopic scenario.

Given the virtual certainty of mean global temperature rising over 2oC and that we will pass important climatic and social tipping points e.g. triggering major sea level rise and regions becoming too hot and humid for humans, any techno-growth triumph will coexist with and may well be constrained by major global unrest and chaos.

This, if for no other reason, should give pause to those advocating nuclear fusion as an answer to the world energy crisis. It is too easy to image a world in chaos with the AI-elite protected by heavily armed militia trying to safeguard themselves from the mob – much as in gated communities now in parts of the world.

However there is another way – **an alternative dream**.

We have a choice – agency is the word often used. We as humanity can decide to adopt a more sensible, a more equitable way and follow a less energy-intense road.

Do people actually to want to be serfs to the emerging AI elite?

Do people want to be on the treadmill of ever accelerating material growth?

Dreams can and do become reality!

Many of these ideas are not new. I am trying to provide a comprehensive framework, clothed with a scientific cloak, to reach for a new reality .

Realising a new reality will require us to work in ways that are largely neglected in this world of individualism, competitive commerce and cut-throat geopolitics.

Ultimately the issues are not scientific, technological or even about our political economy but relate to our human ethics and values.

I can do better than end with Waldo Williams

Bydd mwyn gymdeithas.

　Bydd eang urddas.

　　Bydd mur i'r ddinas.

　　　　Bydd terfyn traha. (*Waldo Williams*)

There'll be gentle community,

　There'll be broad dignity,

　　There'll be walls to the city,

　　　　Arrogance shall fail. (*translation by Anthony Conran*)

Addenda.

1. On Christmas Eve 1967, less than four months before his assassination, Martin Luther King said, "It really boils down to this: that all life is interrelated. We are all caught in an inescapable network of mutuality, tied into a single garment of destiny. Whatever affects one destiny, affects all indirectly."

2. Ponder about these two energy scenarios.

Had this planet not been endowed with vast reserves of fossil fuels, how might global society have evolved in the nineteenth and twentieth centuries if based largely on neo-current photosynthate from forests and farming?

　Had atomic (fission) power, as Lewis Strauss, the anti-hero in the film Oppenheimer, anticipated in 1954, generated electricity 'too cheap to meter', what might have been the outcome for both the global north and south and for our environment?

3. It's worth noting the energy demand of the techno-optimist scenario would be quite enormous.

　1 Bitcoin transaction = 27 days electricity use by average US household 29 kWh; ChatBoxGBT already uses 0.5 million kWh per day which is still largely generated by burning fossil fuels!

3
Ynni a Phŵer:
Hanfodol, Caethiwus a Gwenwynig

Yn y bennod hon cyflwynir cyfieithiad o'r testun (Pennod 2) sy'n sylfaen i'r ddarlith traddodais yn y Saesneg yn y Senedd yng Nghaerdydd ar y 24ain o Fedi 2024. Yn fy sgwrs roeddwn yn ceisio cyfleu hanfodion fy namcaniaeth i gynulleidfa eang, lled anwyddonol. Trefnwyd yr achlysur i ddathlu llwyddiant y safle we, Nation.Cymru, a gweledigaeth ei sylfaenwyr. Mawr yw fy nyled i Jon Gower ac i Ali Anwar am eu cefnogaeth.

Fel yng ngweddill y llyfr hwn rwy'n ceisio esbonio ac ymhelaethu ar gynnwys fy llyfr, "Energy the Great Driver; Seven Revolutions and the Challenges of Climate Change".

Os am ddadansoddiad mwy manwl a wyddonol rhaid troi at y pumed bennod, Energy and Power; the Main Scientific Essay. Ond fe geir ymdriniaeth o'r cefndir esblygiadol yn yr erthyglau a ail gynhyrchir o O'r Pedwar Gwynt gyda chyd-weithrediad parod Sioned Rowlands.

Ynni a Phŵer:
Hanfodol, Caethiwus a Gwenwynig

Mae'n anrhydedd o'r mwyaf cael fy ngwahodd i draddodi darlith gyntaf Nation.Cymru. Yn fwy fyth felly gan ei bod yn cael ei rhoi yn y Senedd.

Rydw i'n ddigon hen i gofio pan nad oedd "Senedd i Gymru" yn ddim ond breuddwyd bell, ac unrhyw un fyddai'n breuddwydio am hynny yn cael ei galw'n granc. Cranc oeddwn i, felly.

Ond gall **Breuddwydion** gael eu **Gwireddu** – da chi, cofiwch hynny wrth wrando arna'i.

Gall breuddwydion yr ychydig droi'n realiti newydd, AC fe fydda'i yn dadlau fod gennym ni, fel dynoliaeth, angen dybryd am freuddwyd newydd ac am realiti newydd.

Nod y ddarlith hon yw dathlu llwyddiant gwefan Nation.Cymru ac anrhydeddu'r sawl fu'n ddigon mentrus ac egnïol i roi bod iddi a pheri iddi lwyddo.

Hoffwn eu llongyfarch hwy ac Ali Anwar a Sefydliad H'mm am roi llwyfan i mi droi fy mreuddwyd yn realiti ffisegol.

Fy mhwnc heno yw Ynni a Phŵer; maent yn hanfodol i fywyd o bob math, ond, fe fydda'i yn dadlau, yn beryglus o gaethiwus ac fe allant fod yn wenwynig.

Maent yn hanfodol i bopeth a wna dynoliaeth; ynni i'n teithio, i wresogi ein tai, i'n hiechyd, i wneud dur, plastig a sment, a phopeth arall yn wir.

Mae arnom angen ynni i redeg ein busnesau preifat a chyhoeddus a'n cyfrifiaduron. Yn fwyfwy felly, meddan nhw wrtha'i, ym Mae Caerdydd, i redeg y systemau awyru mewn fflatiau sy'n rhy boeth.

Mae'r ynni yn ein bwyd yn caniatáu i ni dyfu at atgynhyrchu ac yn wir i baratoi'r digwyddiad hwn: i mi siarad â chi ac i chi wrando.

Mae ynni meddyliol yn realiti ffisegol yn ogystal ag yn ymadrodd.

Mae ynni yn un o gysyniadau mwyaf sylfaenol a threiddiol gwyddoniaeth.

Dywed cangen o wyddoniaeth o'r enw thermodynameg fod popeth – pob digwyddiad – yn dibynnu ar **drafodion ynni**.

Ond nid yma yr ydw i i ddathlu pwysigrwydd ynni ond i awgrymu, er bod trafodion ynni yn hanfodol, y gall gormodedd o ynni a phŵer mewn cymdeithas ddynol fod yn wenwynig – yn wir, eu bod felly.

Os ydw i'n gywir yn hyn, yna yr ydym ni fel dynoliaeth yn wynebu rhai penderfyniadau anodd iawn ac y mae'n rhaid i ni ail-feddwl am rai o'n rhagdybiaethau anwylaf, yn fyd-eang ac yn lleol.

Rwyf yn sylweddoli, wrth gwrs, fy mod yn un o res hirfaith o hen ddynion sy'n darogan gwae; hen farchog sy'n ceisio stopio car gwyllt trwy weiddi – wow! Ond gallaf hefyd ragweld canlyniad cadarnhaol.

Mater i chi yma, y gynulleidfa yn y Senedd ac ar y we i benderfynu ai hen granc ydw i, neu rywun sy'n gwneud achos sy'n eich perswadio.

Bydd rhywfaint o'r hyn fydd gen i i'w ddweud yn swnio'n ddigysur. Ond rydw i'n gobeithio hefyd eich ysbrydoli a breuddwyd: i bledio'r achos dros agwedd newydd at broblemau cynyddol dynol ryw.

Fe gofiwch arwyddair enwog Carwyn James "Mynnwch ddial yn gyntaf". Felly rhag ofn i unrhyw rai ohonoch orfod gadael yn gynnar neu ddiflasu neu ddisgyn i gysgu, gadewch i mi gychwyn, nid yn gymaint â'm dial, ond gyda fy neges i chi.

Datgan yr ydw i fod ein problemau sylfaen **hyd yn oed** yn ddyfnach nag effeithiau posib newid hinsawdd trychinebu, a'r problemau amgylcheddol a chymdeithasol dwys eraill yr ydym yn ymgodymu â hwy. Maent wedi eu dadrithio â democratiaeth, yn ddiofal am ein rhyddid ac yn agored i gam-fanteisio. Mae llawer yn teimlo nad yw'r byd yn deg â hwy yn bersonol.

Rydw i'n dadlau mai sail llawer o'n problemau yw ein perthynas dyngedfennol ag ynni, **waeth beth yw ffynhonnell yr ynni hwnnw**.

Rydw i'n dadlau, hyd yn oed os llwyddwn i wneud toriadau enfawr yn ein hallyriadau nwyon tŷ gwydr a hyd yn oed gyrraedd sero net erbyn canol y ganrif; uchelgais, gwaetha'r modd, y mae amheuaeth ddofn yn ei gylch, wnawn ni ddim datrys rhai o'n problemau sylfaenol os parhawn ar y llwybr o geisio ecsploetio mwy a mwy o ynni.

Rydw i'n dadlau mewn gwirionedd, hyd yn oed os deuwn o hyd i nirvana o ynni rhad diderfyn nad yw'n llygru, y bydd y blaned hon a dynoliaeth yn dal mewn trafferthion enbyd.

Rydw i hefyd yn dadlau fod ffordd arall fydd yn parchu ein dynoliaeth ni ac eraill, ac yn sicrhau planed iachach.

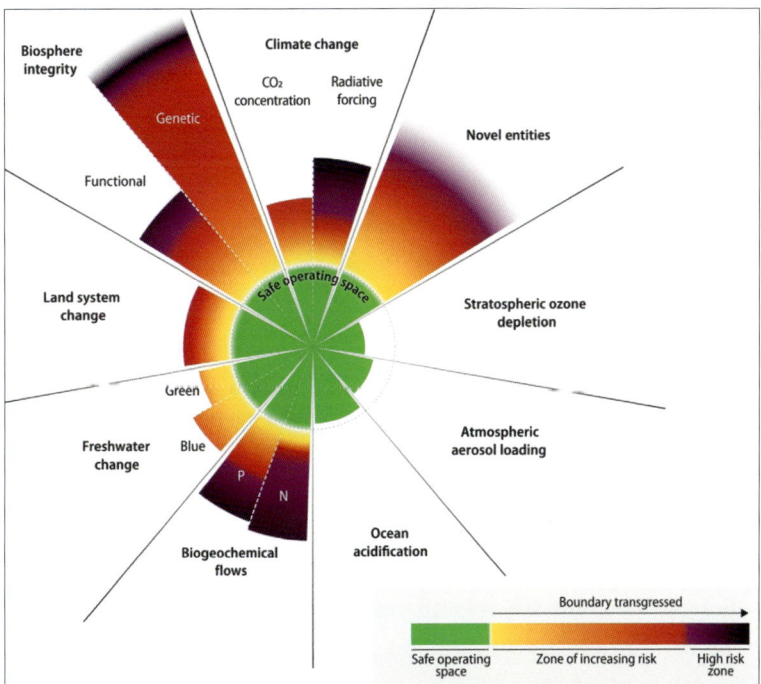

Ffigwr 1. Amcangyfrif o'r graddau y mae gweithgaredd dynol wedi torri terfynau diogel mewn 9 maes.
Dangosir gofod gweithredu diogel mewn gwyrdd a pharthau risg cynyddol yn felyn i biws tywyll. Sylwch mai dim ond un o'r 6 gweithgaredd sy'n mynd dros y terfynau diogel yw newid hinsawdd.
https://www.science.org/doi/10.1126/sciadv.adh2458

Darlunnir dirywiad iechyd ein planed yn ddramatig yn Ffigwr 1. Mae'n dangos, mewn chwech o'r naw prif faes pryder amgylcheddol sydd yn cynnwys cyfanrwydd y biosffer a llygredd yn ogystal â newid hinsawdd, fod y terfynau diogel yn cael eu torri.

I ddechrau, gadewch i mi geisio esbonio – yn gryno iawn – rhyfaint o'r wyddoniaeth gefndir sydd wedi dylanwadu ar y ffordd yr ydw i'n meddwl.

Cysyniad ffisegol yw ynni, a dywed y wyddoniaeth wrthym, po fwyaf o ynni, neu i roi'r enw cywir, **trawsnewidiadau ynni rhydd**, y gellir ei ecsploetio neu ei ddefnyddio yn fwy effeithlon o lawer, mwyaf yn y byd o waith y gellir ei gyflawni o fewn amser penodol a mwyaf yn y byd o bŵer a gynhyrchir.

Y diffiniad o bŵer yw gwaith fesul amser uned a gwaith yw'r hyn sy'n peri i bethau ddigwydd.

Does a wnelo **rhydd** neu **am ddim** yn y cyswllt hwn ddim oll ag arian: mae'n golygu bod yr ynni yn rhydd i wneud gwaith ac i achosi trawsnewidiadau.

Yn gyffredinol, mae ynni, neu yn hytrach ynni a màs, yn ôl hafaliad enwog Einstein $e=mc^2$, yn cael ei gadw, ond yn raddol, pan fo trawsnewidiad yn digwydd, mae peth ynni yn cael ei golli i ffurfiau na all wneud unrhyw waith – meddyliwch, er enghraifft, am y gwres ffrithiannol gwastraff a ryddheir gan beiriant neu'r gwres o'ch cyfrifiadur, ac yn wir, y gwastraff gwres metabolig a gollir gan ein cyrff – llawer ohono o'n pennau.

Entropi yw'r enw ar y gydran ynni "coll" hwn. Ni all wneud gwaith buddiol. Mae rhai'n ei ddehongli fel cynnydd mewn anhrefn, ac eraill fel diffyg gwybodaeth i ddiffinio sefyllfa benodol.

Serch hynny, mae arbrofion a damcaniaethau wedi dangos, mewn systemau sydd ymhell o fod yn gytbwys, y gall llif ynni **yn ddigymell** roi bod i systemau a **strwythurau trefnus a chymhleth**.

Mae systemau o'r fath yn ansad yn eu hanfod a byddant yn dymchwel oni chânt eu bwydo'n gyson â chyflenwad o ynni- meddyliwch am gorwyntoedd yn gwanhau dros dir.

Mae nifer o bobl enwog wedi honni y gellir cymhwyso'r cysyniadau hyn i fioleg ac i gymdeithas ddynol. Yn fy llyfr a'm llyfryn, rwy'n manylu am hyn.

Am y tro – yn syml – gadewch i mi honni hyn: po fwyaf o ynni y gellir ei dapio, mwyaf o waith y gellir ei wneud a phŵer y gellir ei gynhyrchu. Gall hyn wedyn arwain at strwythurau a chymdeithasau â mwy o gymhlethdod a photensial deinamig ond trefnedig.

Defnyddir astroffisegydd, Eric Chaisson yr ymadrodd "**saeth cymhlethdod**" i ddisgrifio hyn.

Ystyriwch sut yr esblygodd bywyd ar y Ddaear dros biliynau o flynyddoedd o un gell gymharol syml i ni – bodau dynol soffistigedig, ymdeimladol gydag ymwybyddiaeth foesol, sy'n byw mewn dinasoedd hynod gymhleth. Ond, gadewch i mi bwysleisio, sy'n dal i ddibynnu'n llwyr ar yr organebau ungellol syml i gynnal yr holl system ryng-ddibynol.

Ategir saeth cymhlethodd gan gefnfor dwfn o ryng-ddibyniaeth.

Mewn bioleg ac yn ein byd dynol, mae sefydlogi'r cymhlethdod hwn yn ein byd di-drefn yn dibynnu ar ffynonellau ynni ac ar hierarchaeth gyflin o fecanweithiau rheoleiddio – a elwir yn aml mewn bioleg yn fecanweithiau homeostatig – i hyrwyddo sefydlogrwydd pob egin-ffurf newydd.

Yn ein cymdeithasau ni, mae'r sefydlogrwydd hwn yn dibynnu ar normau ein hymddygiad, a etifeddwyd, fwy na thebyg, o'n hynafiaid oesoedd maith yn ôl – agweddau o'r hyn yr ydym yn aml yn ei alw yn ddynol natur.

Ac mewn diwylliannau modern a chymdeithasau trefnedig, ar y deddfau a'r rheoliadau a ddyfeisiwyd gennym, yn aml mewn Seneddau fel hon, ac yn ogystal ar doreth o arferion, dylanwadau a chredoau. Heb reoliadau, cyfreithiau a safonau ymddygiad o'r fath, fyddai dim modd i unrhyw gymdeithas ddynol weithredu.

Datblygodd llawer o'r "rheoliadau" hyn dros yr ychydig ganrifoedd diwethaf wedi'r Chwyldro Diwydiannol, er bod gan ein credoau a'n hathroniaethau crefyddol, wrth gwrs, wreiddiau dyfnach.

Gadewch i mi bwysleisio rhai pwyntiau hanfodol.

Yn gyntaf, wrth i fwy o ynni gael ei ecsploetio, **bydd modd gwneud mwy o waith mewn amser penodol**.

Felly bydd digwyddiadau yn dilyn ei gilydd yn gyflymach fyth – hynny yw, bydd popeth yn cyflymu'n gyson.

Yn wir, o gofio'r cynnydd enfawr a wnaeth dynoliaeth o ran ecsploetio ynni yn y 200 mlynedd wedi'r Chwyldro Diwydiannol, does dim syndod fod pobl yn sôn am y "cyflymu mawr" – rydym wedi byw trwy'r profiad. (gweler Ffigyrau 2.3 a 4)

Yn ail, mae cydberthynas rhwng gwybodaeth ac ynni. Mae entropi cynyddol yn golygu cynnydd mewn anhrefn neu ddiffyg gwybodaeth ddiffiniol.

Ar y llaw arall, awgrymodd Erwin Schrodinger mai nodwedd celloedd – uned sylfaenol pob system fyw ddeinamig – yw cronni entropi negyddol, sef bod mwy o ynni a mwy o wybodaeth wedi'u gwreiddio ynddynt ac mewn cymhlethdod cynyddol.

Yn drydydd, o edrych ar hanes bywyd ar y Ddaear sy'n ymestyn dros 4 biliwn o flynyddoedd, mae modd gweld patrwm o ddilyniant o newidiadau allweddol o ran dal a defnyddio ynni.

Does dim dichon hyd yn oed amlinellu'r rhain heno, ond dylwn ddweud fod gan y tri olaf hyn effaith uniongyrchol iawn ar ein dyfodiad ni; esblygiad *Homo sapiens*.

Yn gyntaf, buddsoddodd ein cyndad *Homo erectus* rhyw filiwn a hanner o flynyddoedd yn ôl mewn ynni i gynyddu pŵer ymenyddol. Gosododd y buddsoddiad hwn mewn nerth ymenyddol yn hytrach na nerth bôn braich hominidiau ar lwybr newydd. Un o'r dulliau ymwneud cymdeithasol mwy soffistigedig hyn, fwy na thebyg, oedd iaith, yn ogystal â thechnolegau newydd. Yna rhyw 8 i 10 mil o flynyddoedd yn ôl, golygodd y chwyldroadau amaethyddol fwy o galorïau bwyd ar gael i alluogi i boblogaethau mwy dwys a chymhleth ddatblygu gyda phoblogaethau mwy a mwy sefydlog. Arweiniodd hyn yn ei dro at yr angen i gadw cofnodion, ysgrifennu a rhifedd. A thros y 250 mlynedd diwethaf, bu'r Chwyldro Diwydiannol, a yrrwyd trwy losgi tanwyddau ffosil, yn drawsnewidiol fel y trafodwyd eisoes.

O ganlyniadau i'r camau breision hyn, er nad oes arnom ni, fel ein cyndeidiau ryw filiwn o flynyddoedd yn ôl, ond angen tua 2,000 i 2,500 o galorïau bwyd y dydd o ynni i gynnal ein hiechyd metabolig sylfaenol, mae ein dulliau o fyw, ar gyfartaledd, bellach yn dibynnu ar yr hyn sy'n cyfateb i gan gwaith hyn o ynni y dydd!

Yn olaf, mae'r newidiadau allweddol hyn mewn ynni wedi creu posibiliadau newydd. Buddsoddi ym mhŵer yr ymennydd, lleferydd, ysgrifennu, rhifedd etc. sydd wedi trawsnewid yr hyn y gall creadur ar ffurf epa gyflawni.

Ymddengys hefyd fod pob posibilrwydd newydd yn caniatáu i egin-elite newydd i ffynnu, gan mai hwy all fanteisio orau ar y cyfleoedd newydd.

Yn fy marn i, mae'r ffiseg waelodol – y thermodynameg – a thystiolaeth bioleg esblygiad, anthropoleg a'n hanes cymdeithasol ac economaidd oll yn gyson â'r gosodiadau hyn.

Dyma hypothesis sylfaenol fy llyfr, 'Energy the Great Driver. https://www.uwp.co.uk/book/energy-the-great-driver/ ac yn y llyfryn.

Dylwn nodi fod y gosodiadau hyn, sydd wedi'u symleiddio, yn tynnu ar waith rhai o wyddonwyr mwyaf blaenllaw'r ganrif ddiwethaf: Wilhelm Ostwald, Erwin Schrodinger, Ilya Prigogine, Daniel Kahneman a'r Arglwydd Rayleigh – oll yn enillwyr Gwobrau Nobel – yn ogystal â'r niwroffisiolegydd Antonio Damazio.

Gwelaf fod y dyfaliadau hyn yn sefydlu fframwaith o ddealltwriaeth.

Fy mhrif nod heno yw ystyried beth allai'r fframwaith hwn olygu i ni a'n planed yn awr ac yn y dyfodol agos.

I ail-adrodd fy ngosodiad sylfaenol:

Po fwyaf o ynni y gallwn ddefnyddio neu ddefnyddio lawer yn fwy effeithlon – er enghraifft, oherwydd pŵer ein hymennydd a'n hoffer ac yn awr, fwy na thebyg, DA (AI) – mwyaf yn y byd o waith y gallwn wneud a'r mwyaf y gellir harneisio pŵer fel y gellir creu'r cymhlethdod deinamig trefnedig ond ansefydlog, gyda'r posibiliadau a'r problemau newydd a ddaw yn sgil hynny.

Mewn cyfres o newidiadau allweddol mawr yn y blaned a llawer iawn mwy o gamau llai, mae'r cymhlethdod cynyddol hwn wedi galluogi cyfleoedd newydd i ymddangos.

Yn nhermau dynoliaeth, nodweddir hyn gan ein hymenyddiau, ein lluniadau cymdeithasol, materol ac yn fwy diweddar, lluniadau economaidd, a'n dinasoedd mawr a chelfyddyd. Ond yn anad dim, ein hymwybyddiaeth, ein hathroniaethau a'n credoau.

Ond gyda'r cymhlethdod cynyddol hwn, cynyddu'n enfawr hefyd wnaeth cyfradd y newid.

Mae hefyd wedi creu angen am gynnydd cyfatebol mewn mecanweithiau rheoleiddio er mwyn sefydlogi'r cymhlethdod. Mae popeth bellach yn newid fel mellten. Er, yn ein hanes biolegol, fod esblygiad rheoleiddio homeostatig biolegol mor effeithlon fel ei fod bron yn anweledig o ddydd i ddydd, mae rheswm da yn awr dros gredu nad yw sefydlogi rheoleiddiol effeithiol bellach yn cadw i fyny â chyfradd y newid.

Beth yw rhan y damcaniaethau hyn yn ein byd modern sy'n wynebu cynhesu byd-eang yn ogystal â phroblemau amgylcheddol mawr (gweler Ffigwr 1, heb sôn am heriau cymdeithasol?

Does dim dwywaith fod ein cymdeithas, ein golud, ein safon byw yn llwyr ddibynnol ar ddigonedd o ynni rhad, cyfleus: sef llosgi tanwyddau ffosil – glo, olew a nwy, a bwyta bwyd rhad.

Bu cynnydd enfawr yng nghyfoeth materol y byd fel sy'n cael ei fesur gan GNP (Ffigwr 2), ac yn y boblogaeth ddynol (Ffigwr 3). Cynyddodd yr olaf hwn o ryw 800 miliwn yn y 1770au i 8 biliwn yn awr.

Mae'r defnydd o ynni o ran cyfanswm a fesul y pen wedi codi i'r entrychion. Mae llawer iawn o nwyddau megis dur, plastigion, sment a bwyd yn dilyn y duedd hon. Yr ydym oll yn dibynnu ar fwyd – ar ynni

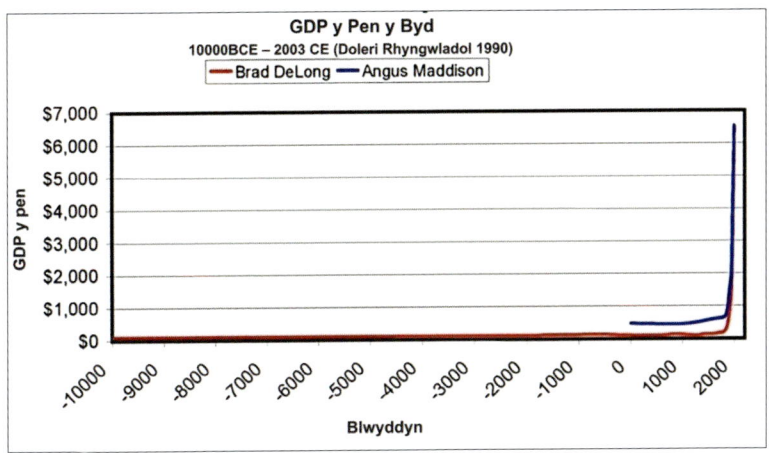

Ffigwr 2. Amcangyfrifon o Gyfoeth Dynol Cyfartalog o'i fesur fel GDP y pen ers Oes yr Iâ ddiwethaf.
https://www.google.co.uk/search?q=world+gdp+history+chart&tbm=isch&tbo=u&source=univ&sa=X&ved=2ahUKEwjTwfqxkL7eAhUDaVAKHdyiCsgQsAR6BAgFEAE&biw=1920&bih=887#imgrc=nDa6wP58RoifAM:

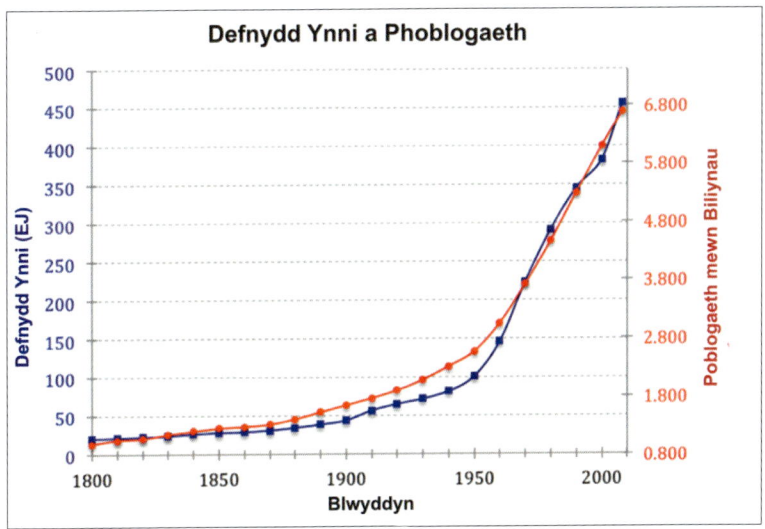

Ffigwr 3. Twf ym Mhoblogaeth y Byd a Defnydd Ynni ers 1800
https://www.e-education.psu.edu/earth104/node/1347

caloriaidd rhad yn ogystal â maethynnau.

Mwy o ynni; mwy a mwy o stwff, yn gynt ac yn gynt!!

Oherwydd bod cydberthynas gref rhwng ffyniant a chyfoeth â defnydd ynni, ymladdwyd rhyfeloedd dros a thrwy gyflenwadau ynni fel yn Wcráin yn awr. Mae adnoddau wedi cael eu dwyn neu eu meddiannu. Gwnaeth rheoli ynni wledydd ac unigolion yn hynod gyfoethog.

Mae gan reolwyr olew Saudi Arabia neu Irac, nwy Qatar neu Rwsia bŵer anhygoel. Bu hyn yn wir unwaith am lo Cymru – glo fu ar flaen y gad am 50 mlynedd – gyda Chaerdydd a'r Gyfnewidfa Lo yn ffynnu (gweler Ffigwr 4).

Ynni yw'r rheswm sylfaenol ein bod yma heddiw ym Mae Caerdydd ac nid ym Machynlleth fel y breuddwydiodd Jan Morris unwaith.

Tanwydd ffosil yn llythrennol oedd wedi tanio ein byd gyda'i ystumio

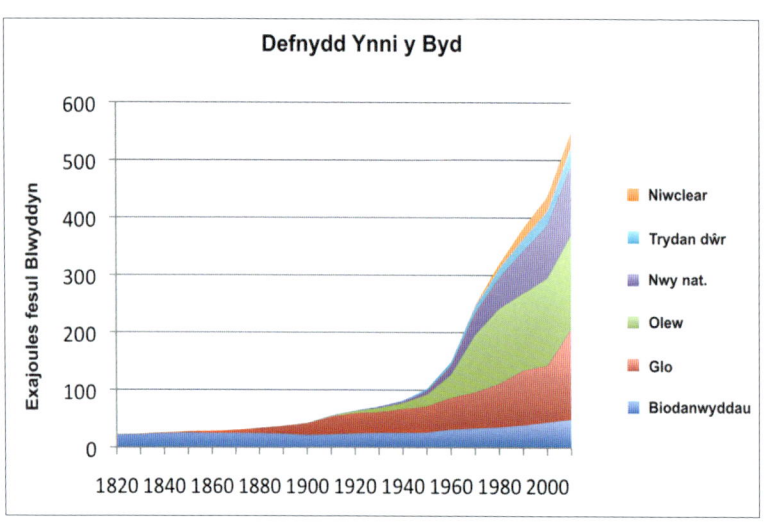

Ffigwr 4. Newidiadau yng nghyfanswm defnydd y byd o ynni ac mewn ffynonellau ynni ers y Chwyldro Diwydiannol cynnar.
https://ourfiniteworld.com/2012/03/12/world-energy-consumption-since-1820-in-charts/

enfawr ar gyfoeth a'i lygru.

Fe wyddom bellach y tu hwnt i amheuaeth resymol fod rhyddhau nwyon tŷ gwydr, carbon deuocsid [CO_2] yn bennaf, methan ac ocsid nitrus o losgi tanwydd ac o amaethyddiaeth a newid y defnydd o dir yn achosi cynhesu byd-eang a newid hinsawdd. Mae hyn yn digwydd ar raddfa a chyfradd sy'n bygwth y drefn economaidd fyd-eang a gwead cymdeithasol sylfaenol ein planed. Ac y mae'r gwead hwnnw ei hun, yn eironig, wedi'i sylfaenu ar gyfoeth o danwydd ffosil.

Mae'n bwysig pwysleisio – ar waethaf yr holl sôn am wrthweithio newid hinsawdd, fod 80 % o'n hynni, yn fyd-eang ac yn y DG, yn dal i ddod o losgi tanwyddau ffosil. Rydym yn chwydu nwyon tŷ gwydr allan yn gynt nag erioed, gan beri newid sydyn yn yr hinsawdd a chynhesu byd-eang.

Yr ydym ni fel bodau dynol yn defnyddio mwy a mwy o adnoddau eraill ein planed gan gynnwys o ffotosynthesis ar draul ffurfiau eraill o fywyd a'u cynefinoedd. Mae hyn yn achosi colli bioamrywiaeth yn ogystal â llygredd cemegol a llychwino gan blastig.

Yr ydym oll yn dibynnu ar fwyd – ar ynni caloriffig rhad yn ogystal ag ar faethynnau – ac efallai y dylem ddychryn o gofio fod y gadwyn amaeth-fwyd yn cyfrif am 30 % o'r holl allyriadau byd-eang.

Mae cynhyrchu bwyd gennym yn dibynnu nid yn unig ar ynni'r haul a ffotosynthesis ond ar danwyddau ffosil i wneud gwrteithiau ac agrogemegolion ac i gludo, rhewi a phrosesu'r bwyd a fwytawn o'r byd benbaladr yn ogystal â chynhyrchu llawer o fethan ac ocsid nitrus, y naill a'r llall yn nwyon tŷ gwydr grymus iawn.

Mae amaethyddiaeth ar raddfa fawr wedi'i galw yn system i droi olew yn fwyd.

Mae'r galw cynyddol ledled y byd am ynni yn golygu mai prin y mae'r gwaith o osod seilwaith newydd i ynni adnewyddol yn cadw i fyny â'r galw cynyddol a diwedd hen weithfeydd niwclear.

Y gobaith yw bod pethau'n newid, ond bron yn rhy araf i osgoi llanast mawr.

Yn sicr, gwrthweithio cynhesu byd-eang anthropogenig yw un o fethiannau mwyaf dynoliaeth.

Yn fras iawn, mae tri math o ymateb ar gael.

Yn gyntaf. Gwadu, dilorni a hau amheuaeth – twyll yw'r cyfan, neu wedi'i orliwio'n fawr. Rydym yn gweld hyn yn arbennig yn UDA sy'n dilyn Trump. Yn y DG, nodweddir hyn gan y Sefydliad Polisi Cynhesu Byd-eang, y lobi gwrth-sero net, Farage a'i gydweithwyr yn Reform, a llawer iawn o'r wasg aden-dde yn Llundain.

Yn ail. Mae'r brif ffrwd wleidyddol, gan gynnwys llywodraeth bresennol y DG, yn cytuno fod arnom angen gostyngiadau mawr mewn allyriadau nwyon tŷ gwydr, cynnydd mawr mewn ynni adnewyddol ac, yn fwy na thebyg, trydan niwclear. Yn y dyfodol agos, trydan o ymholltiad ac yn nes ymlaen, gobeithio, ymasiad. Mae'r brif ffrwd hon yn cydnabod yn gynyddol y gall dibyniaeth ar ddal a storio carbon a geobeiranneg planedol ddod yn angenrheidiol wrth i bethau waethygu'n wirioneddol.

Amcan cyffredinol a phwysicaf y brif ffrwd yw sicrhau parhad y model economaidd presennol seiliedig ar dwf blynyddol, diderfyn mewn GDP – bod hynny yn y DG a ledled y byd. Mae hyn oherwydd bod twf sy'n cael ei arwain gan y defnyddiwr yn cael ei weld yn un o ragofynion sefydlogrwydd cymdeithasol a lleihau tlodi.

Felly ie! Gadewch i ni dorri allyriadau – ond, ond, ond!

Gwnaeth y weinyddiaeth Doriaidd ddiweddar gyn lleied ag sydd modd ar waethaf ymlyniad rhethregol at sero net. Rhaid aros i weld pa mor gyflym y bydd Llafur dan Starmer yn gweithredu. Pwerau cyfyngedig yn unig sydd gan Lywodraeth Cymru. Mae China yn buddsoddi symiau enfawr mewn ynni adnewyddol, ond mae tanwyddau ffosil yn dal yn hollbwysig. Yn India, glo sy'n rheoli o hyd. Yn Rwsia, maent yn gobeithio y gall Siberia, efallai, elwa o newid hinsawdd.

Yn gyffredinol, bu newid arwyddocaol yn rhy araf o lawer. Ond i wneud pethau'n waeth, os bydd Trump yn ennill etholiad UDA, does wybod beth fydd yn digwydd. Bydd UDA, fel y pŵer mwyaf yn y byd, wedyn yn ymuno â rhengoedd y sawl sy'n gwadu ac yn twyllo.

Yr hyn sy'n bwysig i'w gofio yw bod y sawl sy'n elwa ar y sefyllfa geo-wleidyddol bresennol a'r model economaidd sy'n hollbresennol, mewn un ffordd neu'r llall, oll am i bethau fynd ymlaen fel y maent. Maent am gadw eu statws a rhaid i ni ein cyfrif ein hunain yn rhan o'r grŵp hwn.

Yn naturiol, mae'r rhai llai datblygedig a'r sawl a adawyd ar ôl yn benderfynol o ddal i fyny.

Yn drydydd. Efallai y gallaf gategoreiddio'r trydydd grŵp fel y rhai sy'n anobeithio, a rhai yn filwriaethus yn eu hanobaith.

Maent yn syniied ein bod ar y llwybr i gynnydd enfawr yn nhymheredd cymedrig y byd, i fynd heibio trobwyntiau o bwys, a thuag at anhrefn byd-eang. Mae ganddynt reswm da dros gredu hyn. Mae rhai yn filwriaethus; eraill yn syml wedi anobeithio.

O gofio pŵer a dylanwad gwleidyddol ac economaidd y brif ffrwd a'u rhagdybiaethau dealledig am ddefnyddio mwy fyth o ynni , hoffwn i ni feddwl mwy am eu safbwynt a sut y mae a wnelo â'm damcaniaeth.

Fel y crybwyllais, y rhagdybiaeth sylfaenol yw y gall twf, fel y'i mesurir gan GDP, ddal i'n gwneud yn gyfoethocach, yn fwy ffyniannus, yn fwy bodlon – ac y bydd y tlodion yn raddol yn dod yn llai tlawd wrth i dwf rhad sy'n ddibynnol ar ynni godi lefel cyffredinol ffyniant. Datblygiad diferu-i-lawr byd-eang a chenedlaethol.

Petaech am weld mynegiant dilyffethair o'r senario twf hon, buaswn yn awgrymu eich bod yn darllen maniffesto'r tecno-optimistiaid gan Marc Andreessen, un o ffigyrau amlycaf Silicon Valley; mae bellach yn un o brif gefnogwyr Trump. https://a16z.com/the-techno-optimist-manifesto/

I ddyfynnu:

"Rydym ni'n credu y dylai ynni fod yn sbiral tuag i fyny. Ynni yw'r peiriant sy'n gyrru ein gwareiddiad. Po fwyaf o ynni sydd gennym, mwyaf o bobl y gallwn gael, a bydd bywyd pawb yn well. Dylem godi pawb i'r lefel o ddefnyddio ynni sydd gennym ni, wedyn cynyddu ein hynny 1,000x, ac yna godi ynni pawb arall 1,000x hefyd."

(Maniffesto'r Tecno-Optimist, Marc Andreessen).

Mae maniffesto Andreesen fel "doethineb y dorf" ar steroids.

Er bod Andreessen a minnau yn cytuno mai ynni yw'r 'peiriant sy'n gyrru ein gwareiddiad", nid ydym yn gytûn ar fawr ddim arall.

Mae'n hanfodol ystyried sut y mae a wnelo fersiwn remp Andreessen neu'r fersiwn brif-ffrwd fwy cymedrol sy'n cael ei harddel gan bron bob gwleidydd a'r rhan fwyaf o economegwyr, â'm dyfaliadau.

Gadewch i ni ragdybio am ennyd fod problem newid hinsawdd, trwy ryw ryfedd wyrth, wedi ei "datrys". Ond fe ddown yn ôl at hynny yn y man.

Beth felly fydd ystyr y 'sbiral ynni tuag i fyny' Andreessen hyd yn oed os na fydd unrhyw gyfraniad at newid hinsawdd?

Rwy'n awgrymu:

A] Mwy o gymhlethdod deinamig yn faterol ac mewn cymdeithas.

B] Mwy o gynnydd mewn galwadau dynol ar adnoddau'r blaned hon, gan gynnwys ffotosynthesis, fwy na thebyg. Bydd hyn ar draul organebau eraill a'u cynefinoedd. Mae Andreesen hyd yn oed yn rhagweld twf ym mhoblogaeth dynoliaeth yn y byd; pob un â'i alwadau ffisegol a chymdeithasol.

C] Bydd cyflymiad pellach yng nghyflymder y newid yn ogystal ag mewn cymhlethdod deinamig y n anorfod. Mae'n debyg y caiff effeithiau dwfn ac y bydd yn tarfu ar ein bywydau cymdeithasol yn ogystal â'n bywydau economaidd, ac ar ein lles seicoleg a'n hiechyd meddyliol. Mae tystiolaeth eisoes yn bodoli am broblemau iechyd meddwl cynyddol, yn enwedig ymysg y genhedlaeth iau.

D] Fel y sylweddolodd Joseph Schumpeter, wrth graidd cyfalafiaeth a'r model twf parhaus mae 'dinistr creadigol'. Bydd cylch dinistr a chreu yn troi'n gynt a chynt: yn wir, bydd yn rhaid iddo wneud wrth i ynni droelli tuag i fyny ac i dechnolegau, dyfeisiadau a ffasiynau newydd gymryd lle eu rhagflaenwyr.

E] Bydd y cymhlethdod newydd yn creu posibiliadau newydd gaiff eu croesawu gan y lleiafrif sydd yn y sefyllfa orau i fanteisio arnynt – daw elite grymus newydd i'r golwg. Mae'n fwy na thebyg mai byd DA a robotiaid fydd hwn, ac efallai cyborgiaid a hybridiau. Sglodion wedi'u mewnblannu yn ein hymennydd, hwyrach?

F] Wrth i gymhlethdod gynyddu fwyfwy, rhaid cael systemau rheoleiddio newydd, yn lleol a ledled y byd, i sicrhau sefydlogrwydd, neu i geisio gwneud hynny o leiaf. Mae rhesymau da dros gredu nad yw ein systemau rheoleiddio eisoes yn gallu cadw i fyny â chyfradd y newid, e.e., rheoleiddio'r cyfryngau cymdeithasol a DA. Mae'n amheus a all ein systemau democrataidd presennol ar gyfer cynhyrchu cyfreithiau a rheoliadau newydd gadw i fyny a'r newid fydd yn digwydd yn gynt nag y mae yn awr.

G] Mae'n fwy na thebyg y bydd y rheoleiddio ychwanegol ar draul ein hawliau a'n rhyddid, ac o bosib ein democratiaethau.

H] Bydd y grwpiau elite newydd fydd yn ymddangos er mwyn manteisio i'r eithaf ar y cyfleoedd newydd yn debygol o ddefnyddio TG a DA i gadarnhau eu statws.

Efallai mai dyma beth mae Andreesen eisiau ac a fydd yn cael ei ganiatáu, yn ddiarwybod efallai, gan y brif ffrwd, ond y cwestiwn hanfodol yw:

Ai dyma beth mae pobl yn gyffredinol eisiau? Ac a yw hyn er eu lles? Ai dyma'u breuddwyd iddynt hwy a'u plant?

Fel y dywedais, mae fy namcaniaeth yn awgrymu y caiff y byd hwn fydd yn uchel mewn ynni effeithiau niferus, ac yn fy marn i, negyddol, ar ein dynoliaeth, ein rhyddid, yn ogystal ag ar fioamrywiaeth ac iechyd cyffredinol y blaned.

Efallai i chi gofio fod fy namcaniaeth yn pwysleisio fod pob ffurf newydd mwy cymhleth o fywyd yn dal yn ddibynnol ar iechyd ei holl ragflaenwyr.

Fy safbwynt i yw bod yn rhaid i ni, yn y byd go-iawn, ymdopi â dau fater gwahanol ond sy'n gyd-gysylltiedig.

Y cyntaf, wrth gwrs, yw newid hinsawdd oherwydd tanwydd ffosil. Mae'n fygythiad cydnabyddedig, real a chynyddol, er yr ymddengys nad yw'n un y gallwn ymateb iddo. Serch hynny, tynnaf sylw yn awr at fygythiad arall, na chafodd gymaint o sylw. Bygythiad mwy llechwraidd yw hwn sy'n deillio o allu dynoliaeth i gyrchu a harneisio ynni **o unrhyw ffynhonnell** i wneud gwaith, creu mwy o bŵer a chyflymu cymhlethdod gyda'r angen cyfatebol i reoleiddio fwyfwy. Fe erys y bygythiad hwn os llwydda dyfeisgarwch dynoliaeth i fanteisio ar ffynonellau ynni nad ydynt yn allyrru nwyon tŷ gwydr. Y paradocs yw y bydd yn cynyddu os daw dynoliaeth o hyd i'r allwedd i ynni rhad diderfyn, o ymasiad niwclear, dyweder, sy'n aml yn cael ei hyrwyddo fel y llwybr tuag at dwf dilyffethair.

Tybed faint o bobl erbyn hyn sy'n gwirioneddol gredu yn y freuddwyd hon o gynnydd materol anorfod a pharhaus.

Yn fy marn i, daw cyfyngiadau'r freuddwyd *economicus* yn fwyfwy amlwg; felly hefyd natur wenwynig yr allyriadau nwyon tŷ gwydr sy'n deillio o danwydd ffosil, yn ogystal ag effeithiau llai amlwg a thanseiliol yr ecsploetio cyflym ar ynni.

Hoffwn bwysleisio fod y natur ddynol yn fwy cymhleth, ac yn wir yn fwy cyfoethog a gofalgar, na'r hyn a ddywed *Homo economicus*.

Mae ein heconomi yn lluniad sy'n ddibynnol ar ynni ond sydd wedi ei osod mewn Daear gyfyngedig, fel y rhagwelwyd gan Bob Ayres, Kate Rawoth ac eraill (Ffigwr 5.)

A yw'r darlun felly yn hollol ddu?

Wel, yn rhyfedd iawn, nac ydi. **Y mae dewis arall, yn wahanol i TINA ('there is no alternative' Mrs. Thatcher!!**

Nid yw popeth ar ben er fy mod, wrth gwrs, yn cydnabod maint yr heriau.

Fy ngobaith cyntaf yw y bydd y fframwaith dealltwriaeth y bûm i'n ei hyrwyddo yn symbylu dadl fywiog a deallus ac y bydd hyn yn ein harwain, fel y dymunodd yr athronwyr Groegaidd filoedd o flynyddoedd yn ôl, at adnabod ein hunain yn well.

Gobeithio y bydd y digwyddiad hwn yn y Senedd yn help i sbarduno'r ddadl hon.

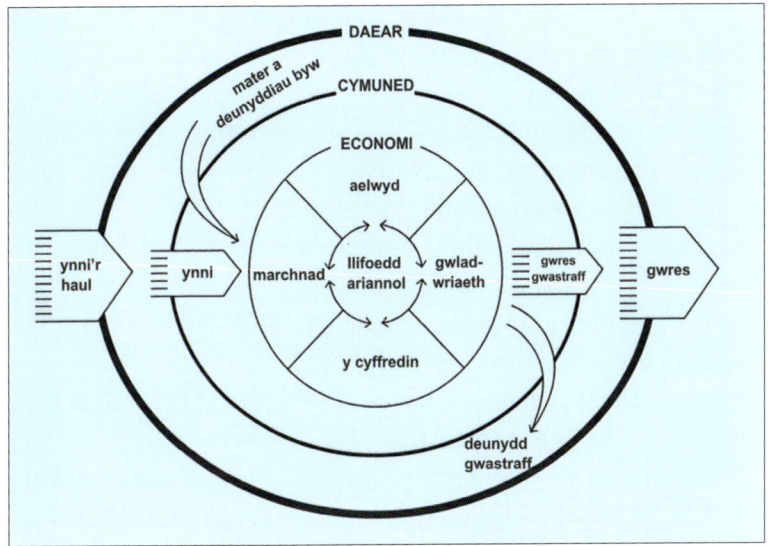

Ffigwr 5. Model o'r Economi wedi'i yrru gan Ynni ond o fewn ffiniau adnoddau'r Ddaear
Kate Raworth. (2018) *Doughnut Economics: Seven ways to think like a 21st Century Economist.* Radom.

Mae llawer bellach yn beirniadu'r model o dwf wedi'i arwain gan ynni o wahanol safbwyntiau, ond hyd yma, lleiafrif ydynt, ac nid oes neb wedi llwyddo i'n hargyhoeddi sut i atal neu arafu cyflymder cynyddol y felin draed economaidd heb greu chwalfa.

Yr hen fodel sy'n dal i dra-arglwyddiaethu, yn rhannol oherwydd ei lwyddiannau hanesyddol go-iawn, ac yn rhannol oherwydd bod llawer yn credu nad oes dewis arall. Serch hynny, ymddengys fod y model ar ddiffygio.

Y paradocs yw y gall agweddau o'r argyfwng newid hinsawdd ddangos y ffordd ymlaen.

Mae'n amlwg fod mwy o ynni isel o ran allyriadau yn hanfodol. Yn benodol, yn y gwledydd mwy cyfoethog, mae angen trydan i wresogi, ar gyfer trafnidiaeth a diwydiant.

Yn anffodus, mae anfanteision a chyfyngiadau i bob ffynhonnell adnewyddol – rhai effeithiau negyddol sy'n rhoi bod i ffenomen 'nimby' ac arafwch cynnydd.

Mae ymholltiad niwclear yn arswydus o ddrud, gyda llawer o anfanteision, gan gynnwys galw trwm am ddŵr i oeri, a phroblem tymor-hir storio gwastraff hynod ymbelydrol.

O gofio amserlen cynhesu byd-eang, ni fydd modd ei gyflwyno mewn pryd yn y DG, heb sôn am ledled y byd.

Yn wir, a minnau wedi gweithio yn Aleppo a llefydd eraill megis tiriogaeth Boco Haram yng ngogledd Nigeria, mae'r syniad o ddegau o filoedd o adweithyddion niwclear, Adweithyddion Modiwlaidd Bach neu beidio, ledled y byd yn rhy frawychus i'w ddirnad.

Breuddwyd at y dyfodol pell yw ymasiad rhad. Y mae potensial i hydrogen gwyrdd, ond bydd hefyd yn mynnu seilwaith newydd anferth.

O gofio'r cyfyngiadau sylfaenol hyn, onid yw'n rhesymol yn economaidd ac yn gymdeithasol i weithio er mwyn adeiladu cymdeithasau lle gall pobl fyw yn dda gan ddefnyddio llai o ynni?

Oni all hyn fod yn ffordd effeithiol o wrthweithio newid hinsawdd yn sydyn ac osgoi elfennau gwaethaf cymdeithas fodern ac effeithiau negyddol 'sbiral ynni' Andreesen?

Pam gwastraffu biliynau ar gynhyrchu o'r newydd os mai gwell fyddai ei wario ar leihau'r galw?

Gwella ac insiwleiddio ein stoc tai a'n swyddfeydd ac ati – oni ddylai POB adeilad newydd fod yn garbon niwtral ac yn isel o ran defnydd ynni?

Oni ddylem roi blaenoriaeth i ddarparu gwell cludiant cyhoeddus a gwneud llai o ddefnydd o geir? Mae gennym ni gar trydan, ond mae creu ceir felly yn dal yn gwneud defnydd gweddol ddwys o ynni a deunyddiau. Annog cerdded a beicio, gwella ein mannau a'n gwasanaethau cyhoeddus, mwynhau gwyliau lleol, ein bywyd gwyllt a'n glan môr a chefn gwlad, a buddsoddi yn ein bywyd diwylliannol.

Rhaid i ni ofyn i ni'n hunain, oni allwn fyw'n well ar lai o ynni?

Mae'n gwestiwn i'w ofyn i'n timau chwaraeon. Allwch chi wir gyfiawnhau hedfan i Dde Affrica mor aml?

Os byddwn yn lleihau'r galw, byddwn yn lleihau problem cynhyrchu trydan ychwanegol, ond nid yn ei dileu. Efallai y bydd mwy o gynhyrchu lleol yn gostwng yr angen am wifrau trawsyrru newydd, ond eto nid yn ei ddileu.

Nid syniadau newydd mo'r rhain. Ond nid dadlau am fanylion yr ydw i heno, ond am ffordd gyffredinol newydd o feddwl.

Breuddwyd, nid am sut i gael mwy a mwy ond sut y galla'i fyw'n well ar lai o ynni?

Mae'r weledigaeth hon yn arwain yn rhesymegol at bwyslais ar y gymuned leol i wrthweithio grym canolog busnesau mawr a dinasoedd mawrion.

Mae'r weledigaeth yn pwysleisio cyfrifoldeb a rhyddid personol, ynghyd a chydweithredu ar lefel gymunedol.

Mae'n brosiectau ynni adnewyddol cymunedol fel Ynni Ogwen ac Anafon nid Amazon.

Nid yw newid meddylfryd fel hyn yn syml ac fe fydd yn costio arian: buddsoddiad cyhoeddus a chydweithredol ar raddfa fawr.

Yn sicr bydd yn golygu mwy o drethi ar y cyfoethog, lleihau triciau i osgoi treth a gwneud i ffwrdd â llochesi treth. A chydweithredu gyda'n cymdogion Ewropeaidd.

Serch hynny, daliaf fod gweledigaeth amgen i'r un o dybio bod lles a ffyniant dynol yn ddibynnol ar dwf diderfyn mewn defnyddio, fel sy'n cael ei fesur yn ôl GDP y pen. Y peth pwysig yw ei bod yn cynnig gobaith am dorri'n sydyn ar allyriadau nwyon tŷ gwydr, yn ogystal â dangos llwybr i osgoi peryglon gwaethaf y 'sbiral ynni'.

Gadewch i mi grynhoi.

Yn ddiamau, ynni yw'r *peiriant sy'n gyrru ein gwareiddiad*, a dyfynnu Marc Andreessen; yn wir, dyna sy'n gyrru bywyd yn ei gyfanrwydd. Yn fy llyfr a'r llyfryn a gyhoeddir heddiw, rwy'n awgrymu bod ffyniant dynol yn ganlyniad i gyfres o esblygiadau mewn ynni ar y blaned dros y 4 biliwn o flynyddoedd a aeth heibio.

Er hynny, mae rhagamcanion a rhagdybiaethau Andreessen, ac yn wir rhai'r brif ffrwd gymedrol, yn sylfaenol wallus ac yn wir yn wenwynig.

Cyfyd y peryglon o ddwy ffynhonnell fras. Y ffaith fod ein cymdeithas yn beryglus gaeth i ynni rhad, cyfleus o danwydd ffosil sydd yn rhoi pŵer enfawr a milain i'r darparwyr. A hefyd y cymhlethdod sy'n gynyddol ddibynnol ar ynni, gan gyflymu newid a heriau rheolaethol, sy'n brofiad beunyddiol i ni ac yn gwneud difrod sylweddol i'r ymwneud a'r cylchdroadau disylw sy'n hanfodol i iechyd y ddynoliaeth a'n planed (gweler Ffigwr 1).

Hyd yn oed os llwyddwn i roi ar frys yn lle ynni o danwydd ffosil ffynonellau eraill sy'n llygru llai – ac nid wyf yn gweld hynny'n digwydd – y bydd yn rhaid i ni o hyd fynd i'r afael â'n perthynas ag ynni, pŵer, cymhlethdod a chyflymder newid. Awgrymu yr wyf, fel gyda llawer o bethau, mai 'gormod o ddim nid yw dda' – yn wir, gall fod yn wenwynig.

Felly rwy'n awgrymu mai ceisio adeiladu cymdeithas sy'n isel mewn ynni ond sy'n ffyniannus yw'r unig ffordd ymlaen o ran cynhesu byd-eang a'r sbiral ynni.

Bydd hyn yn gofyn am flaenoriaethau economaidd newydd fel dewis amgen yn lle GDP, a phrin yr wyf wedi cyffwrdd â hwy.

Yn sicr, byddwn yn wynebu anawsterau mawr wrth geisio gwireddu'r freuddwyd hon.

Crybwyllaf dri ohonynt yn unig.

Caiff y freuddwyd ei gwrthwynebu'n ffyrnig gan y pŵer sydd gan gyfoeth mawr, fel y gwelir yn achos Andreessen a'i gydweithwyr mewn TG. Hwy sy'n cyllido Trump, a gallant ledaenu camwybodaeth a gweld eu hunain fel yr elite sy'n elwa o'r sbiral ynni.

Yn ail, dim ond mewn byd tecach a mwy cyfartal y gellir gwireddu'r freuddwyd. Mae pobl dlawd iawn, hyd yn oed yng Nghymru, sy'n defnyddio ychydig iawn o ynni y pen a heb allyrru llawer o nwyon tŷ gwydr. Mae ganddynt angen dilys am lawer o ynni – mwy o ynni sy'n rhydd o nwyon tŷ gwydr. Heb fwy o degwch a chydraddoldeb lleol a byd-eang, fe awn i'r gors.

Yn olaf, rwy'n cydnabod y bydd technolegau newydd yn hanfodol er mwyn datrys llawer o broblemau dynoliaeth, ond rhaid i ni warchod rhag hubris a bod yn effro o hyd i beryglon parhaus yr hyn sy'n hysbys ac yn anhysbys.

Gall canlyniad rhesymegol y patrymau a amlinellais yn hawdd olygu ymddangosiad tecno-elite neu hyd yn oed cyborgiaid, a all gamfanteisio ar y byd newydd hwn sy'n ddwys o ran gwybodaeth ac ynni, a thra-arglwyddiaethu dros y gweddill.

Senario ddystopaidd teilwng o Orwell yw hyn i mi.

O ystyried ei bod yn weddol sicr y bydd tymheredd cymedrig y byd yn codi dros 2°C ac y byddwn yn pasio trobwyntiau pwysig, e.e., achosi cynnydd mawr yn lefel y moroedd, a rhanbarthau yn mynd yn rhy boeth a llaith i ddynoliaeth, bydd unrhyw fuddugoliaeth gan dwf technoleg yn digwydd ar yr un pryd ag anhrefn a thryblith mawr ledled y byd – a gall hyn gyfyngu ar y twf hwnnw.

Os nad am unrhyw reswm arall, dylai hyn beri i'r sawl sydd o blaid ymasiad niwclear fel ateb i argyfwng ynni'r byd ail-feddwl. Hawdd iawn yw dychmygu byd mewn anhrefn gyda'r elite DA yn cael eu gwarchod gan filisia personol arfog yn ceisio diogelu eu hunain rhag y dyrfa – yn debyg iawn i'r modd y mae cymunedau y tu ôl i glwydi yn gwneud yn awr mewn rhai rhannau o'r byd.

Fodd bynnag, y mae ffordd arall – **breuddwyd amgen**.

Mae gennym ddewis – cyfrwng yw'r gair sy'n cael ei ddefnyddio'n aml. Gallwn ni fel dynoliaeth benderfynu mabwysiadu llwybr mwy synhwyrol, mwy cyfartal a llai dwys o ran ynni.

A yw pobl yn wir eisiau bod yn gaethweision i'r elite DA newydd?

A yw pobl yn wir eisiau bod ar felin droed twf materol sy'n cyflymu fwyfwy?

Gall breuddwydion gael eu gwireddu, ac y mae hynny wedi digwydd!

Wrth gwrs, 'dyw llawer o'r syniadau hyn ddim yn newydd. Ceisio rhoi fframwaith cynhwysfawr yr wyf i, mewn gwisg wyddonol.

Bydd gwireddu breuddwyd newydd yn golygu y bydd yn rhaid i ni, fodau dynol, weithio mewn ffyrdd sydd wedi eu hesgeuluso i raddau helaeth yn y byd hwn o unigolyddiaeth, masnach gystadleuol a geowleidyddiaeth ymosodol.

Yn y pen draw, nid materion gwyddonol, technolegol neu hyd yn oed economaidd wleidyddol yw'r rhain, ond mae a wnelont â'n gwerthoedd dynol.

Alla'i ddim gwneud yn well na diweddu gyda cherdd Waldo Williams

Bydd mwyn gymdeithas.
 Bydd eang urddas.
  Bydd mur i'r ddinas.
      Bydd terfyn traha.

Addenda.
1. Ar Noswyl Nadolig 1967, lai na phedwar mis cyn iddo gael ei lofruddio, dywedodd Martin Luther King, "Hyn ydyw yn y bôn: fod bywyd oll yn gyd-gysylltiedig. Rydym oll wedi ein dal mewn rhwydwaith gyda'n gilydd nad oes modd dianc ohono, mewn un wisg o dynged. Mae beth bynnag sy'n effeithio ar dynged un yn effeithio ar bawb yn anuniongyrchol."
2. Ystyriwch y ddwy senario ynni isod.

Pe na bai gan y blaned hon gyfoeth o danwyddau ffosil, sut y gallasai cymdeithas fyd-eang fod wedi esblygu yn y bedwaredd ganrif ar bymtheg a'r ugeinfed petai'n seiliedig i raddau helaeth ar ffotosynthed cerrynt newydd o goedwigoedd a ffermio?

Petai pŵer atomig (ymholltiad), fel y rhagwelodd Lewis Strauss, y gwrtharwr yn y ffilm Oppenheimer, yn 1954 wedi cynhyrchu trydan 'rhy rad i'w osod ar fesurydd', beth fuasai'r canlyniad wedi bod i ogledd a de y byd ac i'n hamgylchedd?

3. Mae'n werth nodi y byddai swm yr ynni a fynnir dan y senario hon yn enfawr.

1 trafodiad Bitcoin = 27 diwrnod o ddefnydd trydan gan aelwyd arferol yn UDA 29 kWh; mae ChatBoxGBT eisoes yn defnyddio 0.5 miliwn kWh y dydd. Mae hyn yn dal i ddibynnu i raddau helaeth ar losgi tanwyddau ffosil!

4
The History of Energy

This chapter is comprised of eight invited short articles published on the Nation.Cymru web site in the summer of 2023.

Jon Gower wrote: "There are some books which come into a reviewer's hands which feel they are worth far more than a summary account. Gareth Wyn Jones' history of energy was certainly one of those.

He was invited to turn the substance of 'Energy, the Great Driver: Seven Revolutions and the Challenges of Climate Change' into a series of weekly articles for Nation.Cymru and we were delighted when he accepted. He both condensed much of the substance of his book but also found the necessary style to convey the urgency of his thinking to a wide readership."

The History of Energy

In this series of short articles, which update my book I will try and put our current problems in their long-term context and consider how, over many millions of years, the exploitation of energy, more correctly, energy transformations, has changed the face of Earth and our lives and, indeed, ask where might it all be leading us?

1 / We have a problem

Humans are inveterate storytellers, always have and always will be. Nowadays our stories usually reach us on our mobile phones or lap tops, historically they were shared around a fire.

But it is still these stories that colour and drive virtually every aspect of our lives. They can motivate us and help us overcome difficulties or reinforce our worse fears and lead to deep gloom and lethargy.

For a couple of centuries, we've been fed and inspired by the story of unending material growth. We call it called 'progress' – by which we mean acquiring more and more stuff, enjoying a consistent rise in our living standards in the expectation that this will make us all happier.

Historically it's a relatively new story but an exceptionally powerful one. Our forebears might have yearned for salvation and a place in heaven and possibly for personal worldly success and power.

But continuing, unceasing 'progress' as we imagine it today was not on the agenda – not part of human story telling. Now it's what we expect and think we deserve!

Stagnating

However in last decades in many countries such as the USA and UK including Wales, all this 'progress' has stalled. Only the very rich are enjoying 'growth' and 'progress'. Many, perhaps most, are stagnating. For many young people their lives are actually much harder than for earlier generations.

So is the dream dying? Our usual reaction is to put our faith in some new leader or a different political party who will, in all probability, be promising to renew, even enhance growth and that we (be it the Welsh, English or British) will soon become 'world leaders'.

But it's worth asking ourselves, at a much more profound level, can these stories we tell ourselves be true. They are fed to us, continually, by politicians from right and left and by the media but are they really credible?

The UK Government, especially post-Brexit, has seemed singularly inept, but there are far deeper questions, only tangentially related to the government of the day in London or Cardiff.

Solutions

Can we and our children, as citizens of Planet Earth living in Wales, expect to enjoy everlasting material growth? Are there the resources available on this planet to deliver plenty for over 8 billion of us? Or

Image: wikimedia

maybe, do we see ourselves as exceptionally deserving cases although we are living together on "Spaceship Earth"?

Will, and indeed can, our scientific and technological expertise and our amazing human inventiveness and enterprise always come up with solutions? Or, may be, we are facing a real turning point in human history?

Quite apart from the easy and often false promises of conventional politics, could we be, in reality, reaching an impasse, a tipping point?

Even worse, is it possible that the continual growth stories we swallow wholesale, are positively dangerous, even disastrous for us as individuals, for our children and grandchildren and for humanity generally?

Should we be heeding global warming as a canary in the mine warning us of the disasters ahead?

Prosperity

On the other hand poverty is endemic globally and, sadly, all too common in Wales. It is not surprising that many in the West, despite the evidence of stagnation over the last 30 to 40 years, put their faith in economic growth.

They see it, with good reason, as the only way to improve their lives, and those of their children. Even less surprising is the insistence of the less developed world on their right to greater prosperity and more of this earth's resources.

These contrasting pictures highlight a cruel and unforgiving dilemma. One that has, in my judgement, the potential to destroy our civilization. It hides a set of issues scarcely acknowledged or discussed; not part of the political discourse.

Cost of living

Ironically the truly horrific war in Ukraine has had one positive impact. It has brought home our critical dependence on energy. It is forcing us and the rest of Europe to rapidly and drastically reduce oil and gas imports from the aggressor Russia.

The war is causing European countries to decarbonise their economies at a rate not contemplated even when facing all the scientific evidence of impending catastrophic climate change and global heating and Greta Thunberg.

The threat to Ukrainian grain exports also highlighted humanity's dependence on grains as a major source of the energy in our diets and that of our animals.

All this has contributed to the current cost of living crisis, and to people suffering, possibly dying, of cold in modern Britain. Without cheap and plentiful energy, or possibly revolutionary ways of coupling energy to work, the economy stalls.

Is there, even remotely, a silver lining? Clearly we must start with an honest appraisal of where we stand and where we are likely heading.

Inadequacy

Let us therefore start thinking seriously about energy and its many roles and impacts.

While Putin's war has had an immediacy which even Greta Thunberg's brave advocacy failed to produce, it has, by default, highlighted the inadequacy of our responses to climate change and global warming.

Floods

One remarkable aspect of the underlying physics is that both the chemical energy in grain and in oil and gas and the energy we can capture from the sun's radiation or from winds and tides, are all encompassed and quantitively interrelated to each other in this one term, "energy".

Few now actually deny that our society's dependence on energy from burning of fossil fuels, initially coal, latterly oil and gas, and that the CO_2 so released, is the prime cause of global warming.

This is also behind the climate disasters visited on so many communities around the world – the searing heat waves, the horrendous fires, the devastating floods and drought and storms.

But, and it's a very big but, we depend on huge quantities of energy to sustain our society and our life styles. Despite all the talk, 80% of this still comes from burning fossil fuels.

As is well known, we need 2,500 to 3,000 food calories in our daily diet but our life-style requires perhaps 8 times as much energy, daily, to warm our homes and workplaces, to run our cars and move goods, to go shopping or on holiday, and is embedded in the goods we buy and the entertainments we enjoy (including powering the web) etc., etc.

2 / Some background on the laws we learnt in school but never really understood!

In my first article I used Russia's aggression in Ukraine to highlight society's dependence on energy whether as fossil fuels or food calories.

Of course, this dependence also helps explain the reluctance of countries, businesses and us as individuals, to take the necessary steps to avoid catastrophic global warming.

Human quest

We hear plenty about the urgent need to reduce our dependence on the fossil fuels that power our society that are leading to the massive greenhouse gas emissions which cause man-made global warming; often called Anthropogenic Global Warming (AGW).

But we are failing to act decisively. Moreover, we hear little about the underlying and fundamental role of energy in both biology and human culture.

A good place to start is with a quotation from the Canadian physicist, Vaclav Smil, apparently one of Bill Gates' favourite authors.

He wrote: "**All natural processes and all human actions are, in the** most fundamental physical sense, transformations of energy. Civilization's advances can be seen as a quest for the higher energy use required to produce increased food harvests, to mobilise a greater output and variety of materials, to produce more, and more diverse goods, to enable higher mobility, and to create access to virtually unlimited amounts of information**".

Smil is here expanding on a concept which dates from before the World War I. In 1912 the Nobel Laurate Wilhelm Ostwald wrote "**free energy is therefore the capital consumed by all creatures of all kinds and by its conversion everything is done**".

Let us underline the enormity of these assertions – **all processes** – **all actions** – **all creatures** – **all kinds**.

Every damn thing depends on energy transformations. But what of the term **free energy**, why **transformations** and why not just **energy**?

Fission or fusion

This takes us to some of the most fundamental laws in all science – the laws of thermodynamics. The first states 'energy cannot be created or destroyed'.

After Einstein and the discovery of atomic structures – we saw that changes in matter, either by nuclear fission (atom bombs) or fusion (hydrogen bombs), can lead to the release of vast amounts of energy.

This means that we needed to revise the First Law to 'Energy and Matter', taken together, cannot be created or destroyed.

Energy in its many and varied forms is defined by an ability to do work i.e. to make things change; be that to move clod or a tonne of earth, burn wood or coal to generate heat or cook or boil water to create steam or indeed the calories in our own food, which allow me to move my hand and to write and think, or the flow of electricity to turn a motor to run a fridge or a washing machine.

Fraud

In every case such energy transformations, as well as doing useful work, release a small amount of energy which is dissipated as low level heat e.g. from friction in a motor or the heat lost from our bodies (note: wear a hat on cold day as our brains use lots of energy!).

This lost energy is no longer able to do work. So, the term **free energy** refers to the part of the transaction that can do useful work while the small amount that is 'lost' i.e., can no longer do work, is termed **entropy**.

In any closed system entropy gradually increases and useful free energy deceases – this is why perpetual motion machines are always a fraud.

Fortunately, our planet Earth is not a closed system as, daily, the planet receives the radiant energy from thermonuclear reactions in the Sun. Not so much 'give us this day our daily bread' but give us this day our energy fix (which of course also helps provide our bread).

Disorder

The second law of thermodynamics can be paraphrased, as 'everything will run down into disorder unless free energy is put into the system' i.e., entropy tends continually to increase in given closed system.

At the level of molecules, say in a gas, entropy is also a measure of increasing internal disorder and randomness. So, when entropy is high, the system is more randomised and disordered. Conversely when entropy is very low or negative, it means the system is more ordered and more complex.

In our daily lives we intuitively realise a very similar phenomenon. If you don't put in the work i.e., use 'energy', then things tend to fall apart in our private lives, in the repair of our homes and socially.

But, of course, too much can also create a bull in china shop!

Entropy

An additional insight was provided by the famous physicist, Ludwig

Boltzmann.

He conceived entropy as a statistical phenomenon which means that it is closely linked with what we all obsess about now 'information technology'.

Information is defined by the number of simple binary (yes/no) decisions required to specify a unique condition or the state of a system.

Broadly the more decisions that are needed to define a given specific state or object, the higher the information content and the more ordered it is.

According to Boltzmann's formula, entropy can be interpreted in similar terms. But it's an inverse relationship – the better defined and more ordered a system, the lower its entropy.

Contradiction

However, in thinking about how these laws may apply on our world, there is an apparent contradiction.

The laws of thermodynamics mean that in any closed system e.g. the Universe, everything will gradually and irreversibly run down; giving us the arrow of time.

Although our Universe is expanding, it is a closed system in the sense that there is nothing outside beaming in energy or matter.

So, the entropy of the Universe must be increasing, leading, over eons of time, to the so-called 'heat death' of the Universe i.e. an order-less, cold and dead Universe, including planet Earth.

But the Earth we see daily is alive, complex, full of beauty and with a large measure of order. How come?

Scary

The first key point is that the Earth is not a closed system. We are in receipt of solar, thermonuclear-generated energy from our star, the Sun. This allows our Earth to exist without reaching equilibrium with near space or, let's face it, our planet too would be cold and lifeless.

However crucially both physical experiments and theory have shown that, under specific conditions, energy transformations in systems well out of equilibrium with their environment can, **spontaneously**, lead to the creation of complex, well-ordered, albeit transient structures.

Perhaps the best known and often scary example of such a phenomenon on Earth is a hurricane which uses energy and matter from very warm oceans to create a vast ordered structure.

But it is one that soon dissipates when the energy source is lost e.g. when the hurricane makes landfall. Everyone has seen the beautiful photographs of hurricanes from satellites as well as reading of their destructive power.

The energy-driven structures then can be lost, 'dissipated' is the term used, due to both too much and too little energy. A very well explored and simple example of this phenomenon is the Benard cell.

Power

This physical background leads to a simple, perhaps deceptively simple, proposition: The more free energy that can be exploited under specific conditions, the more work that can be done and power applied (in physics, power is defined as work per unit time).

This, in turn, can be coupled to the spontaneous emergence of 'structures' of more and more ordered complexity.

However, in marked contrast to the hurricanes and the Benard cells mentioned earlier, in biology, such complex systems have found ways to stabilize themselves/be stabilized and cope with changes in energy and material flows.

And I will argue later the same holds for human society.

Change

In **Energy the Great Driver; Seven Revolutions and the Challenges of Climate Change**, I explore how these propositions can be applied to and, I believe, illuminate both biological and human history.

This exploration has shown that a number of important, revolutionary step-changes in the energy economy of our planet can be identified, as I will discuss in my next article.

These changes were very gradual in the early millennia but have accelerated dramatically: that is the time between the major revolutions has decreased dramatically.

The energy revolutions have not only had profound effects on life forms but also the chemistry of the Earth's oceans, atmosphere and even geology.

However, I wish to add other new dimensions to this concept.

Firstly, the mechanisms by which stability can be achieved in these energy-dependent structures. Secondly, the other properties that have emerged over time. And thirdly the implications of this analysis both to our combating climate change crisis and how human society might emerge for it.

I am suggesting the stories we are telling ourselves need to change dramatically to avoid damaging our future.

NOTE: Others including John Maynard Smith and Eors Szathmary, Robert Ayres, Eric Chaisson, Tim Lenton and Andrew Watson and Olivia Judson have made similar but not identical proposals about the energy events. In the case of Eric Chaisson he has also applied similar concepts to the evolution of stars.

3 | The big energy steps

It's worth recalling Vaclav Smil's comprehensive statement paraphrased as 'everything damn thing depends on energy transformations'.

Given this basic principle, it is hardly surprising that, during our Earth's 4.5 billion year history, some major step-changes in such energy transformations can be identified.

These revolutions have not only defined the trajectory of life on Earth, but also caused huge and, at times, catastrophic changes in the chemistry of the Earth's atmosphere and oceans, even its geology.

Patterns

I will mention very briefly six major revolutionary steps while acknowledging other possible candidates and, importantly, recognising that a myriad of smaller energy step changes can be identified.

My purpose in these articles is not to analyse the science, be it physical, biological or social, but to highlight the patterns that emerge.

These, I argue, have led to our species dominating this planet and, in all likelihood, we are following a disastrous pathway and embracing a dangerous vision.

The first of these energy revolutionary changes is of course the appearance of life itself. Evidence for the **first living cells** dates to about 3.8 billion [thousand million] years ago.

How the first living cell arose is, as might be expected, the subject of much, fraught controversy and I will only emphasise one, albeit fundamental, aspect.

A cell, the most fundamental unit of biology, is a prime example of phenomenon I've alluded to previously. It is an island of ordered structured complexity, indeed a quite amazingly precise and sophisticated structure, in a sea of a much more limited order.

Quantum mechanics

The famous physicist Erwin Schrödinger – he of quantum mechanics, wave equations and a cat in a box which could be both dead and alive – first proposed that, for any living cell to exist it must tap into a source of energy to create and sustain its internal order.

Erwin Schrödinger

So a cell or indeed any living organism can be said to have gained order – negative entropy and embedded information – but it does so at the expense of its environment, which becomes more disordered i.e. increases in its entropy.

Crucially, according to the physics, this could have occurred spontaneously provided there was a steady flow available energy.

Life, in turn, has transformed the chemical composition of the Earth's atmosphere and seas and even much of its geology. Because of this, the planet itself can be seen as a non-equilibrium 'Gaia'-like, entity.

One exciting outcome is that the search for life in the Universe is, in part, a search of bodies, like Earth, out of equilibrium with their immediate environments e.g. in terms of their atmospheric chemistry.

Energy capture

One big issue is where did this initial energy first come from and how was it captured? It had to be geological as at that time there were no organisms such as small plants able to capture the Sun's radiation i.e. no photosynthesis.

One possibility is the highly alkaline liquids seeping through volcanic vents from deep within the Earth's crust.

These have been found to percolate through micro-porous material into the early seas or oceans. Such vents create natural and relatively stable gradients of pH, which are sources of energy.

Two points are worth stressing. Firstly, at the very outset, the geochemical energy source must have been realigned into being contained in and exploited by a self-contained, dynamic cell, discretely separate from its environment.

Secondly, my emphasis on energy does not diminish the huge importance of other issues such as how the biochemical components of the first cell were assembled, and especially the origin of DNA and RNA. There is a lot to learn!

Simple cells

Quite extraordinarily from our modern perspective, these 'simple', single cells with little visible internal structure were the only life on Earth for about a thousand million years.

These organisms, are termed **prokaryotes**, and are divided into two distinct 'domains' – Bacteria and Archaea. In everyday language we talk of 'bacteria' but in recent decades it's been shown that these simple, organisms, the prokaryotes, come in two types with different internal chemistry.

Both originally depended on geochemical energy and coexisted for about a billion years. Indeed both types of simple celled organisms – 'both prokaryotic domains' – still flourish today, including in our bodies.

Then some 2.5 billion year ago a second revolution. Some cells, instead of using energy from geological sources, developed the ability to harness energy from the sun! These were bacteria – called cyanobacteria (or blue green micro-algae).

They developed a capacity to capture a plentiful source of energy; sunlight. We call this process '**oxygenic photosynthesis**'. It traps the energy from the Sun using it to split water [H_2O] into oxygen gas, protons and electrons.

The last two provide cellular energy and can be used to fix CO_2, which existed even in the primaeval atmosphere and capture nutrients. The oxygen gas escaped into the atmosphere. Interestingly the capacity to fix atmospheric CO_2 predates the emergence of oxidative photosynthesis – a sequence of events which still has consequences today.

Atmosphere

The evolution of photosynthesis started a process which, over billions of years, has created the atmosphere which we now breath. It also radically changed the chemistry of the oceans and much geology.

The escape of O_2, despite its importance to us in respiration, was an early global 'pollution" event as oxygen is very reactive and can be highly damaging.

Cells some 2 billion years ago had to evolve chemical mechanisms to live with the growing levels of this highly reactive gas both in the atmosphere and dissolved in the seas.

Oxygen

For several hundred thousand years until perhaps 1.7 to 2 billion years ago, life on this planet remained prokaryotic, including the photosynthetic cyanobacteria, with a very slowing increasing atmospheric level of O_2 – creeping up to ~5% (cf. current ~20%).

There then emerges in the geological record clear evidence of a third major revolution. Cells are found with much more complex internal structures – this new domain is termed the **Eukaryotes**.

They were formed by what appears to be a one off and amazing fusion of an archaean and a bacterial cell – a process called **endosymbiosis**. The former providing the bulk of a eukaryotic cell and the later, its mitochondria.

Cells in this new domain contained a nucleus and DNA wrapped into chromosomes, and other internal organelles including mitochondria. The latter, small internal organelles derived from a bacterial ancestor, act as power-packs.

Each has retained only a small fraction of its original bacterial genome. Hence such cells have about 1000 times more power per gene and per genome compared with prokaryotes.

'Busy' eukaryotic cells can contain as many thousand mitochondria But, as the eukaryotic cells are also larger and of a great mass and complexity, the processes behind this step change remains controversial.

The step began the formation of complex life forms – but it was a long, and laborious process.

Critically all multicellular organisms, be they fungi, trees, lizards or us humans are formed from eukaryotic cells. This unique symbiotic event has transformed the physical complexity, wealth and diversity of life. Notwithstanding this, the biochemical capabilities and diversity of prokaryotes remains wonderous and important.

For hundreds of thousands of years all eukaryotes were single cells – termed protists. Then some 500 to 600 million years ago evidence of multi-cellular organisms is found in rocks, including the recent discoveries near Llandrindod.

The emergence of these diverse organisms which are the ancestral precursors of fungi, plants and animals (and other lines which have disappeared) coincides with a spike in oxygen levels in the atmosphere.

Yr Wyddfa

As coupling the breakdown of sugars to oxygen-based respiration increases the energetic efficiency, this may deserve to be called another energy-related step change. But as far as I can judge the matter remains unproven.

Over the next five hundred thousand years or so, evolutionary biology allowed life on Earth to flourish and diversify. Life survived at least 5 major extinction events, periods of snowball earth, major tectonic changes in the distribution of land masses and oceans and numerous ice ages.

Gradually it evolved into the Earth as we know her. We tend to think of the Earth beneath our feet as solid, reliable and unchanging but on longer timeframe it is anything but. It's amazingly, even scarily, dynamic and unstable.

Our coalmeasures were laid down when this land was tropical and sea shells can be found in rocks high upon on Yr Wyddfa. Only 20,000 year ago Wales was covered in meters of ice.

Much more recently three energy revolutions have occurred which have led directly to human dominion over this planet and even a new geological era – the Anthropocene – recognising that our human imprint on planet is seen in its geology!

About 1.8 million years ago (note only **million**) additional energy was expended on **brain** development in an early human ancestor, on intelligence and communication skills.

The evolution of this enhanced brain capacity was energised by improved food intake and digestibility, probably helped by cooking, and the assignment of that energy and material to the brain. 'Brain' proving over time a much better investment than 'brawn'.

Brainy hominid

We humans are unique among animals in preferentially devoting some 25% of all energy to our brain functions. In all probability this led to greater intelligence, new social interactions, a capacity to assimilate and use information and of course find more and better food – a virtuous cycle.

Amazingly nearly a million year before our arrival, earlier humans, Homo erectus, the first brainy hominid, spread from its African origin to most of this planet. It became an important predator and started the evolutionary processes of leading to us.

By a complex and tortuous path, **Homo sapiens** emerged about 300,000 years ago: a blink of an eye in 'Earth time'. As late of 40,000 years ago we shared this Earth with several other Homo species including such Neanderthals with whom we, in Europe, interbred.

Homo erectus.
Adult female reconstruction.
Smithsonian, Washington D.C

Even more recently, starting 10 to 8 thousand years ago as the last Ice Age was receding, a number of **agricultural revolutions** occurred around the world around the world.

From a European perspective, the most significant was the dawn of agriculture in the Fertile Crescent in West Asia; now mainly in Syria and Iraq. This led to improved food energy supplies which later supported more dense and permanent settlements.

Enhanced production of food calories through agriculture led to the emergence, albeit fitfully, of more complex societies, to record keeping and writing and to the great classical civilizations not only around the Mediterranean, and in China and India but worldwide.

The Industrial Revolution

This sequence of energy step changes has culminated in the **Industrial Revolution**, starting only a little over 200 years ago. This supercharged

The Industrial Revolution

change by our learning to exploit hydro-carbon fossil fuels for gain more power.

This human exploitation of power combined with global commerce and capitalism and the development of our technical and scientific expertise has been transformational.

It has allowed an extraordinarily rapid increase in numbers of **Homo sapiens** (about ~300 million in Roman times to currently 8,000 million), together with the growth of urbanization and industrialization.

It also catalysed colonisation, as well as changes in the systems of local and international governance and in economic management. Together they created of our staggeringly complex modern societies. We must also recognise and fear the parallel energizing of our capacity to make war, locally and globally.

In the next article I will consider the ramifications of these six major energy step changes before addressing the lessons for climate change and our futures; an issue ever more urgent as so much of this Earth is now not warming – as we over complacently say – but heating to unprecedented levels, as many will be experiencing personally.

4 / Emerging change

The rapid gallop through nearly 4 billion years in the last article obviously simplified and left out many important aspects. These included many other smaller energy step changes such as evolution of warm-bloodedness, bi-pedism, and the colonisation of land. Although not my main theme, it is important to note that there has been a parallel energising of violence in human societies since the time of our earliest ancestors.

Wars and quarrels would have originally involved fists, throwing spears and cudgels. Now it's drones, supersonic missiles and thermonuclear bombs. (The carbon footprint of Russia's invasion of Ukraine must be enormous!) However my objective in this fourth article is to pick up some lessons from this hectic journey we followed in the third article.

New ways of exploiting energy

Clearly the six revolutionary energy-dependent step changes differed from each other in important ways. The second step involved capturing a completely new source of energy, i.e. tapping into sunlight through photosynthesis. Others have initiated different ways in the coupling energy to work and complexity.

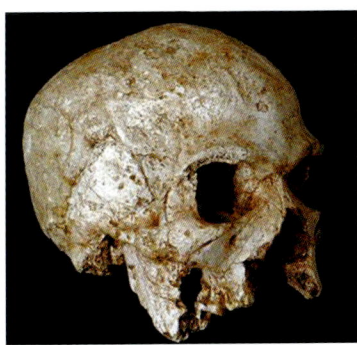
Hofmeyr Skull, Homo sapiens

For example, the acquisition of brain power in early humans depended on better food and then using the extra energy and nutrition to further increase their brain power and their abilities to hunt, gather and likely making food more nutritious by cooking and to work together; a complicated set of positive feedbacks.

Using our brains

The agricultural revolutions were similarly phased and complicated. Humans became able to command for themselves, at favoured locations, more of the energy derived from photosynthesis. This allowed an increase in their population which overtime, and again using their braininess, led to new ways of its coupling of work and power and to the emergence of complex societies and structures (see 'A' and 'B' as sketched below).

In the case of the 6th step, the Industrial Revolution, a new and convenient energy source, fossil fuels, was exploited and, as we all know, increasingly clever and sophisticated ways of using it have been discovered. Indeed, from our earliest times humans have been toolmakers and have use tools to leverage increasing useful work from the resources available to them.

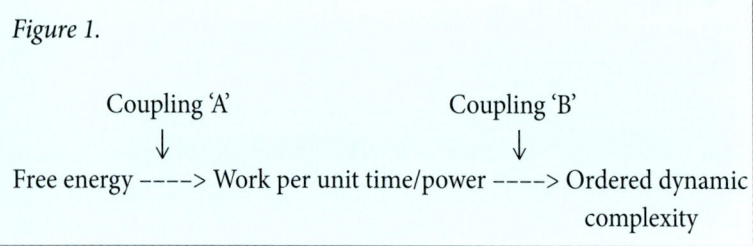

Figure 1.

Coupling 'A'   Coupling 'B'
     ↓   ↓
Free energy ----> Work per unit time/power ----> Ordered dynamic complexity

Stabilising the growing complexity

As I've outlined previously, structures that are dependent on a constant flow of energy are intrinsically unstable. All organisms and human social and material constructs must therefore have evolved mechanisms to stabilize and regulate themselves so coping with irregular energy flows and changing external environments.

One remarkable feature of a cell, and of life itself, is that it has released itself from the second-by-second tyranny of a dependence on a steady flux of external free energy. It does so by storing chemical energy and by an ability to respond to changes in its environment. How the first cell achieved this degree of stability remains a mystery.

In biology the term **homeostasis** is used for the regulatory and stabilizing aspects of metabolism. We all know about the importance of checking our blood pressure and body temperature and keeping these and other factors around given set points or normative values. Some of these factors are remarkable constant over a wide range of organisms, and are good indicators of our health. We check to see if our internal regulatory systems – our body's homeostatic mechanisms – are in good working order.

Similarly, even singled celled organisms have such mechanisms to stabilise their metabolism and can be seen to swim away from stresses e.g. too much heat, and towards food sources. This homeostatic principle can be applied much more widely to all aspects of the relationships between the individual cells or whole organisms and their surrounds or, indeed, humans and their social environments.

Inevitability of regulation

These ideas were developed by a Portuguese American neuroscientist Antonio Damasio. He realised that an arrow of evolutionary change in the hierarchy of such homeostatic, regulatory mechanisms can be recognised from a single prokaryotic cell to a complex multi-organelle cells, to multi-cellular, multi-organ animals and higher plants through to relatively primitive and now much more complex human societies.

Natural selection operating over thousands of generations appears to have ensured the homeostatic stabilization lower down the evolutionary hierarchy is virtually invisible in higher organisms or even in societies unless exposed by disease or other disaster. Day to day we humans are unaware of the importance of maintaining the balance of sugar or sodium or our blood pressure near to the appropriate set-point. Biological homeostasis works exquisitely well but a major deviation implies a medical condition and may result in death.

Antonio Damasio has argued that this homeostatic hierarchy should be extended to human behaviour, especially the ways in which we express and control our emotions and feelings. From our earliest ancestors through to modern humans, these emotions and feelings have evolved gradually to enable a degree of social, sustainable, yet dynamic, cohesion – social homeostasis.

They, as well as our brain power, have permitted our cultural diversity and collective technological prowess. These adaptations have been critical to the evolutionary success of **H. sapiens** compared with other apes in allowing humans to coexist and collaborate in close proximity and relative harmony unlike most our closest primate ancestors. Regulation of our personal behaviour and in humanity's social systems is vital and inevitable.

More power: speeding up change

Secondly, it is striking, when considering the individual step changes, how events have accelerated over the eons of time. The early revolutions were separated at about a billion year but with the emergence of brain power, the changes came thick and fast. At first hundreds of thousand, then thousands of years, now merely decades. It is only a little over two hundred years since James Watt invented the first relatively efficient steam engine and Matthew Boulton proclaimed "**I sell here, Sir, what all the world desires to have – power**".

James Watt by Henry Howard

Power is defined physically, as **work per unit time**. It is unsurprising therefore that more and more energy, or its better coupling to work/power and growing complexity is reflected in this massive acceleration in the rate of change. This has profound implications for us today as I will discuss later.

While Bolton made a fortune selling the new-fangled power from steam engines, his quote also underlines the close relationship of energy and power: physically, economically and politically.

The powerful

Alfred Lotka and Howard Odum and others have advocated the **'maximum power principle'**. That is that self-organising systems, out of equilibrium with their surroundings, systematically maximise their power intake to increase their own success and survival chances. The authors were mainly concerned with biology but it is not far-fetched to see similar trends in human society. Power in human communities is about the ability to command either directly or indirectly the work of others or the fruits of that work.

Before the Industrial Revolution this usually revolved around the ownership of land, especially fertile land, and the production of food and the consequent availability of manpower and horses in peace or war. The relationship of food supply to power was very close. It is no accident that the powerful Marcher Lords controlled the best land.

Havoc: the risks

As described vividly by Peter Frankopan in **The Earth Transformed**, human history before and after the Industrial Revolution is replete with examples of powerful men (usually) or groups creating great empires, but the tyrants then being overthrown by events for which they were ill-prepared. Such events were often environmental disasters – droughts, volcanic eruptions and various diseases and pests. They brought havoc and great misery and death to the ordinary people, largely because of food shortages.

Great emperors and empires fell not only because of competing tyrants but because of unanticipated environmental crises. 'Power may come out the mouth of gun' (Chairman Mao) but always depends on

Megacity

the control of energy – historically this has been calorific energy but now it's mainly guns and explosives! But Russia's current blocking of Ukrainian grain exports shows the old powerplays still count!

Interdependency and elites

Thirdly, while each step change heralded a new level of complexity, – some authors refer to 'the arrow of complexity' – the new emerging forms remain dependent on the previous levels of life. Thus, the eukaryotic world depends on the prokaryotic one, as indeed we, **Homo sapiens**, do. Virtually all life on Earth depends on photosynthetic energy. Modern human megacities depend entirely on the less complex life forms e.g. bacteria in sewage farms, as well as food acquired using skills some dating back to the agricultural revolutions.

However, it also appears that the revolutions have created opportunities for new elites to emerge, certainly after the evolution of braininess, First **Homo erectus**, then **H sapiens**, much later after the agricultural and industrial revolutions, new power-controlling human elites have emerged.

It has also been proposed that at each of energy levels there has been a maximum degree of societal and cultural complexity compatible with the available energy source and its concomitant technologies. For example, it is suggested that the Roman and Chinese Song empires represent the high watermark for societies dependent on, and energised and empowered, by agriculturally-derived calories with only a modest input from classical wind and water power.

Competing and cooperating

Fourthly this analysis also highlights the subtle balance between cooperation and competition in determining the arrow or fate of evolutionary change. You may recall our earlier discussion that all complex cells (eukaryotes) and consequently all multi-cellular life has evolved from the fusion of a single archaean and bacterial cell. This unique combining is known as endosymbiosis.

Other, only a little less dramatic, endosymbiotic events have occurred. Cyanobacteria, the bacteria that evolved a capacity to undertake oxygenic photosynthesis, have also been 'endosymbiotically' captured by eukaryotic protist [i.e. single cells with complex internal structures including mitochondria] resulting in the ancestors of modern photosynthetic plants and trees. Lichens and coral have arisen from the cohabitation of fungi, alga and cyanobacteria, while the colonisation of land depended on the positive, cooperative interactions of lichen, fungi and plants.

The success of our earliest ancestors depended, almost certainly, on their enhanced capacity to cooperate, to successfully hunt, gather and cook food. Greater brain power improved social skills as well as no doubt promoting competition for resources and living space.

This emphasis on cooperation is not to minimise the importance of Darwinian natural selection and competition. The survival of the fittest has clearly played a vital part in evolution. But assessing the balance of cooperation and competition in biology and indeed in human society is not easy. It is clear that raw competition is not the whole story.

Energy and information

Finally, there is the difficult issue of the relationships of these energy step changes and indeed energy itself with information processing. With the possible exception of step 2 (the capture of solar radiation by oxygenic photosynthesis), each step change has involved some major change in the processing of information.

Step 1, the origin of a simple cell, as I discussed, required a reliable energy source but of course also depended the genetic code and a capacity for the intergenerational information transmission. In the case of Step 3 the fusion of a bacterial and archaean cell to form the first complex eukaryotic cell, there were major changes in cell structure, in how genetic selection works and in cell energetics.

Eukaryotes have significantly more energy available to carry out the 'work' of their genes. 'Busy' eukaryotic cells such as our muscle cells or the neurons in our brains have thousands of mitochondrial powerpacks per cell. Step 4, as we've discussed, saw the enhancement of proto-human neural capacity and the emergence of non-genetic information transfer.

New ways of transmitting information and coordinating activity, and likely the earliest type of primitive speech, emerged. This allowed increased direct transmission of learnt experience from parent to sibling and within a tribal group to all its young. And beyond, no doubt, as we are great copiers. Step 5 also led to enhanced information transfer to facilitate urbanised, social interactions. In time record keeping, writing and mathematics and libraries developed.

While the Industrial Revolution [Step 6] has improved our capacity to transmit knowledge, initially, by better printing and distribution, then radio and television. By now we are in a globalized world of digitization, many forms of electronic data transmission and processing, including our current obsession with the social media, "big data" and artificial intelligence (AI).

One feature of energy which I mentioned in an early article in this series, is that, as well as there being 'free energy' i.e. that capable, in transformations, of doing useful work, there is 'residual' energy which has lost this ability e.g. heat generated by friction or by cell metabolism; this energy is called entropy.

You may recall in article 2 that entropy also a broad measure of disorder at the molecular level. Thus, organised systems are defined by detailed information, disordered more random system much less so. So there appears to be a broad relationship between energy (usable free and entropy) and information, which is important as we enter the world of AI. However, it is not an aspect I will pursue further.

Our current crises

What is important right now is the future of us humans and the rest of life on this planet, remembering we are interdependent. I believe the patterns we've been discussing have profound implications on our current human crisis and where we go from here.

There are lessons we can learn about both why we have been so slow to react and how we should react to global warming and the challenges of artificial intelligence? Is it all out of our control? No not necessarily. But we should urgently be thinking about our future in new ways.

5 / The energy revolutions

In the first four articles I've tried to explain the fundamental significance of (free) energy to the long march of life on this planet and to the relatively recent but dramatic rise of **Homo sapiens**. Free energy simply means the energy that is available to useful work and generate power.

Many of the individual ideas I've been discussing have been advocated by others but the synthesis is my own responsibility.

These ideas of the underlying 'big history' may excite some and hopefully increase our understanding, but even more crucial are the

Image by Steve Buissine via Pixabay
Power station

lessons they might hold for our current crises, to the stories by which we live and to humanity's future.

Revolutions

It is obvious that we are now in the midst of at least one, but quite possibly two concurrent, energy revolutions. The first is very different to all its predecessors. We are being forced, because of the imminent danger of climate catastrophe, to give up fossil fuels; the very sources that have energised our economy and brought prosperity to a significant proportion of people.

Secondly, the human technological enterprise and scientific knowhow have developed bioengineering, computing, artificial intelligence and robotics in ways which are fundamentally changing both the relationships between energy, power and complexity and also the rate of change in our society.

Record high

Other than a few diehard deniers, the dangers of catastrophic climate change and global heating, rather than any more benign-sounding warming, are all too apparent and scary. Nevertheless, despite the high flown promises and innumerable conferences, progress in cutting the emissions of the main greenhouse gases (CO_2, CH_4, N_2O etc) is pitifully slow.

As of the last week in April this year (2023), the atmospheric CO_2 was at a record high. Worse, the annual increase in CO_2 was the largest on record in a non-El Niño year. Similarly, methane and nitrous oxide are at record levels. To date we are not even reducing the rate of growth of these dangerous gases, still less achieving cuts.

Tipping points

A few points stand out. While the atmospheric concentrations of the Green House Gases and mean temperatures are widely reported, it is often forgotten that most of the extra heat (>90%) is absorbed by the oceans which cover 70% of the global surface area.

Mean oceanic temperatures are also at record levels and will affect global weather patterns for generations to come. The 2015 Paris Accord committed the global community to limiting the mean atmospheric temperature rise to less than 2.0°C and as near 1.5°C as possible.

These objectives reflected the best estimates of temperatures which would avoid triggering irreversible tipping points such as melting of ice sheets, methane and CO_2 release by the decay of the tundra in Siberia and North America, or the loss the tropical rainforests' capacity to be carbon sinks.

Enormously damaging

Any realistic prospect of limiting the mean rise to +1.5°C is gone – short of a series of huge volcanic eruptions or a nuclear war which will themselves be enormously damaging (be careful for what you wish). Even breaching the +2°C ceiling now seems likely but not quite inevitable.

However, my concerns are not simply about our reluctance to give up fossil fuels or their replacement by renewables and perhaps electrical energy from nuclear fission and fusion. **A huge issue is the societal and environmental impact of using more and more energy from any source.**

These articles are exploring the basic idea that more energy leads to more work and power and to greater complexity, which in turn requires an ability to control and stabilise each new level of complexity.

In biology the term **homeostasis** is often uses to describe these regulatory mechanisms. Both the physics and the historic record also show that doing more work per unit time accelerates the rate of change, as was discussed previously.

Pumping gas

Human excess

From a human perspective, global warming can be viewed as a homeostatic failure. It is altering the subtle global balances needed to support our civilisation not just in Wales but in India with temperature nearing 50°C or the Pacific Islands threatened by sea level rise.

Our actions are endangering the very energy-dependent complexity that has allowed some of us to thrive materially. This reveals a major weakness in our man-made regulatory systems e.g. costs of pollution are not shown on company balance sheets.

On the other hand, from a Gaia-like perspective, global warming can be seen as the planet's response to human excess and a way of supressing an over-dominant species. Either way humanity is clearly in a major hole. And on top of this we are facing the rapid and vast changes being brought about by digitization and new ways to couple work and power to complexity.

Humans are not so much gaining extra brainpower or wisdom as an extraordinary capacity to process vast quantities of data (still using energy) but with an extraordinary efficiency. But it's an ability that may actually be endangering humanity.

Taking the two current revolutionary trends together, it appears virtually certain that more energy (even if low carbon) and/or new ways of more efficiently coupling energy use to work and power e.g. through artificial intelligence, etc. will:

i] Increase still further the material and social complexity of society.
ii] Further accelerate the rate of social and economic change.
iii] Reinforce the need for more 'regulations' to stabilize the growing complexities, with unpredictable consequences to human well-being especially given the rise of surveillance and big data; currently in the hands of a very small number of companies and government.
iv] Increase our demands on all other planetary resources – renewable and non-renewable; e.g. more food and more copper and lithium etc.

These predictions are consistent with the patterns that emerge from Earth's long history as I've sketched them.

In the previous article, I discussed briefly the homeostatic hierarchy required to maintain stability. As life evolved from a simple single cell to first complex cells and much later complex multi-cellular, interdependent organisms, then new layers of regulation emerged. In our society, such social regulation probably has two major sources.

Inherited and man-made regulation

The first is our inherited feelings and emotions. Our physiological and psychological mechanisms for self-control which permit cooperation as well as competition. Unlike other apes we live and usually co-exist in relative peace – imagine a rush hour train packed full of male and female chimps. Some of the ways we regulate our interactions derive from our long history as early humans; from living in small, largely self-sufficient but quite often threatened groups.

Richard Wrangham has argued that we have, in effect, domesticated ourselves by reducing the alpha male aggression found in other apes. We have controlled our **reactive** aggression but possibly at the expense of **proactive** aggression. That is a propensity to create 'cunning plans' to dominate and eliminate rivals to power.

We tend to be tribal, and somewhat parochial while sometimes seeking a strong, charismatic leader – some will remember all the talk of the 'smack of firm government' in the 1980's. And one can now add Trump, Berlusconi and many others to this rollcall. Despite our innate

Climate change

generosity, we also tend to resist giving up our hard won prosperity and resources that support it.

While we often seek strong leaders to protect us, they can prey on our prejudices. A situation made worse by our tendency to make decisions instinctively without really giving too much thought to them and our poor statistical assessments.

All this is poor preparation for dealing with global warming which requires forethought, everyone to pull together and, quite probably, for the better-off to make sacrifices today to avoid calamity tomorrow. As rule of thumb – the richer one is, the higher one's own carbon footprint.

Unfortunately CO_2 emissions from anywhere and everybody impact everywhere and on anybody. In terms of climate change, we are truly globalized.

Secondly since the Agricultural Revolution larger and closer communities and later great cities have emerged. To cope with this increasing complexity, we have had to set up our own human-devised systems of regulations, laws and practices and some form of enforcement, although these varied widely.

Over the last 100 years or so, the world has increasing adopted the consumerist, individualistic capitalist model based on globalised versions of market-based economics which can be summarised as an allegiance to **Homo economicus**. The model is based on the assumption that economic self-interest, enterprise, even greed is paramount.

Nevertheless it is presumed that the efforts of the few, while of course making them wealthy, will lead to an overall increase in the living standards of all. So, although they are winners and losers, in theory, we all gain and become materially wealthier. This then is the basis of the story of 'progress' I described in the first article.

Our contended futures

A very big question is the likely combined impacts of these two types of regulation, together with the other factors I've alluded to?

Given our behavioural inheritance, our man-made regulatory systems, the accelerating speed of change, the unremittent pursuit of power and wealth and our growing demands on the Earth's resources as our population grows and more yearn for prosperity, what might the future hold?

Are our actions compatible, in the short term, with combating climate change and, in the medium term, with human wellbeing and planetary health and limiting our dependence on greater and greater use of energy and resources? These issues will be the focus of the next articles.

6 | Our current dilemmas

The conclusions from the five previous articles are profoundly disturbing.

Both the historic record and theory suggest that accessing more, even low-carbon, energy or ever more efficient ways of using it may not, in reality, be a panacea. Indeed they seem likely to lead to both greater human disfunction and continuing planetary degradation.

The patterns we have been discussing suggest that more work and power will further accelerate the rate of change and generate more and more complex societies. Our demands on both renewable and non-renewable resources will also increase as will the pressures on the rest of the biosphere.

These trends will also likely lead to the emergence of new elites and almost certainly to an increase in the need for more regulation; quite possibly more intrusive and stringent regulation. This, at a time, when our digital expertise is making surveillance more pervasive.

The speed of change

Many troubling issues arise around how we should, indeed whether we can, manage these outcomes and cope with the speed of change and its consequences.

What sort of society might emerge, locally and globally? What of those currently poor and dispossessed? What of the rest of life on this planet? What type of society do we actually desire? Are we trapped by the current story and its actors? Do we have both a coherent vision and the determination to implement it?

Undoubtedly some will argue that our technological skills, ingenuity and entrepreneurism will provide solutions and allow our current social assumptions and the model of unlimited, exponential economic, consumerist growth to continue and spread. But, in truth, the evidence for such a view is not convincing.

Essential systems

I have previously argued that the homeostatic regulatory systems which allow humans to live together in relative harmony are derived partly from our intrinsic, behavioural characteristics evolved over our long pre-human and human past and the laws, rules and conventions that we have developed, many quite recently and post the industrial revolution.

These are essential to sustain our lives both as individuals and members of civilised communities. Such systems, of course, vary widely among human societies and cultures and over time. There is no one inevitable right way.

And, crucially in these matters, we have agency; that is we humans, uniquely, can conceive of and plan for a desired future. Although it is wise not to ignore our hard-wired inherited behaviours or how they can be manipulated.

Self-sustaining technologies

It is important to place our current global heating crisis in this frame work as well as the complex technical and economic issues that must necessarily exercise us.

Indeed we must, over the next decade or so, find sufficient very low carbon energy to maintain human society all over the world without ruining the planet and with it our own societies. But even that will not, of itself, suffice.

Paradoxically, as well as the well-rehearsed scientific problems, our technological prowess is part of the problem in several ways.

First of all, technologies and their capabilities are to an extent self-sustaining. They have their own internal logic and drive.

Outfall from Wylfa Power Station

I would offer nuclear power, such as Wylfa, is an example of this phenomenon. Nuclear electricity has largely been a failure as it has proved too expensive, too accident prone and produces contaminated wastes which will require thousands of years of careful management; an unresolved conundrum after well over 50 years.

Further it also is a poor match for the emerging, and cheaper renewable sources – solar, wind. The latter suffer from intermittency and seasonality but nuclear electrical generation works best when output is constant.

The characteristics of nuclear generation are not complementary to the intermittent renewables. Surely events in Zaporizhzhia should convince everyone that, in a world of growing civic strife, building thousands of atomic power station is too great a risk to be contemplated.

What might have happened to such stations in built Aleppo before the catastrophic Syrian war or Maidurugi, before the rise of Boko Maram: cities of which I have first-hand experience?

Nevertheless the idea is backed by major politicians and players within the military-industrial complex, as first identified by Eisenhower, and by centralised planners, be they capitalistic or communistic or the old CEGB.

Paradoxically this backing is related to commitment to the fearsome power of and preferential use of plutonium in atomic bombs not as timely response to climate change.

Surprises

Our technological history is also littered with surprises; both known and unknown unknowns. James Watts and Matthew Bolton would be staggered to realise not only the increased prosperity their invention, the steam engine presaged but that we are, as consequence, now facing climate catastrophe.

English inventor Matthew Boulton by Carl Frederik von Breda

Globally there is a growing assumption that in order to achieve net zero (or even an approximation thereto), the capture and storage of atmospheric carbon dioxide on a massive scale will be required; that is finding ways to suck CO_2 out of the atmosphere and store it out of harm's way for centuries.

Promoting such a scenario suits the powerful oil and gas interests, fossil-fuel rich countries and right-wing, laissez-faire economists and politicians. It will for certain feature strongly in COP 27 in the Emirates next autumn. However the technology is itself in its infancy, currently very expensive and full of known and unknown unknowns.

Risky interventions

All the scientific evidence shows that the global atmospheric and oceanic circulations that control local weather patterns are sensitive to change in unpredictable ways – chaotic in its scientific sense.

Yet we are contemplating massive, highly risky interventions because of special interests and our reluctance to change our life styles and inability to face reality. Humans would do well to avoid hubris. But such presumption is certainly embedded in Western, individualistic, laissez-faire society and, maybe, in our humanity. Sages and prophets from all religions, cultures and eras have warned of its dangers. Few have listened.

The rate of accelerating change brings its own problems. Joseph Schumpeter famously wrote of the modern economy as a "gale of creative destruction" and a "process of industrial mutation that continuously revolutionizes the economic structure from within, incessantly destroying the old one, incessantly creating a new one".

An obvious question is, as this process is continuing to accelerate as result of AI and miniaturization, how will our humanity fare? In Wales, as worldwide, we are devoted to our own 'milltir sgwar', to our own identities and crave a degree of stability. Can we cope with this ever increasing speeded change and, crucially, being incidental and expendable cogs in an unrelenting cycle of economic destruction and re-construction?

This cycle almost completely out of the control, not just of individuals and communities, but of many nations. Power centripetally accrues to a few locations and individuals. Governments compete to attract investments. I've read that the UK government has offered near a

£1billion to tempt an Indian company to build a factory make EV car batteries in Somerset.

Tensions

A few prosper mightily but many are left behind at great psychological and social cost. In the **Tyranny of Merit**, Michael Sandel argues that this over emphasis on merit and competition not only exclude the many but create a huge tensions and conflict among the successful, giving them both an undeserved sense of entitlement and many social hang-ups.

This I would argue is also a major problem for combating climate change:- so 'If I can afford to hire a private jet for a weekend break in the Caribbean, why should I worry about anyone else?'

Human regulatory systems inevitably take time to evolve and mature and take time to change. It is arguable that we are already not keeping up with the rate of change.

We have so far failed to find an appropriate regulatory balance for the emergent social media to detriment of many, especially young people – social poison, misinformation and conspiracies abound. There is near panic in seeking to control the potential misuse of AI.

James Lovelock in his last book **The Novacene** had an unique perspective. He envisaged the emergence of a hyper intelligent, AI-grounded, cyborg elite which humanity should and would be happy to serve in the interest of Gaia.

I view his vision with dread. It is an outcome which we should use all our human agency to avoid. That said, it is not inconsistent with the trends I am discussing in these articles, if taken to one possible logical conclusion.

Planetary burden

A few points need emphasis. The greater the human population the greater will be the burden placed on this planet's resources. The poorest will quite legitimately seek a better life and richer people will resist any change in their status. If we have plentiful cheap energy, the greater the demand will be for goods of all kinds; each of these will have their own material as well as energy foot print.

The cost and viability of energy also determines which material resources are economic to exploit. So unlimited cheap energy will allow poorer will be the mineral resources we can mine and extract economically and more habitat destruction.

Similarly given the atmospheric physics and chemistry of CO_2, the larger the population, the smaller will be the average carbon amount/ration? available to each of us; noting also that this total has to fall rapidly. Or, alternatively, the greater will be the inequality between those with the capacity to command a disproportionally large share and the rest.

Population growth

It follows that the stabilization of both the total population and decreasing the resource demands of the wealthy are priorities. Fortunately the rate of population growth is beginning to level off. This positive development will necessarily mean an aging population; fewer young and proportionally more old people. In a number of counties the population is already actually decreasing.

Such a voluntary trend should be welcome. But it is not a welcome truth within the modern economic growth model and a source of embarrassment to the UK government both wanting and not wanting youthful immigrants to keep the system functioning.

There are also some welcome signs that some countries are severing the historic link between total energy consumption and economic growth as measured by GDP. However more generally greenhouse gas emissions are closely related to wealth either on an individual or country basis and it follows that the majority of the cuts must inevitably come from the better off.

Global emissions

The core dilemma can be illustrated readily. Currently the global population of about 8 billion emits over 50 billion tonnes of greenhouse gases each year (expresses in terms of CO_2 equivalents, that is combining the impact of all the gases); that about 6 to 7 tonnes per person on Earth per year.

But per head annual country emissions vary from >25 tonnes and more to below 0.1 tonne. [USA and Australia ~15; China ~8, India ~2, many African countries 0.1 and below]. Some 30% of these emissions come from land use, land use change, making agrichemicals and the rest of the food chain (carbon dioxide, methane and nitrous oxide). The global population is expected to be around 10 billion by mid-century.

Even optimistically it seems improbable that we will reduce emissions to below 1.5 tonnes per head per year and still feed the world. We must also create the 'energy-space' to allow the very poor to have a better life and maintain a viable society. So we may be left, even if countries really press hard for rapid emission's reductions, with residual annual global emissions of around 10 to 15 billion tonnes.

Carbon capture runs currently at about $100 to $200 per tonne so only 10 billion tonnes annually (i.e. about I tonne per head) would cost, at the lower estimate, $1 trillion per year. Such capture and storage processes (CCS) are as yet at the pilot state and, as I've noted highly risky and full of known and unknown unknowns and will be difficult to implement in time.

The core dilemmas are quite apparent with no easy answers but I suggest that saving energy – seeking positively to reduce energy use – seems a good bargain as well as consistent with the concepts I've been exploring.

7 Can less mean more?

The global challenges of climate change and heating, biodiversity loss and the other aspects of environmental decay that confront humanity are unique.

Our species is now so powerful, so dominant and so resource-demanding as to be despoiling the wildlife, soils, oceans and atmosphere on a global scale. We, humans, are even commanding about 20 % of all terrestrial photosynthesis but, as much as 80%, in areas of high population density.

This is, in addition, to burning billions of tonnes of fossil fuels annually to energise our life styles. Small wonder all other wildlife is being squeezed out.

Fossil fuels

Despite the conferences, the promises and the gradual rise of renewables, some 80% of our energy, globally and locally, still comes from burning fossil fuels.

As we've been discussing, these emissions are endangering the very fabric that sustains us. Yet these emissions underpin the exceptional material prosperity of close on a billion people. Even so, some 650 million live on less than $2 a day.

Almost a quarter of the global population, about 23 percent, live below the poverty line, defined as $3.65 a day in the poorest countries, and almost half the world's population, nearly 4 billion, live below the $6.85 poverty line set for the middle income countries. (To put these numbers in a Welsh/UK context, currently $12.3 per day = £10 per day = £3,650 per year).

The poorer half use much less energy and release far fewer greenhouse gases, as we discussed in the last article, and the latter mainly from their food. Prosperity and poverty may be fellow travellers but with drastically different profiles.

Pollution

Human food chain

Fortunately the rate of population growth is slowing. Nevertheless we face, by the middle of the 21st century, the daunting problem of giving hope and sustenance to some 10 billion humans without wrecking the Earth.

This in a context, let me note again, that about 30% of all GHG emissions come from the human food chain.

Methane emissions from ruminants are but one issue, although a very important one in Wales. Land loses CO_2 when ploughed, so increasing arable areas will create a spike in emissions unless there is a parallel drop in animal emissions.

Also crop yields are near-linearly related to available nitrogen. Much of that comes from the energy-intensive Haber process for converting atmospheric nitrogen to ammonia. But about 1% of the nitrogen that is applied to increase yields of food crop is lost to the atmosphere as N_2O; a greenhouse gas x 300 as effective as CO_2.

Much food is also wasted by the prosperous; while food is spoilt by the poor, largely because of inadequate storage; they have neither electricity or fridges. The post farm-gate food chain contributes about a third of the sector's total emissions; transport, refrigeration, processing, storage, display in supermarkets and of course cooking etc. at home.

Hill farmers

As Albert Bartlett wrote, 'Modern agriculture is the use of land to convert petroleum into food'. I can hear the cries of Welsh hill farmers in my ears but lamb is not going to provide the billions with their daily 2,500 food calories and, unfortunately, even upland grazing is far from carbon neutral.

We may enjoy our lamb but cereals are the key energy commodity as is being demonstrated yet again by Russia's war in Ukraine.

Of course, feeding such grains and soya to animals in feed-lots to satisfy the world's growing demand for meat, does nothing to solve our problems. Globally some 220 million hectares are devoted to wheat, 206 million to maize and 165 to paddy rice which implies that much vaunted vertical farming, that is growing crops in doors using hydroponics and artificial light, is at best a distraction [cf. Wales ~2.1 million hectares].

These numbers summarise very scary issues. we are warned that poor harvests in some cereal producer-countries may well precipitate a global food problem well before warming exceeds 1.5°C. We are riding our luck!

Social upheavals

Our cruel dilemma deserves emphasis. The Industrial Revolution and its attendant secondary agricultural revolutions, including the displacement of many indigenous peoples from their lands, allowed agribusiness to flourish and the human numbers to soar, albeit often in poverty and near servitude.

The consequential moral and practical responsibilities now fall on this generation.

Greenhouse gas emissions from all humanity, including from such basics as food production, must be reduced to near zero in 25 years. In doing so we must seek to avoid mass starvation and mass migration.

Failure will leave important and populous regions such as the plains of northern India from the Indus river to the Bay of Bengal as death traps. This will ensure huge social upheavals and mass migration. In all likelihood it will lead both 'reactive' and 'proactive' aggression as we've discussed.

The directly affected will react as their livelihoods are threatened and the authorities and external powers will proactively seek to minimise the impacts of the unrest on their status or patch.

As occurred in Syria some will see an opportunity to make serious mischief. Such strife cannot and will not be confined within neat geographical limits.

Extinctions

Life on our planet has lived through many crises; at least five major extinctions and periods of Snowball Earth and of great heat and, more recently. a sequence of Ice Ages.

Living organisms have survived, diversified and indeed flourished. The extinctions were caused by external or major geological events.

The dinosaurs were not the agents of their own demise. Given a bit of asteroidal luck. they might have reigned for a few more million years. Now we, after but a few centuries of industrialisation, are the agents of our own possible downfall, even destruction. Fortunately we also have agency; we are the only potential source of our own salvation.

Energy transition

I have been arguing that a fundamentally salient problem arises from our use of energy. To reiterate "every damn thing depends on energy transition".

Broadly, more energy allows more work and greater power leading to greater complexity and an accelerating speed of change, now reinforced by digitization and AI.

Both the current climate change crisis and the AI revolution have brought these issues into sharp focus. I argue the way forward to retain our humanity, our well-being, our "souls" to use the old religious term, and our sanity is deliberatively to seek to limit our energy use.

The onus to do so must lie primarily with those now using the most, as the energy-deprived poor deserve better and their share. Furthermore these changes must happen quickly.

Nimbies and deniers

This core conclusion chimes with the time-line in the 2022 6th IPCC Report. The rapid reducing GHG emissions to avoid a highly dangerous increase in the mean global temperature of over 2°C must take place in the next decade or two if we are to avoid dangerous tipping points.

However installing, globally, whole new energy systems powered by renewable electricity and maybe hydrogen will take time and a substantial investment not only of money but of natural resources including lithium, copper and rare earths.

Such massive change will inevitably meet with resistance as all renewable energy resources and electricity storage schemes have their downsides.

The nimbies and deniers, as well as voicing some justifiable concerns, will peddle disinformation. I would argue therefore that both the practical short-term priority to reduce greenhouse gas emissions and my broader energy-based analysis are mutually reinforcing. Cutting emissions through reduced energy use should be the overwhelming priority.

Can this be achieved? Why are we so reluctant to face reality?

8 Our future?

It is not so difficult to see why progress in cutting greenhouse gas emissions to date been so slow. The deniers were ploughing and sowing doubt in fertile ground.

Our society, politics and economy are based on cheap, convenient energy, as we saw when first discussing Ukraine and the Russia invasion in article one.

Fossil fuels might have fuelled wars and given status and power to unsavoury tyrants and tycoons but they have been the source of many benefits and lesser fortunes.

Cheap energy has enabled a conveyer-belt of new innovations and gadgets of varying utility and helped sustain economic/GDP growth. Consumerism is the source of much of our own well-being and satisfaction. Shopping is what people do for fun and for entertainment as well as status.

Votes are gained by promising world-beating growth. The whole system has been underpinned by a quasi-religious belief in an individualistic, highly competitive but subtly rigged, free market economics and its associated social policies.

The worship of Homo economicus is on display, daily, in many London papers and politicians. In contrast, the pressing global environmental and social challenges requiring a recognition of our joint responsibilities and of a global, never mind a local, common good, are downplayed.

Hard-wired behaviour

I have argued in my book *Energy the Great Driver* and briefly in Article 6 of this series that our inherited, hard-wired behaviours tend to make us parochial, and easily swayed by the promises and the rhetoric of charismatic, authoritarian leaders.

We have also adopted socio-economic regulatory systems which favour resource exploitation and pay little heed to equity and unfairness.

But all this is not a given! One of the achievements of the hard right wing is the idea of TINA – there is no alternative! There has to be or we all face disaster.

Seeking to prioritise using less energy will not logically lead to greater social justice but it implies a major shift in our ethics and value systems. In democracies we have agency and could exercise it.

Addiction to power

One of my greatest concerns is our long relationship, even addiction, to power. In an article updating the book , I outlined the ideas of Alfred Lotka and Howard Odum.

They suggested that biological organisms seek to maximize their flow of energy per unit time i.e. their power. Those that do so, survive best in Darwinian competition, provided always they can also access other essential resources e.g. water and nutrients (processes which themselves depend on energy).

This evolutionary process has however punctuated by major crashes due mainly to external events, e.g. asteroid strikes or other forces e.g. tectonic drift or volcanic eruptions.

Nevertheless, in each case it has then re-established itself. New opportunities have arisen from the rubble, much as the assent of early ancestors of humanity can be traced back to new situation that emerged from the extinction of the Dinosaurs.

Crashes

Something very similar appears to apply in human societies as described recently by Peter Frankopan. He shows history to be full of examples of individuals and empires gaining more and more power so controlling the labour of many peoples, enforcing their will and making themselves and their inner cohort richer. (If this were not so, why bother with conquering an empire!!)

But these triumphs have always been limited and inevitably followed by crashes. Nevertheless the next aspirant to 'power' is undeterred. Historically power more or less directly revolved around control of land and food resources and consequently the control of the labour of other humans and animals.

Today power can still be seen in similar terms, but as the control of embodied energy and information in other humans. It is now expanded to include all the consequential manufactures, devices and services. Wealth can be seen as a modern extension, even an embodiment, of traditional resource-based power.

In human society the pursuit of power, as recognised over two centuries ago by Matthew Bolton, has been a key driver of change and of a Schumpeterian acceleration of 'creative destruction', as we've discussed.

Thus it appears that both our innate hominid behaviour and the regulatory regimes established in last century reflect are related to and have, indeed, reinforce this tendency to crave power.

Dark ages

In these articles I am suggesting that we must aim to live on far less energy if we are not to undermine the very earth-systems that have allowed **Homo sapiens** to flourish. Maybe the pursuit of power in this sense has run its course!

The logic of the analysis is that the specific mix of psychology, sociology, technology and economics that has been so successful, especially in Western countries, for about 250 years has indeed run its course.

Carl Linnaeus, who coined the term "Homo sapiens", portrait by Alexander Roslin, 1773

If we carry on seeking more and more cheap energy to allow more and more work and power to be exploited by a growing population, we are doomed, in my view, to experience a catastrophic crash: the darkest of dark ages.

Such crashes have happened before, regionally, in human history and in biology and can happen again. In the past such crashes in human history have been triggered, regionally, by major volcanic eruptions, climatic extremes, plagues or external enemies and a mixture of hubris and a lack of foresight by powerful leaders.

Now we are, **globally**, the authors of our own crisis. The hubris and lack of foresight characteristic of ancient emperors, remains. Unfortunately, one must harbour grave doubts whether the powerful and all too often authoritarian elites in this modern world will listen.

This sad state of affairs is illustrated the frustrations of 'extinction rebellion', by the blindness of Putin's the invasion of Ukraine and by environmental cowardice of both Conservatives and Labour in Westminster.

However I would also argue, paradoxically, that the current crises could be an opportunity.

Simpler lifestyles

The climate change and AI are making (some) people sit up and notice. The pressures to reduce GHG emissions can only grow. Given that all renewable energy resources have their own difficulties. and nuclear is an irrelevance in next 15 to 20 years, by which time it will be too late!), our responses could generate the means of avoiding even worse catastrophes in the future.

I've no doubt that we could rapidly reduce our GHG emissions **and** our energy use by adopting the best practices and well-established technologies found around the world. Some relatively simple technologies and the acceptance of some modest constraints could make a rapid difference.

We could and should promote a more cooperative rather than a naively competitive agenda. We could and should adopt simpler lifestyles. In passing let me note that, closing heavy industry, only to rely on importing goods with heavy carbon footprints from abroad, does not qualify as a success.

Holidays in the Emirates or Las Vegas are not essential. Sports should not require regular flights to and from South Africa. Importing skills and carers to look after an aging population is not an adequate response to the **good news** that population growth is slowing.

Sustainability

We need above all a new story – one of human flourishing on our singular planet in which 'less can mean more'.

Our current story describes progress and happiness in terms of unending material growth. Such growth conventionally is measured by national or regional GPD or variants of it. These are widely recognised to be poor and distorted measures of well-being and human flourishing. (A simple example – the longer your commute to work the greater your contribution to GDP).

These measures ignore our dependence on planetary health, on economic externalities and the whole biosphere. In our economic system global threats such a climate change, biodiversity loss and pollution are largely disregarded.

Over the years many have developed better indexes such as HDI (Human Development Index), ISEW (index of Sustainable Economic Welfare) which evolved into GPI (Genuine Progress Index) to measure our wellbeing.

Economists such as Herman Daly have proposed clear criteria for real sustainability and newer models such as doughnut and no growth economics have emerged. They are highly critical of mainstream economics but sadly have gained little traction.

Our current economic management is based on hypotheses such as everyone acting so as to maximise their personal economic welfare, on the automatic equilibration of supply and demand, on actors in the economy having equal knowledge and an appreciation of all possibilities etc.

These are at best farfetched, at worst downright misleading. Yet extrapolations from mathematical models derived from such assumptions drive many government policies and investment decisions – see here.

In terms of the patterns and concepts explored in these articles, all this represents a major failure of man-made (homeostatic) regulation. However my energy-based theory does not lead to one specific regulatory regime; only that appropriate regulation is essential to avoid collapse.

A greater capacity to do work per unit time and emergent complexity and an acceleration of change will need a balancing increase in regulation if it is not to become and more unstable. The theory provides a **framework of understanding**, not a policy blueprint.

In Wales we have our Future Generations Act which formally recognises some of these dilemmas. This has emboldened Welsh Government to reject the M4 across the Gwent levels: a matter of great practical and symbolic significance and pride. Of course and characteristically, the decision is still criticised by acolytes of **Homo economicus**.

The primary issue that emerges is not the science or our technologies (although both are relevant) but how we view of our place in this world. The stories we live by. Do we have the imagination, will and determination to change these and do so urgently? Success is far from assured.

Failure

These changes will indeed cost money – lots of money – as rampant inequality will ensure failure. But in truth the world is extraordinary rich.

Image by Geralt via Pixabay

Disproportionally and deliberately in the last half century especially in the USA and UK wealth has accrued to the rich – the so-called 1%. Thus well over $20 trillion is secreted in tax havens, often dependent British Overseas Territories; in international law these fall under King Charles III's Privy Council and, in reality, they are offshoots of the City of London.

This treasure could be used to transform the world BUT it will be tightly guarded.

(For a perspective on a trillion $; a thousand [103] seconds = <17 minutes; a million [106] = 12 days; a billion [109] = 31 years; a trillion [1012] = 31,688 years; 20 trillion = ~600 thousand years!! – $20 trillion is big number!!!)

The aim of these articles is to start a process of writing a better story about ourselves and our place on this planet. I base it on our fundamental dependence on energy transactions, one of the most important but mysterious concepts in science and on this 'energy' enabling work and power.

Without such transformations nothing can be done but in excess, as in the explosion of a hydrogen bomb, the eruption of super-volcano or even a massive hurricane, havoc.

My conclusion is not new. Many great sages and religious leaders have said very similar things, although using different vocabularies. We must reject the story based on continuous exponential growth in energy and resource use driven by greed and the pursuit of power. There is clearly an urgent need to address in detail how such changes might be achieved and achieved quickly but that would require more articles.

In my book I wrote 'creating a Planet Earth that is a viable home for humanity and all the other organisms with which we share her is surely the greatest project ever contemplated'. Hopefully it is an aspiration that we can share and an inspiration on which we can build, despite the daunting difficulties.

The alternative, I fear, is a crash which humanity may well survive but at very great cost.

5
Energy and Power

This chapter elaborates and extends the hypotheses and conjectures. It is based on material I've accessed since the first publications and the advice I have received from experts more versed in the various individual disciplines than I.

While I believe it presents a coherent hypothesis, it also raises many questions which lie beyond my capacity to answer satisfactorily. For example, the relationship between free energy entropy and information and their implications for humanity's future, the nature of and stabilisation of social complexity, our capability as human to cope with ever more rapid change, the harnessing of science and technology for the public good and fundamental ethical even religious questions about the 'good', free will and our values.

Energy and Power

1 | The Hypothesis:

1.1. Introduction.

Energy is one of science's most remarkable concepts. It can occur in many, apparently unrelated forms that, amazingly, can be quantitatively inter-related to each other with great precision. It is implicit in gravity, in the flow of electrons or water or heat, in radiation, such as light from the Sun, as well as in the food we eat and in the coal, oil and gas we burn. It is contained in mechanical motion as well as in sound.

The common denominator is that energy, in its many guises, can makes things happen. In physics and chemistry – as well as in everyday language – 'do work'. Again this work can take many and varied forms – heat an electric fire or a blast furnace, light a blub or run an electric motor or steam engine and allow people to run and procreate, talk and communicate.

One consequence of this is that energy and work can be defined in a confusing plethora of units; joules, watts, electron-volts, calories ([2] and see Appendix 1).

Crucially one of science's most fundamental laws is that energy cannot be created or destroyed. It can only be changed from one form to another. However, with each transaction, some energy is lost as waste heat which can no longer do useful work – this 'loss' is termed **entropy**. So, although the total universal quantity of energy is constant, the **free (i.e. useful)** energy is gradually declining while entropy is increasing. Entropy is related to increasing disorder, or perhaps better a lack of information. and embodied energy or negentropy with order. Many believe that the trend of gradually increasing entropy gives rise to **the arrow of time**.

To add to the problem, Einstein famously showed mass and energy are interchangeable, as described by his famous equation; energy equals mass times the speed of sound squared ($e = m.c^2$). Consequently, we must now say it is energy and mass combined that are conserved in the Universe.

Energy is a very slippery but fundamental concept.

If you are puzzled you are not alone. One of the most famous physicists, Richard Feynman, said;

"**It is important to realize that *in physics today, we have no knowledge of what energy is.* We do not have a picture that energy comes in little blobs of a definite amount. It is not that way**"[3].

In the context of this essay and these hypotheses, it is important that ,despite the universal trend towards increasing entropy and 'disorder', under certain circumstances in discrete systems and structures, free energy transformations can lead to enhanced 'dynamic order'.

Both empirical observations and theory show that accessing new *free energy* sources or revolutionary changes in the ways such energy sources can be exploited, have allowed more work to be carried out and power generated. In turn I will argue that these capacities have resulted in the sequential development of physical, biological and, latterly, human socio-economic and material structures of growing dynamic complexity. As I will elaborate, Eric Chaisson called this trend, **the arrow of complexity**: a powerful, but not I think, an entirely appropriate metaphor.

Importantly however, such structures and systems are intrinsically unstable and require specific mechanisms to ensure their sustainability.

In Earth's multi-billion-year history a series of major step-changes in the above relationships can be identified while, undoubtedly, a myriad of other smaller but important steps have occurred in the planetary energy economy. Over time, I am arguing that these have led to human planetary dominance and our remarkable material prosperity, even though the latter is distributed very unevenly. Given the issue of intrinsic instability, these trends have also necessitated the parallel evolution of a hierarchy of homeostatic mechanisms to stabilise the emerging ordered structures. Such structures, in terms of thermodynamics and energy flow, are poised far from equilibrium with their surroundings and will would collapse should the energy supply be curtailed. I argue this is as much the case in modern megacities as in the earliest single living cell several billion years ago. However in biology and in socio-economic elaborate systems have evolved to stay and circumvent these underlying trends.

In this paper the evidence for these propositions is outlined and their implications for our current global crises and their possible resolution explored.

Increasingly our fossil-fuelled prosperity is recognized to contain the seeds of its own destruction and is forcing us, albeit reluctantly, to seek to drastically reduce our dependence on such fuels to avoid catastrophic anthropogenic global warming (AGW). This revolutionary change is occurring while humans are also trying to adapt to the emerging digital age of artificial intelligence (AI) and nanotechnology. These latter tools are themselves also are drastically altering the relationships between energy use, work and complexity.

Many of these issues are widely discussed. However I am suggesting that there is another arguably fundamental relationship which receives little attention. Namely the implications for both human welfare and planetary health of an ever increasing exploitation of energy **regardless of the source of that energy**. Let me emphasise, even if the source of the energy does not generate pollution or other environmental damage, that free energy use and its coupling to work and power is itself a serious issue.

It is widely assumed in the global political and business communities that, provided GHG emissions can be minimised and catastrophic AGW avoided (e.g. by putative 'net zero' emission by mid-century), the current trajectory of exponential consumer-based economic growth and of increasing total global energy and resource use remains valid. Historical observations and physical, biological, and behavioural evidence explored in this paper suggest this assumption is not valid and, highly unlikely, to be compatible with human or planetary wellbeing.

1.2. The Underlying Science.
Few can have lived through recent years without becoming aware of the centrality of energy to human society. The sudden spike in energy costs in the UK caused real hardship as well as political consternation. The Russian invasion of Ukraine is an 'energy war' against both Ukraine itself and Europe generally. The blocking of Ukrainian grain exports (energy as food calories) threatened hunger in some poor countries and was seen by Putin to create additional leverage.

The devastating impacts of AGW from our continuing reliance on fossil fuel energy has contributed to the global prevalence of fires, floods, excessive heat, droughts and extreme weather events.

While the global community gathers annually at UNCOPs to talk about reducing, even eliminating, this reliance on fossil fuels, too little is being achieved. Our efforts remain sadly inadequate. It is a tragic commentary on our capacity to plan intelligently that the Russian military threat has catalysed a much more rapid decarbonization of the European energy economy than ever envisaged in response to climate change itself.

The centrality of energy in our daily lives reflects some of science's most important principles[4,5]. To quote the respected physicist, Vaclav Smil[4]:

"All natural processes and all human actions are, in the most fundamental physical sense, transformations of energy. Civilization's advances can be seen as a quest for the higher energy use required to produce increased food harvests, to mobilise a greater output and variety of materials, to produce more, and more diverse goods, to enable higher mobility, and to create access to virtually unlimited amounts of information".

As noted in the Introduction the fundamental concept is "**free energy**" i.e. useful energy. This must be expended to do any work. As mentioned, the resulting transformations release low-grade thermal energy, termed **entropy**, which has lost its ability to do useful work. Overall energy cannot be created or destroyed. So, in closed systems, the entropic component gradually rises and capacity to do useful work declines. At a cosmic level, outlined in Appendix 1, this leads to the concept of Universal Heat Death. That is the Universe, as a closed system, is gradually losing order or a lack of specifying information and useful energy – leading to nothingness – "heat death".

Despite this universal trend, it has been observed and theorised that an expenditure of free energy can lead, under prescribed conditions, to the spontaneous emergence of entities of physical and material complexity.

The work of the Nobel Laureate Ilya Prigogine and his colleagues has provided a theoretical understanding of these phenomena, demonstrating that the ***localized***, ***spontaneous*** creation of ordered, dynamic complexity – in so-called ***dissipative structures*** – is consistent with the overall cosmic and thermodynamic trend to higher entropy and greater disorder[6]. Such dynamic physical structures depend on a continuous flow of energy and a turnover of matter, but dissipate if subjected to too rapid or an insufficient energy flux. This phenomenon has been explored in detail in a simple well-explored model system – the Rayleigh-Bénard cell[7].

In physical terms these emergent structures can be conceived of as dynamic, ordered entities of low or negative entropy (see Appendix 1) but high embodied-information and energy content i.e. am ordered entity requires more information to specify and prescribe its state.

These phenomena have been widely discussed and seen as relevant to both biology and human society as will be elaborated. Indeed, a living cell may be viewed is a prime example of this phenomenon. In 1944, well before Prigogine's work, Erwin Schrödinger[8] proposed that a cell should be viewed as a temporary island of increased order [***negative*** *entropy*] within an overall system fated to descend into disorder.

Schrödinger first used the term 'negative entropy' to describe this localised ordered, but dynamic state; one 'feeding' on the increasing entropy of its external environment. Well known non-biological examples of such dissipative structures include hurricanes and eddies in fast-following streams. These 'structures' will collapse if their energy supply changes e.g. if the thermal gradient arising from the warm oceans feeding a hurricane is lost over land.

In an analogous way, any cell, organism or society may be seen as also dependent on a flow of external energy to maintain its structural, dynamic integrity. When external free energy is no longer available (or the internal stores of food energy are exhausted), death and decay will ensue. Consider how long a large city would survive if all its energy supplies were cut off. Nevertheless, in ecosystems, the embedded chemical energy potential of the dead components may still be the source of energy and chemicals for other organisms lower down the energy/food chain.

A critical factor for any organism or society is its ability to regulate this flow of free energy and, concomitantly, of material – and to survive glitches!

1.3. Major Energy Step Changes.
As I've mentioned, an examination of Earth's planetary history suggests that new sources of free energy, or major step changes in the ability to convert available free energy to work, have led to revolutionary changes, initially in biological and, latterly, in socio-economic, structural complexity.

Each step has also offered new evolutionary opportunities and divergence. The newly emergent entities, nevertheless, have remained dependent on the activities of the existing less complex forms; that is on functional, dynamic ecosystems and a healthy global system.

In '*Energy The Great Driver*'[1] I identified and discussed at some length six major historical step-changes in the planetary energy economy, which profoundly impacted on Earth's atmosphere, oceans and geology, as well as on life. Consequently in this essay I will only offer a very brief summary of the evidence, noting the approximate time of each step. This time line is illustrated in Figure 1.

These are:

1.3.1. Step Change 1. Evolution of first living cell. (~3.8 billion years ago)

As noted, a cell is an energy-dependent, localized, complex, organised and dynamic entity existing far out of equilibrium with its surroundings. The original energy source to allow such an entity to arise and how it was exploited is the subject of much speculation. The currently favoured sources, which might have allowed a proto-cell to be established, are natural long-lived gradient of pH [H+] that occur in alkaline oceanic vents.

Evidence for such prokaryotic life, which occurred in two distinct forms Bacteria and Archaea, dates back to over 3.5, may be to 3.8, billion years ago. I course recognise that any cell also requires its chemical building blocks and, if it is to survive from generation to generation, a mechanism for coding the essential information and its reliable and consistent largely transfer. The genetic code embodied in DNA and RNA provides such capability. ([9] [10] [11] [12]) explore variants on the possible sources of these primordial energy fluxes.

1.3.2. Step Change 2. Solar energy capture and oxygenic photosynthesis. (~2.7 billion years)

For perhaps over billion years life depended on existing but limited planetary sources of chemical energy e.g. highly reduced organic or inorganic compounds, as well as various terminal electron receptors to create an exploitable energy gradient. From about 2.7 billion years ago there is evidence of the evolution of mechanisms to allow the capture of solar radiant energy by photosynthesis.

This crucial step allowed organisms to exploit an abundant, extra-planetary energy source and remove the energetic constraints of a limited supply of chemical electron donors and acceptors in redox gradients. Life on earth, with few exceptions is now energized by the photosynthetic capture of the sun's radiant energy although localized ecosystems dependent on chemical energy still exist in lightless caves and mines.

Very gradually photosynthesis also lead to the oxygenation of the atmosphere and a complete change in planetary (atmospheric, oceanic and geological) chemistry. (See[13] for more comprehensive account.)

1.2.3. Step Change 3. An energy revolution with appearance of the first eukaryotic cell. (~2- 1.7 billion years ago)

For several hundred thousand years until some 1.7 to 2 billion years ago, life on this planet remained prokaryotic. It was confined to single cells with little internal visible structure e.g. no nucleus or internal organelles and capable of the ready exchange of strands of DNA. At about that time evidence emerges in geological record of the evolution of a more internally complex eukaryotic cell – eukaryotes. These were formed by a unique endo-symbiotic fusion of an Archaean and a Bacterial cell; the former providing the bulk of the eukaryotic cell and the later its mitochondria which act as power packs retaining only about 13 of their original genes[14].

Eukaryotes exhibit greater cellular and genomic complexity, presaging multicellularity and the evolution of complex plants, fungi and animals with defined organs carrying out specific functions. They have greater metabolic power per cell and per gene and per genome[15] but as eukaryotic cells are larger and of a great mass and complexity, it is not proven that that the step change can be unambiguously attributed to crossing an energetic barrier (see also[16] and especially[17]).

Recently Schavemaker and Munoz-Gomez[17] have shown that larger and faster-dividing prokaryotes would have a shortage of respiratory membrane area and a need to direct more energy into DNA. Thus, they argue that, although mitochondria may not have been required by the first eukaryotes, eukaryote diversification was ultimately dependent on presence of mitochondria.

This diversification first led to multicellular organisms and ultimately to the natural world we recognise full of varied plants, from trees to bryophytes and alga, huge variety of fungi and animals as different as beetles, lions and humans.

1.3.4. Step Change 4. Energy expended on brain development, hominid intelligence and communication skills.
(~1.8 million year ago)

While there are a number of candidate energy step changes in the next hundreds of millions of years, such as the evolution of multi-cellularity and the extension of photosynthesis to the terrestrial sphere, the evolution of a more intelligent Hominins is an unambiguous revolutionary change.

An early Hominin, likely *Homo erectus*, took the revolutionary path of investing additional energy and associated nutrients in their mental abilities. The evolution of this enhanced brain capacity was energised by much improved food intake and digestibility, likely enhanced by cooking. The assignment of that energy and material to the brain as opposed to brawn set one species on a new evolutionary course.

Homo species, including of course ourselves *Homo sapiens*, are unique in assigning 20 to 25% of their metabolic energy to their brains. This investment has led to superior intelligence and greater social interactions, quite possible supported by growing language skills. Such changes would be mutually reinforcing by promoting collaborative hunting and tool making and inter-generational and inter-tribal information transfer.

In hundreds of thousands not millions of years, hominin species spreading from their African origin to most of this planet. Over time Homo species become the dominant predators securing more of the Earth resources to sustain their growing dominance and capabilities [18][19].

1.3.5. Step Change 5. The agricultural revolutions with improved energy supply and more dense and permanent settlements. (~10 to 8 thousand years ago)

Remarkably recently, as the last Ice Age ended, there is clear evidence of discrete agricultural revolutions in a number of regions including Mesopotamia, China and the Americas. In all cases crops were domesticated and grown leading to settled community and enhanced, more reliable but more demanding food supplies. Enhanced human energy (food calories) captured through agriculture led to emergence, albeit fitfully, of more complex societies, record keeping and writing [20][21][22].

1.3.6. Step Change 6. Industrial revolution based on fossil fuel utilization. (~250 years ago)

Human exploitation of power embedded in fossil fuels and the development of technical, scientific and latterly digital expertise. Rapid increase in numbers of *Homo sapiens*, the growth of urbanization industrialization and changes in the systems of local and international governance, in economic management. Parallel increases in the capacity to make war, locally and globally e.g.[23][24].

Aspects of step changes 4, 5 and 6, including the role of energy transactions, will be discussed more fully in later Sections (see especially 1.6.

A number of other authors including John Maynard Smith and Eors Szathmary[25], Robert Ayres[26]. Eric Chaisson[27] Tim Lenton and Andrew Watson[13] and Olivia Judson[28] (see also Lovelock[29]), has made similar but not identical proposals. In some cases, other possible step changes are highlighted. Chaisson developed a comprehensive hypothesis based on changes in the local free energy rate density which he applied to cosmic as well as planetary evolution[27].

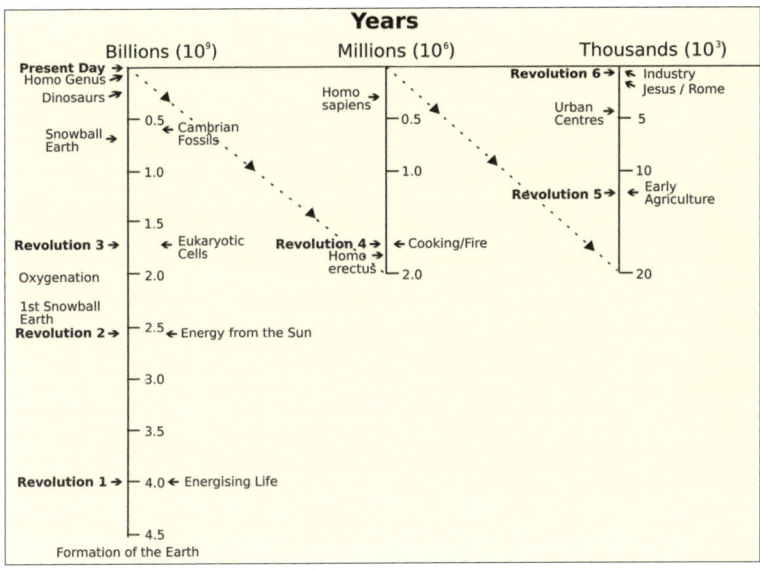

Figure 1. Timeline of Major Events

1.4. Other Possible Step Changes.

Undoubtedly many other smaller but crucial 'energy' events have punctuated biological and social evolution.

Within biology, possible examples include the evolution of bipedism in early pre-hominids making movement more energy efficient. The emergence of warm-bloodedness in mammals with a resulting increase in energy demand but allowing work in cooler conditions. This, it is suggested, and increased the value of investment in offspring likely enhancing the value of empathy and care. The putative relationship of multi-cellularity with the peak of atmospheric oxygen and greater metabolic efficiency preceding the Cambrian explosion when many progenitors of animals appear in the fossil record may be another highly significant revolution. Similarly the endo-symbiotic evolution of photosynthesis in eukaryotic cells leading to plants and later the colonization of land by plants, increase in the global capacity for photosynthetic energy and carbon capture. Unfortunately a more critical examination of such steps lies outside the scope of this paper.

A wealth of relevant examples of changes in the energy economy throughout human history are explored in Smil's work[4], including the improved exploitation of wind and water power by humans prior to the Industrial Revolution.

Conversely major, but usually localized, declines in the availability of calorific food energy to human societies have played a crucial role in human history and the fall of many great empires (see Peter Francopan later). Such harvest failures have been due to droughts, volcanic events and long term climatic changes and frequently wars and social upheavals (sometimes catalysed deliberately, or by their culpable neglect, by external powers e.g. Holodomor, Irish and Bengal famines).

Major changes in the 'energising' of the weapons of war can be interpreted in similar light. Starting from spears depending on human strength and its mechanical supplementation by bows and arrows and various siege devices, human ingenuity has led to the use of gunpowder and other chemical explosives. Now we have highly energized projectiles from guns and rockets of remarkable and deadly precision [see also[26]]. Also relevant to the energetic of war is the mobility afforded first by horses and chariots and now by fossil fuel-driven aircraft, missiles and tanks. We must also recognise we now face the potentially catastrophic energy-release from a nuclear war and fission and fusion.

1.5. The Imminent Challenges.

In section 1. 3 I outlined six historic step changes in the global energy economy culminating in the Industrial Revolution dependent on the burning fossil fuels. While the CO_2 emitted is the prime cause of AGW, additional CO_2 has been and continues to be released due to land use changes, dominantly forest clearance. Methane, a highly potent but short-lived gas, and nitrous oxide, both high potent and long lived, are also indirect products of the Industrial Revolution. But latter gas is present in less a thousand times the atmospheric concentration of CO_2.

Methane emission are a major by-product of gas and oil extraction. Often as a result of poor practice. However the food chain is also a major source of these gases. Methane is a product of the fermentative anaerobic metabolism of fodder by ruminants as well as the anaerobic breakdown of resultant slurry and manure. Globally most pastoral systems dependent on ruminants – cattle, buffalo, goats and sheep. Historically and still today they remain a crucial to human societies in many parts of the world. Methane is also produced during paddy rice production, again because the system is largely anaerobic. Nitrous oxide is released by soil bacteria as a proportion of applied nitrogen fertilizer as well as by combustion of fuel in vehicles.

The intimate relation of the human food chain with GHG emissions is a major problem for future generations.

In sharp contrast to the earlier revolutions humanity is now engaged in a deliberative seventh revolution. Namely seeking to drastically curtail these emissions while sustaining an economy based on cheap convenient energy, including food calories. This at a time when, on one hand, a large proportion of the world's energy-deprived poor are bent on emulating the material abundance of the rich. And, on the other hand, the advent of digitisation and information technology [AI] is altering the relationships between energy and work and certainly social complexity.

These challenges will be the subjects of the next two sections.

1.5.1. Step Change 7. Energy sources allowing the discarding of fossil fuels. (current/next 10/20 years)

The scientific establishment, embodied in the International Panel on Climate Change (IPCC), has presented compelling evidence that the increasing atmospheric accumulation of the greenhouse gases (GHG) (including CO_2, CH_4 and N_2O), is leading in a few decades to catastrophic climate change and global heating. As noted this is due mainly, but not exclusively, to the burning of a fossil hydro-carbons.

The greenhouse gases lead to the retention of an additional increment of incidental solar radiation in the oceans (>90%), ice caps and the atmosphere ([30][31][32]). However this exploitation of fossil fuels was, and remains, pivotal to the Industrial Revolution, to our current liberal-market economy and to prosperity.

The 7th revolution is therefore unique. Human society, alerted by science to a growing risk, is seeking to discard the energy source (fossil fuels) which is fundamental to our prosperity and central to business and global geopolitics. New energy sources are being sought which do not threaten AGW. But the nature of the market economy, global power structures and human behavioural, all militate against the rapid action necessitated by the IPCC data and tend to support the continuing use of fossil fuels. This conflict is being played out, annually, in the UNCC COP events as well as in the machinations of national politics.

The mainstream in all countries, irrespective of political orientation, hopes and expects a gradual switch to low emissions energy sources to be made with minimal disruption. Indeed the GDP growth model of the economy anticipate and necessitates a continuing rise in global energy demand and availability.

It cannot be over emphasised that we are 'living' the unique **seventh revolutionary** step-change. We are seeking to energise and feed a *H. sapiens* population approaching 10 billion within the next 20 to 30 years[33] in an energy-dependent economy with 'net-zero' GHG emissions. Moreover, AGW is but one of number of areas in which human activity is exceeding safe planetary boundaries[34]. These include various forms of pollution, including by phosphorus and nitrogen, plastics and other chemicals, land use changes – some involving the loss of tropical forests as carbon sinks. There are threats to fresh water habitats aggravated by human demands on water resources often for irrigated agriculture as well as for industrial processes and drinking.

The political consensus is that major changes can be achieved while avoiding both catastrophic global warning and a breakdown of human society. In reality, it is not clear that this will be achieved.

It is assumed by virtually all mainstream economists and politicians that:

[a] the current globalised free-market economic model, possibly with minor modifications, remains valid and

[b] renewable, low-C fuels (wind, solar, tidal, hydro etc.), with likely additional contributions from nuclear fission and possibly nuclear fusion, and appropriate energy storage will allow a near seamless transition.

[c] Some recourse to Carbon Capture and Storage (CCS) is increasingly assumed. Some also consider that forms of global geo-engineering will be necessary as the use of fossil fuels will not be phased out completely or sufficiently quickly. Certain products – steel, cement, plastic and industrial nitrogen fixation for food production – are notably energy intensive and are especially problematical.

How the food chain will achieve near net zero as well as coping with climatic extremes is a very real issue but not as widely debated. Digitization and computation are also themselves making rapidly increasing demands on the global electricity supply.

[d] Underpinning this attitude is a faith that human ingenuity, primed by capitalism, will produce timely technological fixes, as will be discussed later.

Three points deserve emphasis.

Firstly, the official target, agreed in the 2015 UN Paris Climate Accord, is to keep any increase in the mean global atmospheric temperature (currently +1.1 to 1.2°C above preindustrial levels) near to +1.5°C. However atmospheric CO_2 concentrations continue to rise, indeed to accelerate. Even the most optimistic projection of a decline in anthropogenic emissions indicate that a mean temperature increase of well over 2°C. Many fear that tipping points will be breached causing reactions which will further accelerate warming e.g. methane release from the Tundra.

Given that this increase is unevenly distributed (greater over land than over the oceans; greater nearer the poles than in the tropics), this implies mean increases of well over 5 to 6°C in some places. Despite the dangers of catastrophic global warming being apparent for at least 30 years (and warned of for well over half a century), a mean rise near to +2.5°C is not unlikely before the much-acclaimed, but disputed, global net-zero emissions are achieved. Currently, the global economy and our living standards still depend for over 80% of their energy on the burning of fossil fuels.

Secondly it is also important to note that larger the human population then, given the long residence time in the atmosphere of additional anthropogenic CO_2, the smaller will be the per capita ration of GHG gases available to each human in the future, Or seems probable the greater will be the required CO_2 capture and storage per head and cost.

Thirdly, in democratic countries the promise of constant material growth is deemed essential to electoral success. In the popular imagination a limitless supply of cheap pollutant-free energy will allow undreamt of prosperity – streets paved with gold. Business and government policies assume and covet huge increases in energy use into the next century. Not only, or even necessarily, to help the poor but to further enrich the already well-off. Vast sums are being invested in harnessing nuclear fusion and technological initiatives for nuclear fission e.g. Small Modular Reactors in pursuit of such a vision with little thought given to its human or planetary implications. With no due diligence.

1.5.2. Possible Step Change 8. New coupling of energy to work through Digitization, Artificial Intelligence [AI] and Miniaturization.
Efforts to replace our primary hydro-carbon energy sources are taking place in the midst of other revolutionary changes which themselves impact on the relationships between free energy, work/power and complexity. Electronics, digitization and major advances in AI, robotics

and miniaturization are revolutionising the coupling of energy transformations and work to complexity[36] as sketched in Fig. 1 below. To date our primary dependence on fossil fuel energy has remained undiminished (~80% of world's energy still coming from fossil fuels). Indeed an increasing % of the global electricity supply is being devoted to computing and data storage[36]. Nevertheless these activities can be interpreted as a concurrent **8th Revolution**; possibly in the long run, comparable with the early hominid investment of energy in brainpower and information processing.

Some of the leaders of this 'digital' revolution also cherish a vision of a technological utopia, based on limitless energy. This is given unbridled expression in Marc Andreessen's Techno-Optimist Manifesto[37] His argument will appeal to many committed to a particular version of 'progress', to free market economics and technical fixes.

1.6. Some Characteristics of the Step Changes and the Coupling (at 'A and B').

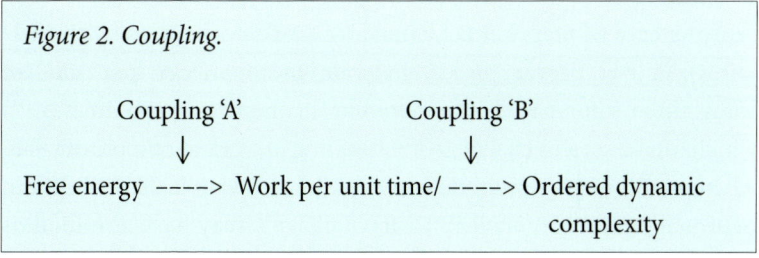

Figure 2. Coupling.

Coupling 'A'   Coupling 'B'
↓   ↓
Free energy ----> Work per unit time/ ----> Ordered dynamic complexity

The major and minor step changes outlined in 1.3 and 1.5 clearly differ in important ways and may be subdivided (see Fig. 1) as:
 a] The accessing of additional energy sources by favoured groups,
 b] An alteration in the relationship between free energy and its ability to do work and promote power – Couple 'A'.
 c] A change in the capacity of work and power to generate complexity – Couple 'B' – or

d] Some subtle combination of the above.

The evolution of oxidative photosynthesis and the capture of solar radiation (Step 2) is clearly an example of the first category. The same may appear to apply to the Industrial Revolution (Step 6) and the exploitation of geologically-stored hydrocarbon energy, although such fossil fuels are ultimately derived from historic photosynthesis. However major changes also took place in the coupling of these energy sources to work with the invention of, first, the steam engine and later many mechanical and electrical devices. All still ultimately powered by coal, oil and gas.

These technologies, perhaps especially the realization of potential of electricity, owe much to experimentation and ingenuity and have impacted on coupling at A and B. These changes in coupling efficiency dramatize the long term impact of our greater brain power (Step 4). Remarkably the modest additional energy investment in hominin brains has had, overtime, huge impacts by enabling new ways of coupling energy to work and to social cooperation, material complexity and civilization.

In the case of Step 3 it is claimed that eukaryotic cells, with their mitochondrial power packs, generate more energy per unit of transmitted information (i.e. per gene to the first approximation[14]) which can be seen as change in the coupling of work to complexity – 'B'. All human cells have multiple mitochondria (1000 to 2000 per cell) but neuronal brain cells, critical to Step Change 4 may have a million or more[38].

The Agricultural Revolution (Step 5) increased local food production and supported larger and denser human populations in specific regions with the gradual emergence of elite castes, the division of labour and enhanced organizational skills. These led to social and economic feedbacks whereby the new changes in primary food production allowed more goods to be produced. In turn promoting more innovation. These changes are exemplified in harnessing horse and bullock power, in more power exploited in manufacturing e.g. smelting, ceramics, in transport and in social organization and politics.

Crucially, it created a demand for record keeping, ultimately leading to writing and libraries. All were also depended on the harnessing of collective, cooperative, as well as coercive, work for complex tasks such as temple and palace building as well a food production and storage. It is apparent that, while the simple model in Fig. 1. may be said to illustrate the possibilities, in virtually all cases the primary energy change had multiple and complex consequences whose impacts took many years to become apparent. Furthermore each step change depended on and built on the potential inherent in its antecedents.

It is also possible to recognise pre-revolutionary changes which potentiated each of the revolutionary steps (see also 1.4). Warm-bloodedness is a possible precursor to enhanced empathy and child care, crucial to early hominids and humans. Bi-pedism is seen as an energy-efficient precursor to the mobility of *Homo erectus*. The gathering and exploitation of useful seeds, fruits, nuts etc. was important prior to the agricultural revolutions. The emergence of new attitudes to commerce and science and technology occurred prior to the industrial revolution.

Earl Cook[39] in the early 70's made an initial quantitative estimate of energy flows in an industrial society compared with the basic human physiological energy requirement. His data were expressed as food calories (kcal) per person per day. A much fuller analysis has been produced by Yadvinder Mahli[40] using the terms metabolism and sociometabolism. He defines biological metabolism as 'the energy exchanged between an organism and its environment' and sociometabolism as 'incorporating biomass resource comsumption from food and fuel' and 'energy generation and consumption from other sources including fossil fuels, and nuclear, solar and wind power'. Therefore 'metabolism' in Mahli's usage is identical to my short-hand 'energy' and Cook's energy flow.

To summarise Malhi's finding. The metabolism of an active human adult produces energy at the rate of about 120W or the power of two old-fashioned tungsten bulbs. This can be equated to a requirement for about 2,500 food calories per day. As emphasised, uniquely in humans 20 to 25 % of this, but only 15 to 20W, is devoted to our brains; that is to information processing with its social, technical and intellectual consequences.

Likely the hunter-gatherer-cooker life style, which persisted for well over a million years and through major climate cycles, required some 300 W or about 6 to 7,000 food calories per day. This number seeks to take into account the energy needed of hunt, cook, create shelter, make tools and weapons. There must have been significant variation from locality to locality so such numbers can only be indicative. However, as Malhi documents, this life style made minimal demands on the natural energy and resource flows and permitted only limited population density and growth. Even so it seems related to the demise of many mega-fauna in many parts of the world.

In contrast, after the agricultural revolution the energy flow required to support human life styles rose to some 2000W per head, equating to some 30-50,000 food calories person per day. This led to significant local ecological change ,described by Malhi as the human colonisation of the landscape and its biodiversity to support humans in greater numbers and density. Variants on this life style remain in many parts of the world although the world of hunter-gatherer-cooker communities have largely been overwhelmed.

As recorded by Malhi, and earlier Cook, the fossil-fuelled Industrial Revolution catalysed another leap in the energy flow or social-metabolism per head. As might be expected from the earlier revolutions, it has permitted a huge increase in human numbers and density. Perhaps less obviously it also caused more photosynthetic energy and carbon tobe harvested to meet our needs at the expense the rest of the biosphere.

The leap is exemplified by the coterminous USA where around 12,000W per head or around 240,000 food calories equivalents per day per person are use. It's worth noting that, even within the US, great poverty and great wealth coexist and this is a mean figure. Undoubtedly the very rich depend on several times the 12,000W per person of energy. Although relatively few enjoy this very high energy life style, it is presented, socially and politically, as the one to which we should all aspire.

This brief summary adds a degree of quantitation to the three historic human step changes and shows a hundred fold increase in energy flow.

Two important points are emphasised by Malhi. Of the total solar energy flux reaching the Earth's surface only about 0.2 % is captured in the biological metabolism – biosphere energy flow – 265 TW (10^{12} W). In terms of this current photosynthetic energy capture (265 TW) about 60% occurs terrestrially and some 40% is marine. This serves to emphasises that the importance of the colonisation of land; a very long drawn out process which started with lower plants before the evolution of higher plants perhaps some 400 million years ago.

Probably several orders of magnitude less energy flow would have been available to support life on Earth prior to the evolution of oxygenic photosynthesis, given the paucity of geological electron donors and receptors in that period.

It must be emphasised that the great wealth and depth of biology, history and human experience is enshrined in both the couplings and emergent complexity. Nevertheless the contention is that, at the deepest level, the emergence of each new step change in the energy economy enabled or catalysed a revolutionary planetary event with fundamental biological, social and even geological impacts. The emerging, dynamic, multi-component structures have made growing demands on planetary resources – initially photosynthate, essential nutrients and water; latterly also the wide range of materials used by humans. These energy-driven events and processes have inescapably made greater and greater resource demands (e.g.[41]).

1.7. Homeostatic Hierarchy.

1.7.1. Physical theory (see 1.2) implies that dissipative structures are unstable without an appropriate and continuous supply of free energy and, in most instances, of material. Such structures exist at the expense of increasing the entropy (crudely 'disorder') of their external environment[6 8 26 27].

Crucially, organisms and human social and material constructs have evolved mechanisms to stabilize and regulate themselves despite irregular energy flows and changing external conditions. These are an essential aspect of cellular metabolism. Indeed, all aspects of the relationships between the individual cells, whole organisms, and human societies, and their external environment have to be regulated and stabilized. One remarkable feature of a cell, and of life itself, is that it has released itself from the second-by-second tyranny of a dependence on a steady flux of external free energy by storing chemical energy and by a variety of metabolic stratagems to survive e.g. spore formation, even hibernation. How the first viable cell achieved this degree of stability remains a mystery, as far as I am aware but a requirement on a consistent energy flux cannot be doubted.

In biology the term *homeostasis* is often used for these regulatory aspects of metabolism. Including, for example, the mechanisms leading to the regulation of specific functions e.g. blood pressure, body temperature in mammals, cell ionic composition; each around given set points or normative values. Some of these are remarkable constant over a wide range of organisms[42]

A hierarchy of such homeostatic mechanisms can be recognised from a single prokaryotic cell to a complex multi-organelle, multi-cellular, multi-organ animals and higher plants through to relatively primitive and now much more complex human societies (see Fig. 3) (and[1] and [43]).

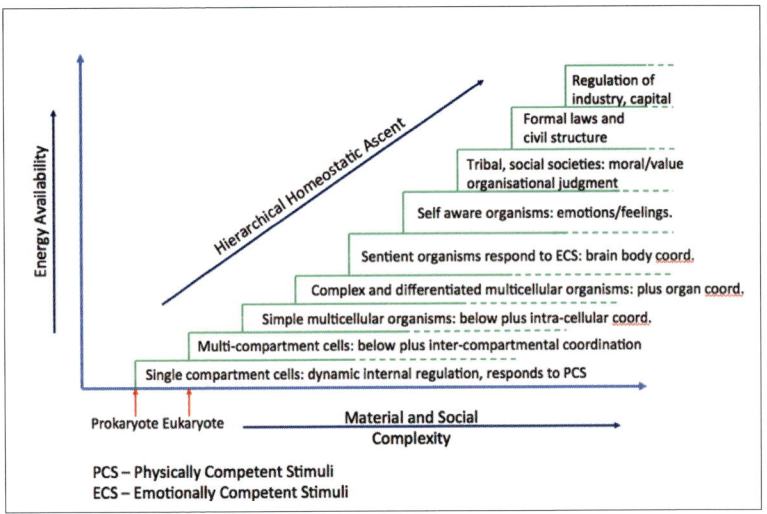

Figure 3. Homeostatic Hierarchy

While each step has posed new regulatory challenges, natural selection operating over thousands of generations appears to have ensured the homeostatic stabilization lower down the evolutionary hierarchy is virtually invisible in higher organisms, or in individual humans, unless revealed by disease or some other disaster. Day by day we are unaware of the importance of maintaining our electrolyte balance or our blood pressure near to the appropriate set-point, still less the detail of our cellular metabolism. Biological homeostasis works exquisitely well. Any major deviation denotes a medical condition and may result in death.

Antonio Damasio[43] has argued that this biological hierarchy should be extended to human behaviour, especially as expressed in and through our emotions and feelings. In our near hominin ancestors and in modern humans, these emotions and feelings must have evolved to enable a degree of social, sustainable, yet dynamic, cohesion. This, in turn, has permitted our socialising, our cultural diversity and our technological prowess. It may be argued these adaptations have been critical to the evolutionary success of *H. sapiens* as compared with other apes. Changes in behavioural and social interactions have allowed humans to coexist and collaborate in close proximity and in relative harmony unlike most of our closest primate ancestors.

Recently, light has been shed on one such mechanism of physiological and psychological adaptation in mammals including *Homo sapiens*.[44] At both a molecular and behavioural level, domestication has been shown to involve the modification of the migratory patterns of neural crest cells during embryonic development. These cells are responsible for both physiological and behavioural changes favouring cooperative behaviour and a decrease in "reactive aggression" i.e. the instinctive aggression characteristic of Chimpanzees. Richard Wrangham has suggested, supported by recent molecular biology[45][46] (but see also[47]), that such domestication has occurred in *Homo sapiens* over the last 300,000 years. Wrangham[44] has further postulated that in humans the culling of the over-aggressive and violent, usually males, by their communities (often their own family) has provided the selection pressure to bring about this trend. However, he suggests that replacing the single dominant 'alpha' male by a group of dominant "male elders", able to rule and kill at little personal risk, has enhanced the human propensity for "proactive aggression". He is referring to premeditated, organized aggression to maintain the status and resources of an elite. Our human history and modern geo-politics are replete with example of such behaviour.

Arguably this is an example of a broader socio-regulatory phenomenon. Over some 1.5 million years the evolution of the innate, behavioural and social characteristics in *Homo erectus* and later *H. sapiens* (and related species) must have achieved a degree of dynamic regulatory/homeostatic stability within small quasi-independent, but also partly inter-dependent, groups of hunter/gatherer/cookers (population density of 0.1 human per km2 or less ;cf. current population density of Singapore over 8,200 people per square kilometre).

Over the last 12,000 years a growing proportion of these also became small part-time farmers. However it is questionable how suitable the homeostatic characteristics that helped stabilize such hunter-gatherer communities and even small farmers are appropriate to those living in a modern, globalised, dense, urbanised society. It is also unlikely that sufficient time has elapsed for any new social, regulatory equilibrium to have evolved biologically in modern humans in the relatively short periods since the agricultural and, especially, the industrial revolutions. Consequently social and political stability now depends on the social norms, regulations and laws developed by human agency and often imposed by authorities to maintain social and economic order; be the latter well thought out or not.

It must be emphasised that:

a] In humans, homeostatic regulation has moved, significantly, from the biological-inherited to the social sphere.

b] Our biological inheritance, including our propensity for rapid, instinctive "system 1" decision-taking (see Kahneman[48], Haidt[49]; see also [44] and [50][51]) remains (albeit much less reactively aggressive than say in chimpanzees) and may have evolved little in a last thousand years.

c] Even a cursory knowledge of prehistory and history shows that many socio-political and economic systems have been tried (e.g. see Graeber and Wengrow[52]. Some have proved more successful than others but none lasting more than a relatively short period. Indeed perhaps the dynamic rise and fall and resurrection of polities may represent a form of long-term dynamic stabilization in the evolution of human society.

d] The rate of change implicit in the acceleration in information processing may itself be a major challenge. i.e. the development of effective regulatory systems may be lagging and thus contributing to social instability.

And e] especially at highest organisational levels, the limited, global control of "proactive aggression" (e.g. invasions and oppression), is of profound significance to human welfare in the nuclear age[39] as well as in our ability to respond cooperatively to climate change.

This hypothesis does not tell us which types of social regulatory systems might be or have been most successful even in specific communities. Indeed, subject to local social and biological selection pressures and given the wide range of human geographical and climate habitats and differing population and resource pressures, different systems could well have evolved and be appropriate in different locations[52]. These conjectures do not proscribe any specific regulatory mechanisms to meet our the current needs; only that social homeostatic regulation is essential and will grow with any increasing energy-driven complexity.

It might be envisaged (e.g. in Fig. 2) the stability of great conurbations depends on the sum of all the homeostatic activities sustaining each one of its denizens (indeed in each of the cells in each person). However, as noted, the lower levels of homeostatic regulation appear largely invisible and subsumed in the regulatory systems operational at the higher communal levels. It seems reasonable to assume that strong selective pressures have been exerted, at all levels, for homeostatic regulation to be achieved efficiently and at as low as possible energy and material cost. Otherwise resources would be diverted away from growth or replication in all organisms and in the social and cultural life of humans.

To the first approximation in modern society, a degree of regulatory stability has been achieved by high-level political and economic interventions; by dialogue, by trade and taxes, by the control of supplies and by the imposition of order by regulations, decrees, laws and treaties; all supported by legalised force. Great cities seem relative robust although any serious break in the energy supply would rapid turn to catastrophe. The impact of Coronavirus revealed both the fragility and the resilience of such high-level regulatory systems in human societies. However, the ruins of Aleppo, Gaza and Mariupol are also evidence of the catastrophic failure of human cooperative regulation and of the continuing threat of 'proactive aggression'. This, in turn, may catalyse localized 'reactive aggression' and horrific barbarity. Similar I would argue that AGW and its increasingly damaging impacts as well as other environmental transgressions exemplify catastrophic regulatory failure.

1. 7. 2. Current 'homeostatic' regulatory mechanisms in human societies are postulated to be a product of both the hierarchical socio-biological mechanisms inherited from our long hominid and biological ancestry and those conventions, regulations and laws that have emerged largely since the agricultural revolution[1 43]. The latter are the product of conscious human cognition, culture and experience and have emerged to permit co-existence and cooperation in varied human aggregations. These appear to have permitted more intense, sometimes more competitive, but often communal and cooperative, social and economic activity. However such social systems and regulatory regimes are not in themselves sufficient.

Profound changes in *institutionalised* welfare, in social support, and in health & housing provision, have occurred. These were necessitated by the energy-induced upheavals of the industrial revolution to compensate for loss of previously 'local' patterns of social reciprocity and care. There have also been technological revolutions in health care, in life expectancy, in education and consequentially in our expectations. Each has demanded a degree of government intervention and energy and resource allocation. A more detailed analysis of these aspects lies outside the scope of the current paper.

1.8. Other Emergent Properties.
An examination of the major revolutionary energy-driven step changes reveals, in addition to the homeostatic hierarchy discussed in Section 1.7, a number of other important consequential and emergent properties [see[1]).

1. The **rate of change**, as revealed by the time elapsed between the major step changes, has accelerated dramatically from the hundreds of

millions of years between the first four revolutions to a few tens of years inherent in humans successfully responding to AGW and adapting to digitization and AI. The energy flow per unit time through the selected components has increased many-fold; this speeding-up accords well with underlying physics i.e. with more energy transformations leading to much more work per unit time. More available energy or major changes in the efficiency of coupling must be anticipated to result in a further acceleration of change, which has major implications for, and raises important questions about, humanity's future welfare and stability as will be addressed later.

2. In most cases there appears to be a **ceiling to the biological and socio-economic complexity compatible with a given energy-resource regime** e.g. all multicellular organisms are eukaryotic. Truly complex prokaryotic, multicellular organisms have never arisen. Many animals and indeed plants communicate in sophisticated ways but speech and other more cultivated social interactions are limited to brainy hominids. Pre agricultural communities would have been limited in their population densities, in their ability to support specialization and did not, as far as is known, evolve systems for recording or writing. Ian Morris[53] claims the Roman and Sang empires, respectively, represent the peaks on cultural and material developmental compatible with the energy resources available from the agricultural revolution. The industrial revolution has, of course, itself enormously transformed the material and social complexity of human society. A process which appears to show no sign of abating.

3. **Cooperation** has been as much a feature of evolutionary history as **competition**. Briefly, endosymbiosis and the cooperative relationships of cells in all multi-cellular and multi-organ organisms are prime examples. Lichens, composite organisms of fungi and either algae or cyanobacteria, are essential to colonizing bare ground[54]. The colonization of land by plants, themselves a product of the endosymbiotic creation of chloroplasts from cyanobacteria, depended

on cooperation with fungi and lichens[55]. The success of various Homo species must have depended on cooperation including the capacity to exchange and transmit information. Often cooperation could gain a competitive advantage. Interactive coordination and some degree of cooperation appear to be features of complex systems.

4. Each step change appears to have empowered a subset of living organisms such as eukaryotes, then hominins and, later, **elite groups** of humans to thrive as they are able to take advantage of the potential created each new revolution; in the case of hominins the 4th, 5th and 6th revolutions. The implications of this conjecture will be discussed later.

5. Nevertheless, overall, there is a network of system-wide **interdependencies**, which include the welfare of the emergent elites. Eukaryotes are still dependant on the prokaryotic world just as humans depend on both prokaryotes and other eukaryotes, including of course other humans. Even the most elite of humans depend on the prokaryotic world in their intestines! [see[56]]

6. Despite our dependence on the exploitation of fossil fuels, humans still command a large % of global photosynthesis[40 57 58], doing so at the expense of biodiversity in general.

7. Major step-changes in the relationship between **exploitation of free energy transformations and information processing** can also be observed.

These are: (note possible exception of Step 2, oxygenic photosynthesis)

Step [1]: Emergence of the genetic code and intergenerational information transmission (see also Lorenz later).

Step [3]: Mitochondria in eukaryotic cells, as noted, make available significantly more free energy per unit of genetic information to carry out the work prescribed by those genes,

Step [4]: Enhanced neural capacity and emergence of non-genetic information transfer which created new ways of transmitting information and coordinating activity, likely a type of primitive speech and increased direct transmission of experience from parent to sibling and within a tribal group to its young (e.g.[18][59]).

Step [5]: Enhanced information transfer by urbanised communities by social interactions and emergence of record keeping, writing and mathematics[60],

Step [6]: Improved 'scientific' data handing leading to initially enhanced printing and information distribution, then radio and television and latterly many forms of electronic data transmission and processing, including currently, artificial intelligence.

Steps [7] and [8]: While the AGW-induced crisis has reinforced the need for global cooperation and information exchange, it does not appear to have catalysed new forms of information processing. In contrast, Step 8 is all about information processing facilitated by AI and the rapid rise of Artificial General Intelligence (AGI). Data are being handled at a pace and scale previously unknown so changing the relationships/coupling between the energy and work and between work and material and social complexity. The consequences of this new capacity to humanity is highly contended.[61][62][63]

In summary:
In the first six revolutions broad and consistent patterns can be discerned. Major incremental changes in harnessing and exploiting free energy has resulted in increased dynamic complexity in living systems and their constructs. Each with new potentials and instabilities and growing resource demands. Increasing the utilizable energy flow and resultant power has dramatically accelerated the rate of change.

Homeostatic adjustments have occurred in parallel resulting in the regulated, quasi-stabilisation of each step change and the sequential plateauing of a new dynamic order. Each of which is characterised by a higher degree of complexity, new power relationships and differentiations and greater embodied information and energy content.

Each new order remains dependent for its success on the continuing health of each of the preceding orders.

These observations led Eric Chaisson[27] to postulate an arrow of complexity but this may not be a good analogy. Rather these processes may be seen as deep current of gradual evolutionary change which, sequentially, catalysed by energy step changes gives rise to series of increasingly rapid sub-flows supporting enhanced complexity and new potentials while remaining dependent on the integrity of the whole system. The flight of the 'arrow' is jerky, accelerating but accumulative.

1.9. Information and Entropy.
In addition to the relationships of energy and information processing noted above in Section 1.8. point 7, the scientific literature explores the profound relationship between energy, especially entropy and negative entropy, and information. In this, the word 'information' is used in a number of distinct senses;
  a] accumulation of data,
  b] knowledge, both in an everyday and more intellectual sense, and
  c] in the formal Shannon sense, as a specified decrease in doubt (Appendix 2 and see also Robert Ayres[26]).

Information in this latter sense is a function of the a *priori* probability of the selection of a given state from all possible states. The number of 'Bits' of information required to solve a problem e.g. decode or transmit a message, corresponds precisely to the minimum number of distinct binary decisions necessary to make the correct final choice. This, in turn, is a function of the number of possible binary choices available and the probability of each outcome. The more exactly a physical state can be defined the more information is embodied in a given selection sequence. The Information Gain can be defined as the difference between the uncertainty before an event and the uncertainty remaining afterwards.

Given the relationship of entropy, which can be ascribed to the absence of information, and information *per se*, as explored in Appendix

2, the energy revolutionary step changes can also be interpreted as step changes in information input into specific entities. This is, in addition, to the changes in information processing noted in Section 1.8 above.

The arrow of complexity as describe by Chaisson (see also Fig.2) can thus be interpreted as an arrow of increasing embedded information, partly in relation to the genetic code but also the free energy flux/information flow through such entities. Chaisson [27] uses "free energy rate density," i.e. the free energy flow per second per gram of material in a given system as his complexity metric. Thus the Sun generates in its nuclear interior 4 x 10^{33} ergs/second and has a mass of 2 x 10^{33} grams, so Chaisson's complexity metric for the Sun is 2 (ergs/sec/gm). However it is not clear if this is appropriate to biology or the social sciences.

While the concept of complexity seems relatively obvious in daily banter, it is not easy to define rigorously.

1.10. The Puzzle of Complexity.

From the time of Darwin, many but not all, have recognized biological evolution as furnishing an arrow of increasing complexity in the life forms on this planet. Few would disagree that a multi-cellular, multi-organ organism is more complex than a single prokaryotic cell. Similarly, human life in, and the daily functioning of, any large city, is more 'complex' that life in small, partly or completely self-sufficient, rural community. Nevertheless, there appears to be no agreed universal definition of the term 'Complexity'.

Complexity is construed as a characteristic of a system or model whose components interact in multiple ways and follow local rules, leading to nonlinearity, randomness, collective dynamics, hierarchy, and emergence. The term is generally used to characterize something with many parts where these parts interact with each other in multiple ways, culminating in an emergence of a higher order that is greater than the sum of its parts.

Complexity, whether of an inert material or a biological or social organization, may be a function of the embedded energy and information required to maintain its integrity of an entity overtime i.e. out of equilibrium with its surroundings (and, as appropriate, its capacity to reproduce). This also allows the exchange of energy and material with its external environment.

A possible definition might read: 'Complex systems (be they biological, socio-economic or non-living) are spatially-constrained, dynamic multi-component entities maintaining a low entropy and embedding a high information content. Some in biology are capable of self-replication. All are multi-component systems which maintain non-random, internal dynamic interactions and regulate the exchanges (fluxes) of energy, information and matter between the entity and its external environment'.

However, it also appears that, as alluded to briefly, mechanisms can evolve in such systems to 'simplify' and smooth complexity. For example, the functionality of single animal or city is remarkably resilient, with some components of redundancy and can be unexpectedly stable. The former can withstand the loss of limb and the latter a major bombing or disease while retaining most of its capabilities. But such complex entities can also suffer from unexpected, catastrophic collapse.

1.11. Energy and Power.

As noted, the energy step changes appear to create new opportunities for specific groups to emerge, exhibiting new potentials and properties and a potential to dominate. Nevertheless, as stressed, the emergent forms remaining dependent on existing forms and on a functional planetary system. Any emergent 'dominant' appears to arise from the ability of new form to command energy and exploit more power i.e. do more work per unit time.

This 'power' concept was advanced a century ago and is primarily associated with Lotka's principle or his law of maximum energy. This states that '*in every instance, natural selection will so operate as to increase the total mass of the organic system, to increase the rate of circulation of matter through the system and to increase the total energy flux through the system so long as there is present an unutilised residue of matter and available energy*'[64 65]. In similar vein Ilya Prigogine[6] argued for a '*principle of minimum entropy*' in such non-equilibrium systems (Appendix 1 and 20 for relationship between free energy, entropy and dynamic order).

Closely related to Lotka's principle is Howard Odum's 'maximum power principle'[66] which states: "*During self-organization, system designs develop and prevail that maximize power intake, energy transformation, and those uses that reinforce production and efficiency*". He postulates that systems that maximize their flow of energy survive in competition. In other words, rather than merely accepting the fact that more energy per unit of time is transformed in a process which operates at maximum power, this principle implies that systems organize and structure themselves naturally to maximize power. Over time, such systems are selected for, whereas those that do not are selected against and eventually eliminated. Odum argues that the free market economy effectively does the same thing for human socio-economic systems and that our economic evolution to date is a product of such a selection process[67].

Chaisson[27] on the other hand, prefers the concept of the optimisation of energy fluxes "w*ith best suited organised systems employing moderate energy-flows large enough to sustain them yet not small enough not to destroy them*". As noted, Malhi's use of the terms biological and socio-biological metabolism[40] are alternates to energy flow in biology and human society, without to my knowledge, any implication of optimisation. Nevertheless he emphasises that the whole of biology and human society depends on these flows which must ascribe to them a central role in evolution.

The ability of any specific energy source to do work for humankind depends on its specific qualities as well as quantity e.g. human metabolism can utilised certain carbohydrates and fats but cannot directly use solar irradiation or burn/eat coal. Odum introduced the term **emergy**, which measures the amount of energy of a lower 'quality' grade required to develop the higher grade in **human terms**. The scale of emergy goes from the 'dilute' energy in sunlight up to plant organic matter e.g. wood, to coal, from coal to oil, to electricity and up to the high quality efforts of computer and human information processing.

Without adopting any of Odum's nomenclature, I wish to suggest that the concept of 'power' – i.e. work done per unit time at a specific location – can be extrapolated and expanded from physics and biology to human society.

Power is, of course, major concern to the social and political sciences.

Success in human terms seems to depend on the mixture of our skill sets as well as competition and cooperation. Together they can give rise to economic and social power, in the three senses employed by Lukes[68] (see also Foucault[69] and Russell[70]). Direct physical power, of course, remains significant. Individual physical power exemplified by a great warrior defending a tribe or securing extra food and other resources for them, has featured strongly in all human history and mythology.

With the energising of brain power (post *H. erectus*; 4th energy step-change) social power, tribal cohesion, layers of influence, including charismatic leadership, must have become critical factors in additional and often complementary to physical power. Now power in human society often arises, sometimes obliquely, from social interactions as exemplified by "influencers".

It is, nevertheless, a consequence of the 'control' of the **work** of 'others'; that is using their production of goods and services and weaponry and money, as a conduit of power, at the behest of and to the satisfaction of the 'powerful'. Our influence and tools lever more power but all remain dependent on free energy transactions.

In a modern state and the corporate world with its pervasive influence (often control) over the media and the message, legitimacy and power depends on popular approval and at least acquiesce. However physical violence and compulsion are never far below the social horizon. If, as I suggest, Odum's postulate is applicable to human behaviour, then the pursuit of power can be traced back to 'systems ecology' and is deep embedded in our evolutionary history. It is expressed in personal and group competition and the sometimes violent attempts to dominate other groups including women by men and some races or castes by others.

In most cases individual humans exercise limited personal social power within a small circle. But some seek much wider domination. Some for purely egotistical reasons others fired by an ideology. Such behaviour may also be related to Wrangham's trade-offs between reactive and proactive aggression in human evolution. Some small groups, inspired by political or religious ideologies and just demagogues, use proactive aggression to seek almost unlimited power. I would assert that both powerlessness, arising from extreme poverty and slavery, and absolute power are corrupting pathologies.

(Note: Should Chaisson's concept of the systematic optimisation of energy flows be preferred to Lotka/Odum maximisation principal, any extrapolation to human behaviour would still imply an evolutionary pressure to secure energy and power.]

In applying his power principle to the economy, Odum was also elaborating on the ideas of Georescu-Roegen[68] which have led to various strands of ecological economics. These consistently highlight the central role of energy. Robert Ayres and others [26,71,72,73,74,75,76,77,78] have analysed the primacy of energy, and the impact of its cost and availability on national and the global economic system. He and others have sought to analyse the economy from a thermodynamic perspective [e.g. Fig. 3 below]. Ayres has also postulated that, in evolution, there is a trend towards maximize the capture and processing of available free energy, (subject to the 'impedance' imposed by the availability of other requisite factors at a specific location, e.g. water and nutrients). To quote, "*most importantly, to converting some of it [free energy] to morphological information embodied as dynamic structures and organizations. To the extent that intelligence enhances this ability, evolution seeks to maximize intelligence".*

As alluded to earlier, we have a propensity for both rapid, instinctive "system 1" decision-taking (Kahneman, Haidt and others, see also[48] and[49]) and a reluctance to engage with deeper more analytical "system 2" thinking. Fascinatingly it has been observed when experiencing "brain fatigue" as we tire or are distracted, we lose some of our capacity to engage in deeper system 2 thinking and to exercise greater self-control. The work of Roy Baumeister[79] on 'mental energy', literally and metaphorically, suggested that our brains can become low in glucose (energy) when distracted or trying to carry out multiple, complex tasks. Our brains can be stimulated to engage with system 2 thinking more effectively when an individual is given a glucose drink. That thinking hard takes extra energy in the Baumeister sense of 'ego depletion' has not been confirmed[80]. Nevertheless there is clear biological evidence that, in humans, the energy supply to our brains is protected and conserved even in times of great need. Malhi observes that a trade-off is found. We make an exceptionally large investment of energy in our brains but the growth rate of children is very slow compared with animals of similar size. I would speculate that our well-established dependence on mental short cuts may be a reflection of ancient hominin trade-off between food supply and investment in intelligence.

Modern legal and regulatory systems are largely the products of analytical system 2 thinking but their execution is heavily influenced by the biases and heuristics characteristic of system 1[48,49]. Even the most sophisticated of human social constructs must be influence by aspects of our instinctive behaviour and possibly our physiologal condition as evidenced by judges handing out heavier sentences in late morning when their 'energy' is depleted[81]. However reliable evidence for such psycho-physiological effects appears difficult to pin down

Overall I suggest our proclivity to seek power can be related to our evolutionary dependence on energy embedded in our neurophysiology as well as to our socio-economic system.

Figure. 4. Modified Socio-economic Model
Kate Rowarth's modified variant of Bob Ayres's original model[75]; emphasising both the importance of energy and resource flows (i.e. the thermodynamic base) and the capture of those flows to maintain human society and its economy.

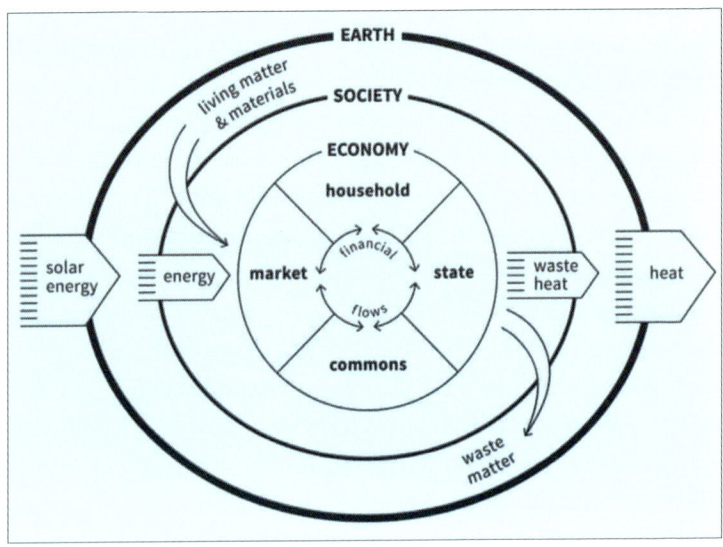

1.12. Overlapping Concepts.

Konrad Lorenz, the father of animal ethnology, wrote[82]: *"Life is an eminently active enterprise aimed at acquiring both a fund of energy and a stock of knowledge, possession of one being instrumental in the acquisition of the other. The immense effectiveness of these two feedback cycles coupled in multiplying interactions, is the pre-condition, indeed the explanation for the fact that life had the power to assert itself against the superior strength of the pitiless inorganic world".*

His words summarise the biological situation succinctly, although couched in very different terms to current molecular genetics, thermodynamics or information theory. His formulation remains highly relevant to this essay and to the human social and economic spheres.

The concepts of coordinated energy and information flow have broad equivalencies in both the socio-dynamics and material structures of human society. As emphasised, the individual and the societal interactions that both make our lives worthwhile and maintain our material prosperity, are dependent on these energy and information flows and their material and social material consequences. It appears in a sense, as information and energy as embedded negentropy are intertwined all is dependent on information processing.

More and more energy and information has been acquired and collated by human cognition since the agricultural and industrial revolutions. These allow the creation of the 'blue-prints' for our various actions, technologies, philosophies and social priorities. Our social interactions and wellbeing, as humans, depend on the myriad of informational signals we receive, process and recall personally, on a second by second basis throughout our lives. These are also sustained by the documented socioeconomic and technological blueprints, laws and regulations that together map out the structures of our society and the goods and services on which we depend.

It is this interlocking scaffold of energy and information interdependency that provides the stage on which humanity must address our profound environmental and social problems.

2 / Implications and Conclusions

2.1. The Framework.

Drawing on a wide range of disciplines, the evidence presented underscores the central role of energy transformations in biological and human history, adding substance to the earlier quotation from Smil[4] "*All natural processes and all human actions are, in the most fundamental physical sense, transformations of energy. Civilization's advances can be seen as a quest for the higher energy use required to produce increased food harvests, to mobilise a greater output and variety of materials, to produce more, and more diverse goods, to enable higher mobility, and to create access to virtually unlimited amounts of information".*

The analysis has also highlighted a sequential series of energy step changes in biology and human history. By exploiting new energy sources, or greatly enhanced energy use efficiency, these have given rise to sequential strata of greater systemic and structural complexity; each with new biological or social potentials (e.g.[1][27][40]) These step changes have also revealed a number of emergent properties which together create a framework for understanding the problem of AGW and many other global issues in a new and hopefully more revealing but no less fraught light.

Fundamental to this understanding is that emergent complex dynamic systems and structures require (homeostatic) regulatory mechanisms to ensure their stability and sustainability. In this light AGM represents a catastrophic failure of these mechanisms after the Industrial Revolution – the 6th energy step change. We are experiencing a potential ruinous destabilizing which has yet to be checked.

In addition this analysis highlights the extraordinary acceleration in the rate of change in energy use and societal complexity since the appearance of Homo species, again most dramatically after the Agricultural and Industrial Revolutions.

This acceleration is challenging the ability of our globalised, digitized society to adapt and devise new and effective forms of regulation at a rate commensurate with the rate of social change and of technological innovation. In terms of human wellbeing, complexity, although not easily defined, is itself a significant issue. The horizon is further darkened by our craving for power which Lotka's and Odum's principles suggest is a deep and essential component of all evolutionary success.

In the 7th Revolution and responding to Anthropogenic Global Warning (AGW), most governments across the political spectrum are seeking, some very reluctantly, to replace cheap and convenient fossil fuels by a mixture of low carbon energy from renewable sources and, possible, contributions from forms of nuclear power. Many assume that a recourse to geo-engineering and carbon capture and storage (CCS) will allow some continuing fossil fuel use and *in extremis* lower atmospheric CO_2 levels. The fundamental policy presumption is that the current economic model, based on a continuing and continuous growth in global energy supply and, concomitant, increasing demands on planetary resources, remains valid.

Economic growth, as measured by GDP and dependent on energy, if not necessarily on more GHG emissions, is deemed essential to mitigate many of humanity's problems including poverty. It is seen to be politically inescapable. It is assumed new game-changing technologies will emerge which will increase human resilience to inevitable climate change, as well as allowing the perpetuation of the current socio-economic model. The extreme expression of this view is found in the Techno-Optimists Manifesto[37].

This consensus and the assumptions that lie behind it are, I contend, deeply flawed. In first place it ignores our core 'homeostatic' failures. These failures arise both from aspects of our inherited, instinctive responses (human nature) and our more recently adopted, socio-economic regulatory systems. We are addressing the symptoms without considering either the cause of the failure or the impacts of our addiction to energy *per se* and increasingly an expanding capacity for information processing.

The global consensus seeks to replace our current source of energy, fossil fuels, with others, some of which are highly problematical. All sources, including all renewables, have some environmental impacts and drawbacks. Some will make heavy demands on non-renewable mineral resource's e.g. Li, Co, Cu, some Rare Earths[41], others effect on aesthetics and/or on wildlife. The underlying aspiration for more and more energy underpins the drive for nuclear fission and fusion with their attendant dangers in a highly uncertain world.

Given the data from the IPCC and the improbable rates at which GHG emissions must be cut annually to avoid mean warming of over 2°C and highly damaging local climate change, time is critically short. It is highly improbable that the nuclear options and indeed other mooted technologies can be implemented on the required time scale. Thus a tacit reliance on geo-engineering is increasingly assumed.

In assuming continuing growth in the economy and energy-use, the fundamental relationships between energy, work/power and complexity and rate of change are disregarded. As is the inevitably increasing of demands on this planet renewable and non-renewable resources.

Sadly:

[a] The evidence that humanity's activities are already transgressing safe planetary boundaries in a number of areas is largely ignored[34].

[b] The evidence that the rate of change is, in and of itself, a serious issue is discounted.

[c] The growing concerns that effective regulation is unable to keep up with the rate of change in areas such as the social media and AI/AIG are passed over.

[d] Issues around human wellbeing are deemed of little concern. As noted in relation to the 8th revolutionary step change, AI and AIG are changing the relationship between energy and information processing and in doing so are not only accelerating the rate of change as well as creating new but unknowable potentials, seen by some as existential threats[61].

[e] The dangers inherent in global geoengineering are often trivialised.

Taking the two concurrent revolutionary trends together, it appears virtually certain that more energy (albeit low carbon) and/or new ways of more efficiently coupling energy use to work and power e.g. through artificial intelligence, nanotechnology and advanced electronics, will:

[a] Increase the material and social complexity of society.
[b] Accelerate the rate of social and economic change.
[c] Reinforce the need for enhanced homeostatic 'regulation' to stabilize the growing complexities, with inherently unpredictable consequence to human freedom and well-being. All likely aggravated by the use of AI and surveillance. It is far from clear that these trends are compatible with democracy and the personal freedoms we currently enjoy.
[d] Increase human demands on all other planetary resources – renewable and non-renewable – especially if population growth continues and poverty is to be diminished.
e] Lead to the adoption of policies and technologies with both known unknown and unknown unknown impacts on our planet and on human welfare.
[f] Likely lead to the emergence of new powerful, controlling, technologically-adept elites.

2.2. Hubris.

Behind the consensus assumptions lies hubris. While some scientific and technological advances are essential, inevitable and to be warmly welcomed, e.g. better methods of storing low carbon electricity, low energy transport systems, the dangers of others (e.g. the 'known' and 'unknown' impacts of new technologies, especially of geoengineering) are immense.

Ironically, AGW is itself a prime example of an 'unknown unknown'. When James Watts invented the first relatively efficient coal-powered steam engine and Matthew Boulton proclaimed [see[1]]'*I sell here, Sir, what all the world desires to have – power*". No one anticipated, or indeed could anticipated, have the impacts their innovation would have on humanity or on Planet Earth. When solid evidence emerged over half a century ago that fossil fuel burning was a serious climatic threat, the response was minimal. An 'unknown unknown' became a 'known' but was still ignored. Indeed ordinated attempts were and still are being made to minimise the problem and dismiss the messengers as cranks.

Another concerning example of human over-confidence is a nuclear fission. Strauss' early conceit[82] has been disproved but many counties are contemplating a very expensive technology with well attested, dangers. It produces highly toxic wastes with life-spans of over a thousand years to try to shore-up the current socio-economic model and the status of certain countries. The technology is intimately linked to nuclear weapons. This in a world that is almost certain to become more and more restive and likely violent as climate change bites.

Some regions will become less and less viable for humans and migration will almost certainly increase. Indeed in much less than a hundred years, not a thousand or more years, major cities are likely to be inundated and global mass migration a reality.

As observed, the ultimate version of the growth consensus and 'econo-tech' hubris is revealed in the Techno-Optimist Manifesto[37]. It is hailed by those wedded to and the beneficiaries of the current dominant economic system as well the climate change deniers.

"We believe energy should be in an upward spiral. Energy is the foundational engine of our civilization. The more energy we have, the more people we can have, and the better everyone's lives can be. We should raise everyone to the energy consumption level we have, then increase our energy 1,000x, then raise everyone else's energy 1,000x as well." (Techno-Optimist Manifesto).

Many who espouse to these views, also believe in the infallibility of the market, and in resisting most forms of regulation assert that markets are perfectly self-regulating. Many find massive wealth inequality an acceptable outcome, regard themselves as 'deserving'. Some see the domination of nature by humans in Biblical terms. Any down sides are deemed a price worth paying for economic growth and the super wealth of a few.

The evidence presented in this essay shows this set of presumptions is not just to be misguided but toxic. Energy may indeed be the 'foundational engine of our civilization' but our exploitation of energy has many profound implications and impacts. The evidence clearly shows the fallibility of our technologies and our judgements and how often we have been caught out by over-confidence (cf.[83] and [84]).

The contrast between the conclusions presented in this essay and those of the 'techno-optimists' defines perhaps the most important choices ever faced by humanity.

2.3. Power and Inequality.
In the work of Lotka and Odum and their disciples, energy fluxes in ecological system and the acquisition of energy and power by organisms are seen as critical to evolutionary success. Foucault and others have theorised that power is equally critical to all human social networks, economics and politics. I have suggested these phenomena may, in reality, have a close evolutionary link.

Social power depends, as elaborated by Lukes[68], on subtle social interactions and on an ability to harness both the work and the outputs of others, including their industries and weapons. Physical compulsion and violence remain an essential part of the equation. Mao concluded, *'power comes out of the mouth of a gun'* but such violence is usually the last resort. Power is now often exerted by influencing, by agenda setting or denying and, increasingly in complex interdependent world, by the financial control exercised by a small elite. They have framed a world and its regulatory systems to serve their own interests. Inequality is the servant of power. To quote Frédéric Bastiat in 1847[85], *"When plunder becomes a way of life for a group of men in a society, over time they create for themselves a legal system that authorizes it and a moral code that glorifies it"*. To which Kevin Anderson added – *'and an expert base that justifies it'* and, I would also add: 'and clever PR and a subservient media to convinced many, maybe most, people is no alternative –TINA'.

Power can be bought and traded. Money is increasingly

interchangeable with power although Smil's emphasis on the ultimate reality of energy transactions remains valid.

Recently the sources of wealth and power has diversified as the world economy has grown more complex. New resources of power have emerged including 'information' itself as seen in Musk, Gates, Bezos and Zuckerberg and other tech giants. Nevertheless Saudi Arabia, the Gulf States, Brunei, the US states and the 'Oil Majors' and, of course, Putin's Russia, stand testimony to the enduring leverage and power of primary energy. And food energy remains an importatn bargening tool.

Throughout history individuals and groups have sought power by controlling the supply of energy, be that food calories in the pre-industrial world and consequently manual and animal power or, after the industrial revolution, fossil fuel power. The status and advances of many civilisations can be seen in their quest for the higher energy use required to produce increased food harvests, and to mobilise armies to underpin the wealth of the rulers. However, as illustrated in Peter Francopan's wonderful and dense volume, *The Earth Transformed*[86], there is also a litany of failure.

Until the last two hundred years, power was virtually always associated with agricultural food and, to lesser extent, commodity-derived wealth e.g. sugar, pepper. Such power, acquired by the control of food-energy for human labour and animal power, was subject to the vagaries of the weather, volcanic eruptions and disease (the latter often catalysed by the former) as well as threats from external competitors and internal socio-regulatory failings. Many great empires and leaders fell as their grip on this acquired power collapsed, often due to hubris, lack of forethought or over-competitive greed, and environmental mischance. Local disasters ensued and countless millions of the powerless, met their, often, horrific deaths.

Whereas historically this exercise of power and the consequences of its over-reach were limited in its geographic reach, our world is now more globalised, homogenised and interconnected than ever before. We are connected second by second electronically and hour by hour aeronautically. AGW, pollution and biodiversity and habitat loss are global issues affecting all mankind, although not equally. Human overreach is now a source of global woe.

Reconciling the pursuit of personal and group power, which I believe has deep evolutionary roots, with the imperatives of climate change is a major concern. Undoubtedly power – physical, social and political – is the major stumbling block restricting the global efforts of combating AGM in many ways.

I've suggested that each step change especially since the evolution of hominids has offered a new elite an opportunity to thrive by making best use of the new energy economy. Indeed we, *H. sapiens*, may well be examples of this trend.

Powerful elites have emerged sequentially from the exploitation of coal, oil and gas. They are a major cause of the delay in addressing climate change. They may well feel entitled to buy their way out of any crisis and to be dismissive of the fates of the masses. Heavily guarded gated communities already exist in many countries.

The Techno-Optimists manifesto reflects this attitude and raises the prospect that the powerful will only relinquish their grip after a total catastrophe. A further constraint is that this group also controls many of the economic resources required to combat AGW.

2.4. Economics and Time.

I've emphasised that the hypotheses and framework outlined do not favour any specific homeostatic mechanisms. It asserts only that regulatory systems must evolve in parallel to growing power and complexity in order to maintain stability.

However AGW and the exceeding of the global environmental boundaries speak clearly of our failure to achieve such stability. To a significant extent, socio-economic data also confirm this trend.

Since the industrial revolution, startling advances have taken place in global wealth, as measured by global and per capita GDP and energy use and in the expectations of both rich and poor Many, but certainly not all, now enjoy a much higher standard of living. And in some cases, greater personal freedom than could have been imagined a few hundred years ago. However, both globally and in-country, wealth, energy and resource availability and poverty are distributed very unevenly in time and space. Indeed they are inversely related. Power and wealth have accrued preferentially to a few who are also, by a very large margin, the main emitters of GHG e.g. CO_2e per head, Qatar >30t; Democratic Republic of Congo, < 0.1t per head[87]. To a degree a few countries have decoupled per capita emissions from GDP growth but this is not the norm. However reducing in country, production emissions at the expense on increasing imports of energy intensive goods (consumption) emission is no solution. As might be expected, the countries that pioneered the industrial revolution have accrued the highest total per capita emissions over the last two centuries.

These facts are important for several reasons:

[a] The atmospheric physics and chemistry of the anthropogenic CO_2 mean that additional anthropogenic emissions have a residence time exceeding a century. Their current impact is a consequence of the total accumulated atmospheric concentrations. Some poorer counties, such as India, argue that fairness requires they can exploit a greater share of the remaining planetary emissions ratio to allow their rapid 'catch-up' development. Cuts in this view should be weighted heavily towards the long term emitters.

[b] As current GHG emissions predominately arise from the wealthier citizens and countries, meaningful cuts can only and must be made by them. But they are likely to be resistant and socially, economically and psychologically have the most to lose.

[c] Establishing the infrastructure for a low GHG emissions world will make redundant some current industries and eliminate the livelihoods of individuals and communities as well as creating stranded

communities and assets. This is a very apparent in Port Talbot now.

[d] The investments in new infrastructure will be expensive and the source of that funding critical.

[e] In the historically better-off cuntries e.g. USA and UK, relative affluence has created the expectation of continuing economic growth. Although, for the many, their living standards and household economies have stalled for some 40 years. Nevertheless there exists a politically potent sense of entitlement which can be exploited by popularists.

[f] During this period, huge wealth has been acquired by a small group, much of which, amounting to >$20 trillion, is held in tax havens[88].

(Note: $10 trillion expressed as 1 second per dollar equates to over three hundred thousand years. This is approximately the time that has elapsed since the emergence of *H. sapiens*.)

Given these factors it is reasonable to concluded that, despite its successes, the current political-economic model and its regulatory regimes are failing socially quite as seriously as environmentally.

Time is not on our side. The evidence, which will not be elaborated in this essay, shows that a >+2°C world will be socially and environmentally distressed. A number of critical tipping points will, in all likelihood, be passed. Sea level rise, heat and humidity and drought and dwindling water resources can be expected to make regions and cities no longer fit for human habitation. Consequently urgent, major changes are required in the regulatory regimes, including the tax system, that are core features of the current economic model. Such changes are now discussed widely by heterodox economists and philosophers [89,90,91].

It appears that peace, social stability and, in the medium term, even modest global prosperity are improbable without a major reappraisal. The effects of AGW can only exacerbate the tensions that are already apparent.

It is moot whether the advances in energy-use efficiency will compensate for the legitimate energy and resource demands of the global poor (especially in Africa and Asia). Their low comparatively *per capita* GHG emissions, and high rates of population growth make it improbable. Paradoxically many of these disadvantaged regions are also in frontline to experience the negative impacts of AGW on food production and poverty. Such constraints may well also hinder the run-out of new technologies as well as reinforcing social tensions and both proactive and reactive violence. The dreams of the techno-optimist are, in reality, the source of nightmares.

Overall I am arguing our planetary transgressions, including not global warming but gross inequalities, are not incidental side-effects or minor facets of our current culture and socio-economic system but outcomes of long, deep trends. They are exacerbated by the industrial revolution but arise from embedded traits; some apparent over many centuries, others millennia.

Two thought experiments commended themselves.

First, consider how the modern world might have developed for the last, say, 300 years if cheap hydrocarbon fossil fuels had not been available. No doubt the rate of change would have been slower. We would not today be facing catastrophic global warming but our scientific knowledge and technologies would have developed apace, albeit in some different directions. Likely our desire for goods and power would be undiminished. Some aspects of environment destruction might even have been worse. But in all probability the proportion of the annual global photosynthetic carbon captured by humans would be even higher at the expense of other land uses and biodiversity. Our population would have grown albeit less rapidly but our caving for power and tendency to dominate and exploit would remain. Food production, in the absence of nitrogen fixation by the fossil-fuel intensive Haber process and other oil-driven activities, might well be at a premium. Forests would be great geopolitical assets and global politics very different. But, I suggest, the fraught relationship of humans with the biosphere would still be apparent and in part, catalysed by both science and capitalism from the 17th century.

Secondly, how might our world have developed had Lewis Strauss' famous prediction[83] that atomic electricity would be "too cheap to meter' proved true? Again while the threats from climate change might be restricted to emissions from land use and food production (perhaps 20 to 25% of the present), most other issues would remain and those derived directly from nuclear technology and radioactive waste increased enormously. Nuclear disasters would likely be more common. Having worked in Aleppo, Syria and Maiduguri, Nigeria, the dangers posed by nuclear proliferation are all too apparen Resource, pollution and population pressures would likely remain and in some cases could well be worse.

2.5. A Renewed Vision.

Human ascendency and prosperity have many strands. From the perspective of this essay, they are product of a sequential accumulation of dynamic but unstable complexity, driven by accelerating currents of energy transactions and information processing. Such a formulation may capture some fundamental trends but, of course, ignores much that is unique in human existence and in our ethical and cultural lives.

Even so *Homo sapiens*' material and cultural success has deep evolutionary roots. Especially, in our ancestral hominins' investing energy in greater intelligence, in imagination and ingenuity. In a new capacity for cooperation and learning as well as competition and pro-active violence.

Over tens of thousands of years our faculties have been honed by competition and cooperation and the imperative to survive in hostile environments, including many external threats from nature and, in all likelihood, fellow humans. Circumstances in which having dominion over the local environment must have seemed, indeed was, highly desirable.

Survival must have depended on an array of traits. Some of which seem to us admirable, such as personal and group loyalty, compassion, companionship and non-violent cooperation. Others more neutral such as ingenuity and bravery. Still others such as a propensity to misogyny and murder, violence and cowardice, greed and jealousy much less so. Nevertheless all these complex and sometimes conflicting traits and how they are expressed or even amphified, in the modern world by our socio-economic constructs, are crucial. They are the basis of our moral universe as well as effecting our ability to respond the great problems now facing humanity.

Many of these traits and constructs, including our inability to devise effective and timely ways of stabilising complexity and change, appear ill-suited to our finding timely solutions to many of our modern problems. Sadly they also provide psychological, social and economic ammunition to those seeking to deny the issues and thwart any progress towards their resolution.

The current crises can thus be seen as the result of deep-rooted trends in the evolution of human society dedicated to the pursuit of energy and power; a culmination or apotheosis of a long history.

Our very successes, and the characteristic that helped bring them about, are now threatening to undo us.

Seeking to stay or reverse such trends would seem a fool's errand.

However such a conclusion is highly flawed. If life, as Schrödinger suggested, feeds off entropy, then humanity feeds off hope. Despair is counter productive.

Over hundreds and thousands of years humans have been story tellers and a new version of the story of human flourishing is essential. Without an alterative vision with wide appeal, the current story and its trends will be maintained until a catastrophe forces a painful change.

While I suspect my essay will have little impact, it may be worthwhile making a few points and suggestions, which may just contain the germ of an alterative approach. This may, in time, lead to the realization that we can live better on less, and flourish within the embrace of this planet.

This approach might be based on the common ground between the urgent need to make drastic cuts in GHG emissions in the next decade, the problems associated with rolling out low carbon energy at the scale and time required and the dangers discussed in this essay of ever accelerating energy use and it consequences.

An alternative dialogue may appeal, especially in developed countries which are the main sources of our over-reach, for several reasons.

First, it is doubtful that ever acceleration energy-driven change, socially and economically, is popular. Is it, indeed, even now seen to be conducive to wellbeing? People obviously wish to live in comfort and a degree of prosperity but too rapid change is deeply unsettling and unpopular.

Secondly, such changes, as I've sought to explain, will, likely, result in a decrease in human freedom and potentially, with the rise an AI elite, may reduce the many to a second class status.

Thirdly, reducing demand will limit the need for investment in renewable energy and storage and its many, often highly unpopular, impacts.

Lowering the gross energy demand, mainly for low carbon electricity and heat, will make it easier to meet that demand with the least environment impact (cf. generation, storage and transmission).

Fourthly, it may be possible to reduce emissions much more quickly by reducng energy demand than by rolling new generation technology world-wide. Indeed the whole impetus towards nuclear energy, and CCS and plant geoengineering should be reduced.

Finally it may be more cost effective to save. Hinkley Point C is now estimated to cost £34 billion (at 2015 prices) and not be generating until 2030. There are about 28 million households in UK; at current prices this one investment equates to about £2,000 per houshold.

In short prioritising energy reducion e.g. improving the housing stock, better public transport, more attractive public spaces, local culture events may have social as well as economic merit.

The turbulence in society, both locally and internationally, seems to imply a growing dissatisfaction with the current model. To date this dissatisfaction is mainly captured by the populists but this may not be inevitable.

It's worth asking, do people, specifically the better off, actually want their lives and those of their children committed to, condemned to, life on an ever accelerating treadmill?

As Joseph Schumpeter[92] famously observed, in capitalism '*creative destruction*' is a "*process of industrial mutation that incessantly revolutionizes the economic structure from within, incessantly destroying the old one, incessantly creating a new one.*" This fundamental characteristic of our technology-driven society is ever more apparent. The great acceleration shows no sign of slowing. It has been remarkably effective in creating new and unexpected opportunities but since the 1980s wealth have been channeled to the 1%, and has taken a heavy toll from individuals and communities.

Since these processes can only further accelerate with pursuit of more and more energy, it's worth asking at what point do the disadvantages of 'destruction' outweigh any advantages of the 'construction'? As I've emphasized each "new" also carries its known and unknown risks and its regulatory challenges.

Inevitably our more recent regulatory systems derive largely from the attempts to stabilize the liberal free market system within its own value system. In this system individuals tend to be treated as utility-maximising machines, as *Homo economicus*. The system is based on rewarding enterprise and lauding individuality which readily evolves into promoting selfishness and greed and accommodating "Phishing for Phools"[93]. But has this run its course?

We are not prisoners of either our inherited or designed homeostatic systems. Our socio-economic system, in its totality, is open to being reconsidered and changed. Perhaps an important aspect of my hypothesis is that it informs us about the underlying currents that we must navigate to avoid shipwreck. It underlines the basic dilemma by highlighting the human characteristics and systems that we must adapt to, and learn to cope with, if to successfully avoid catastrophe.

One crucial human characteristic, physiologically a consequence of our astonishing neural network, is our free will. We have a capacity to make reasoned judgements as well as instinctive choices [e.g.[94]] although some will deny this. Related to our free will is a capacity to create a social milieu; a contextual terrain which then influences the decisions and expectations of any society as well as individual decisions.

This terrain maybe open to change.

Early in the essay I referred to the Bénard cell as a well-studied model system for the spontaneous emergence of a complex, dynamic structure in an energy flux. Such structures can be destroyed by both too little or too great a flux. Could this be a parable for our condition?

A critical question is whether humans, collectively, will exercise their free will to avoid the calamity which our overshooting of safe planetary limits threatens. This, when that overshoot is the product of the values and lifestyles of a minority.

The less privileged majority are fully aware of this paradox. Nevertheless they aspire to these living standards and are increasingly adopting the life-style and diet given an opportunity. The more affluent majority, not unexpectedly, seek to protect their advantage. One of insights from Kahneman's Prospect Theory is that humans are much more reluctant to accept a perceived loss that to embrace a novel but uncertain gain; a trait central to the tactics of the AGW-denier lobby.

Global, regional and even local gross inequity is a central issue. It is impossible to see how the toxicity of energy excess and overshoot can be avoilded with much greater fairness. Huge disparities in living standards and opportunity have often lead historically to and are reversed by uprisings and warfare [e.g.[95]]. This suggests that enlightened self-interest should lead to serious efforts to curtail GHG emissions and other planetary insults and efforts to redistribute wealth. But this has not occurred. The main financial flows remain from the poor counties to the rich. Tax havens flourish at the expence of the poor. GDP per capita remains the economic measure of choice and its growth promoted as key to progress, despite all the evidence to the contrary.

Underlying this discussion are deep philosphical and ethical even religious issues; namely how we envisage our lives, what are our values and our relationship with the rest of both humanity and the natural world. These issues have exercises some of the greatest mind for thousand's years. Even a minimal exploration of how these conjectures should be interpreted in the light of these works is well beyond the scope of this papere and the capability of this author.

Nevertheless it seem fair to state that plentiful cheap energy and the resultant cheap goods and services, boosted by a pervasive consumerist ethic, dominates our culture. The results are not sustainable environmentally, socially or economically [see[86]].

A significant minority are rejecting this pattern. The thrust of my hypotheses is that humankind, specifically the currently prosperous should deliberately seek to live on less energy and by doing so will likely live better. The thesis also shows that exhortations, as in Genesis and Marc Andreessen, to have dominion over nature are both dated and dangerous.

We must realize that we are part of nature and dependent on the natural capital and services provided. Ironically the problems associated with the essential switch to renewable energy and the huge costs and uncertainties of nuclear power may offer a deeper hope. Both this energy-based hypothesis and the urgent pressures to cut GHG emissions drastically suggest that reducing energy use should be both a pressing priority and common sense.

There remains the need for a compelling alternative vision of human wellbeing; the eternal concern of the great thinkers and religious leaders of history. This hypothesis, while offering a frame of understanding and an integrated scientific cloak, does not reveal any new ethical insight.

The hypothesis does not provide any ready answers but asserts that the pursuit of more energy and power will have consequences that, with a very high probability, will be highly damaging. Deliberately seeking to live on much less energy, curbing the growth in complexity and need for more regulation should be liberating. Perhaps by doing humans can achieve greater fulfillment and be at greater ease with themselves, their fellow humans and the rest of the natural world, the task of convincing humans of this remains daunting.

3 / Conclusions

Simply replacing fossil fuel energy with renewable energy, supplemented possibly but dangerously by nuclear fusion and fission, will not resolve many underlying issues.

Many of the initiatives contemplated to attain net zero have both known and unknown risks attached. Consequently, for humankind and other life to flourish within this bounded planet, we must lighten our tread and find fulfilment without energy-driven consumerism. The better off must use much less energy to allow more use by the poorest as well as make fewer demands on the Earth's other material resources. This implies that we embrace a more equitable lifestyle and re-assert the value of the whole natural world.

This may seem an improbable dream but it offers not only solace but profound joy. The alternative requires humans to attempt not only to dominate but to totally re-engineer this planet. This may appear superficially attractive but presages catastrophe.

Albert Schweitzer: "*Man has lost the capacity to foresee and to forestall. He will end by destroying the earth*".

Appendix 1.
- **"All natural processes and all human actions are, in the most fundamental sense, transformations of energy"**. See Smil[4].
- Energy exists in many different quantitatively interchangeable forms; examples being: light, heat, kinetic/mechanical energy, gravitational energy, electrical energy, sound energy, chemical energy, nuclear or atomic energy and so on. Each form can be converted or changed quantitatively into other forms.
- In its many forms, energy is a well-attested but 'virtual', 'abstract' concept nonetheless one based on detailed quantitative experimentation.
- Smil likens energy to a 'currency' that can be considered underpinning global, indeed cosmic transformations and exchanges, but one which comes in many forms; these 'local' forms – equivalent to local currency – US dollars or euros or naira etc. underpin our society and may be harvested to do work and that work can yield various 'products'. (See Fig. 1)
- The First Law of Thermodynamics (the conservation of energy (now energy and matter) states that energy cannot be created or destroyed, but it can be changed from one form to another one. However, in any energy transformation a small proportion of the total initial energy is lost as low-grade heat and is no longer available to do useful work. A limitation of this First Law is that it does not quantify the amount of energy transferred in this process. The inevitable loss of potential to do useful work is encapsulated by the Second Law of Thermodynamics, where the concept of entropy is used to define the quantity of lost useful energy.
- One simple statement of the Second Law of Thermodynamics is that in a closed system heat always moves from hotter systems to colder ones. Only for an open system, where energy in some form can be added, is it possible to reverse the direction of heat flow. Another simple interpretation of the Second Law is that in a cyclic process

not all heat energy can be converted into useful work. The term **free energy** is often employed for the 'useful' energy component which can do work.
- Another consequence of the Second Law is that the total work done by a system in changing between two states depends not only on the nature of these two states, but also on the path taken between intermediate states.
- The component we term entropy can be related to an increase in molecular disorder. Thus, as energy transformations take place, over time, any closed system will run down as more disorder (entropy) is created and the 'free energy' component declines.
- Boltzmann has defined **entropy** as a statistical mechanical property of a system – i.e. the sum of all the microscopic configurations of molecules, say of a gas, compatible with its macro-properties of volume, pressure and temperature. The fewer the compatible microstates the greater the order and the lower the entropy so linking order, information and entropy.
- The Second Law of Thermodynamics states that the entropy of the entire universe, as an isolated system, will always increase over time. i.e. the Universe will ultimately decline into uniform disorder and death. The second law also states that the changes in the entropy i.e. increasing disorder, in the universe can never be negative.
- Free Energy enables work to be carried out and power to be applied and is defined in terms of this work. As noted, energy occurs in many forms leading to both a plethora of units (e.g. Joules or kWatt hours or electron-volts) and great number of possible energy conversions; each leading in turn to a variety of work and power outputs (power being work per unit time). So kinetic energy may be harnessed for electricity generation or chemical energy may drive biological metabolism and all human actions and/or fuel cells. But any energy transaction will lead to a decrease in free energy, with some energy 'lost' as low grade heat e.g. from friction, and an

increase in overall entropy. Thus, any closed system, including our universe, will gradually equilibrate and trend to a higher entropy and an increase in molecular disorder.

(Summarised in any standard text book; see also[3] and[4]. For detail on units, see[2]. I am grateful to Emeritus Professor Ron Pethig for his advice and inputs.)

Appendix 2.
Entropy, Negentropy and Information

The thread connecting these three concepts can be found where statistical mechanics, probability theory and information theory intersect. The common origin is the equation, chiselled into Ludwig Boltzmann's tombstone in Vienna's Zentralfriedhof cemetery, that provides an exact quantitative connection between disorder and order:

$S = k \log W$

The symbol S represents the entropy, k the Boltzmann constant (the proportionality factor that relates the average relative kinetic energy of particles in a gas with the thermodynamic temperature of that gas) -1.38×10^{-23} joules per oKelvin or 3.3×10^{-24} calories per degree Celsius) and W is a quantitative measure of the disorder of an isolated (closed) system of particles. This disorder is specified by both the random heat-induced motions of the particles and the degree to which they are mixed and spatially coordinated. As noted in 1.2. and Appendix 1, Schrödinger postulated that living organisms maintain themselves as organised entities of negative entropy by 'continually sucking orderliness i.e. energy, from its environment'. On Earth the most powerful source of energy giving rise to 'negative entropy' is the radiant energy of the sun, itself the result of the Einsteinian conversion of matter into energy.

Schrodinger explained that, if W is a measure of disorder, its reciprocal $1/W$, can be regarded as a direct measure of order. Since the logarithm of $1/W$ is just minus the logarithm W, Boltzmann's equation can be written as:

-(entropy) = k log (1/W)

The term 'negative entropy' is not to be confused with 'negentropy'. The thread connecting 'entropy' and 'negentropy' involves the treatment of random variables by probability theory. Consider the random kinetic energy of each particle in a given system at any precise moment. For a sufficiently large number of particles, the distribution of kinetic energies can be represented as a continuous curve or function. Out of all possible distributions, characterised by its mean value and variance, the Gaussian probability distribution is the one that corresponds to the highest entropy (greatest disorder) attainable by the system of particles. In information theory, the quantity 'negentropy' is used to measure the entropy difference between an observed distribution of random variables and the Gaussian distribution corresponding to the same mean and variance. Negentropy never has a negative value and is zero only for the specific case where the information signal is Gaussian.

Shannon Entropy is a central concept in information theory, and quantifies the amount of uncertainty involved in either the measured value of a random variable or the observed result of a random process i.e. higher entropy equates to greater uncertainty i.e. in both information theory and statistical mechanics the equations describing a decrease in order or uncertainty are formally analogous.

Quantifying information is the foundation of the field of information theory. The intuition behind quantifying information is the idea of measuring how much 'surprise' there is in an event. Those events that are rare (low probability) are more surprising and therefore have more information than those events that are common (high probability).

- **Low Probability Event**: High Information (*surprising*).
- **High Probability Event**: Low Information (*unsurprising*).

The amount of information in an event, depending on the probability of the event, is termed "*Shannon information*," or simply the "*information*," and can be calculated for a discrete event x as follows:

- information(x) = -log($p(x)$)

Where *log*() is the base-2 logarithm and *p(x)* is the probability of the event *x*.

The choice of the base-2 logarithm means that the units of the information measured are in bits (binary digits). This can be directly interpreted in the information processing sense as the number of bits required to represent the event.

In Brillouin's 'negentropy principle of information'[96], the changing of an information bit value (*e.g.*, yes/no decision, erasure, display, measurement) requires an energy expenditure of at least kT log$_e$ 2, where k is Boltzmann's constant and T is the absolute temperature.

(I am grateful to Prof Ron Pethig for his advice and inputs)

Footnotes

1. R Gareth Wyn Jones. (2019) *Energy The Great Driver: Seven Revolutions and the Challenges of Climate Change*. UoWP https://www.uwp.co.uk/book/energy-the-great-driver/
2. For a formal explanation of energy units see: https://www.aps.org/policy/reports/popa-reports/energy/units.cf
3. Richard Feynman (1964) *The Feynman Lectures on Physics: "Conservation of Energy"*; vol I, Lecture 4, p 2-4
4. Vaclav Smil. (2017) Energy and Civilization: A History. MIT Press.
5. Wilhelm Ostwald. (1912) Der energetische Imperativ. Leipzig Academische Verlagsgesseiahaft. https://archive.org/details/derenergetische00ostwgoog
6. Ilya Prigogine. (1972) La thermodynamique de la Vie. La recherché. Also: https://www.nobelprize.org/nobel_prizes/chemistry/laureates/1977/prigogine-lecture.pdf.
Also (1980) *From Being to Becoming*. W.H. Freeman San francosco quoted in [26]
Also Rene Lefever. (2018) The Rehabilitation of Irreversible Processes and Dissipative Structures 50th Anniversary.
https://royalsocietypublishing.org/doi/10.1098/rsta.2017.0365
7. For Benard cells see https://www.sciencedirect.com/topics/physics-and-astronomy/benard-cells also
https://psl.noaa.gov/outreach/education/science/convection/RBCells.html).
8. Erwin Schrödinger. (1944) *What is life*. Reprinted as Erwin Schrodinger. (1967) What is Life? and *Mind and Matter*, CUP.
9. Nick Lane and William F. Martin. (2012) The origin of membrane bioenergetics. Cell 151, 1408-1416
10. Freeman Dyson. (1999) *Origins of Life*, CUP.
11. Armen Mulkidjanian et al. (2012) Origin of first cells at terrestrial, anoxic geothermal fields. Pnas 109 (14), E821-830
12. Jeremy England. See https://www.amazon.co.uk/Every-Life-Fire-Thermodynamics-Explains/dp/1541699017
13. John Lenton and Andrew Watson. (2011) *Revolutions that made the Earth*. OUP, Oxford
14. Nick Lane. (2015) *The Vital Question: Why is Life the Way it is?* Profile Books, also William F. Martin (2017) Ancient Cells: Going back in Genes. The Biologist, 64:2 20-23.
15. Austin Booth and W Ford Doolittle (2015) in Pnas. see https://philpapers.org/rec/BOOEHS
16. K. Chiyomaru and K. Takemoto (2020) Revisiting the hypothesis of an energetic barrier to genome complexity between eukaryotes and prokaryotes. https://royalsocietypublishing.org/doi/full/10.1098/rsos.191859
17. Schavemaker, Paul E ; Muñoz-Gómez, Sergio A; The role of mitochondrial energetics in the origin and diversification of eukaryotes: Nature ecology & evolution, 2022, Vol.6 (9), p.1307-1317
18. Suzan Herculano-Houzel. (2016) *The Human Advantage: A New Understanding of how our Brain became Remarkable*, MIT Press.
19. Richard Wrangham. (2009) *Catching Fire; How Cooking made us Human*, Profile Books
20. James C Scott (2017) *Against the Grain: a deep history of the earliest states*. Yale University Press
21. Jared Diamond. (1998) *Guns, Germs and Steel*, Vintage.

22. Peter R Shewry, Wheat. J. Exp Bot. 60.1637-1663 and M. Feldman (1995) Wheats. In *Evolution of crop plants*. eds Smart J and Simmonds NW. pp.185 -196 Longman Scientific and Technical, Harlow. https://climate.nasa.gov/climate_resources/24/graphic-the-relentless-rise-of-carbon-dioxide/
23. Hugh Thomas (1981) *An Unfinished History of the World*, Pan London.
24. David Lands (1969) T*he Unbound Prometheus,* PS of University of Cambridge.
25. John Maynard Smith and Eors Szathmary. (1997) *The Major Transitions in Evolution.* OUP, Oxford.
26. Robert Ayres (1988) Self organisation in biology and economics https://pure.iiasa.ac.at/id/eprint/3068/1/RR-88-001.pdf
27. Eric Chaisson (2001) *Cosmic Evolution: the rise of complexity in nature*. Havard also https://lweb.cfa.harvard.edu/~ejchaisson/reprints/nasa_cosmos_and_culture.pdf
28. Olivia Judson. (2017) The energy expansions of evolution. Nature Ecology and Evolution https://www.nature.com/articles/s41559-017-0138
29. James Lovelock (1988) *The Ages of Gaia*. OUP and (2010) *The Vanishing Face of Gaia.* Penguin
30. IPCC 6th Assessment (2021/22) *Synthesis Report* https://www.ipcc.ch/assessment-report/ar6/
31. John Houghton. (2015) *Global Warming: The Complete Briefin*g. 5th Edition, CUP.
32. Kevin Anderson and G. Peters (2014) The trouble with negative emissions. Science. 354 (3609), 182-183
33. For global population projections see: https://www.un.org/development/desa/pd/sites/www.un.org.development.desa.pd/files/wpp2022_summary_of_results.pdf.
34. Katherine Richardson et al. (2023) Earth beyond six of nine Planetary Boundaries. Science Advances 9. 37. https://www.science.org/doi/10.1126/sciadv.adh2458
35. https://www.europarl.europa.eu/RegData/etudes/BRIE/2019/637967/EPRS_BRI(2019)637967_EN.pdf
36. For energy use in computing etc see: https://www.iea.org/reports/data-centres-and-data-transmission-networks
37. Marc Andreessen https://a16z.com/the-techno-optimist-manifesto/.
38. https://www.scientificamerican.com/article/could-mitochondria-be-the-key-to-a-healthy-brain/
39. Earl Cook (1971) The Flow of Energy in an Industrial Society. Scientfic American 225[3], 134-147
40. Vadvinder Malhi (2013) The Metabolism of a Human-dominated Planet. In *Is the Planet Full?* ed. Ian Goldin p.142-163. OUP
41. Henry Sanderson (2022) Volt Rush; T*he Winners and Losers in the Race to go Green*. OneWorld Press.
42. Mario Giordano (2013) *Homeostasis: an underestimated focal point of ecology and evolution*. Plant Science, 211, 92-10
43. Antonio Damasio. (2004). *Looking for Spinoza*. Vintage Books, London: also Antonio Damasio. (2018) T*he Strange Order of Things: Life Feeling and the Making of Culture*. Knopf Doubleday.
44. Richard Wrangham (2020) *The Goodness Paradox: How evolution made us both more and less violent*.
45. AS Wilkins, RW Wrangham and WT Fitch (2014) The 'domestication syndrome' in mammals: a united explanation based on neural crest behaviour and genetics. Genetics

198(4), 1771
46. Zanella M, Vitriolo A, Andirko A, Martins PT, Sturm S, et al. (2019). Dosage analysis of the 7q11. 23 Williams region identifies BAZ1B as a major human gene patterning the modern human face and underlying self-domestication. Sci Adv. 5:eaaw7908. [PMC free article] [PubMed] [Google Scholar]
47. Martin Johnnson et al. (2021) The neural crest hypothesis; no unified explanation for domestication. Genetics see: https://www.ncbi.nlm.nih.gov/pmc/articles/PMC8633120/
48. Daniel Kahneman. (2011) *Thinking Fast and Slow*, Penguin.
49. Jonathan Haidt. (2013) *The righteous mind: Why Good People are divided by Politics and Religion*. Gildan Media
50. Simon Gächter and Jonathan Schulz (2016) *Intrinsic honesty and the prevalence of rule violations across societies*. Nature. 531, 496-499
51. Oliver Scott Curry (2019) *What's wrong with Moral Foundations Theory*. https://behavioralscientist.org/whats-wrong-with-moral-foundations-theory-and-how-to-get-moral-psychology-right/
52. David Graeber and David Wengrow (2022) *The Dawn of Everything*, Penguin
53. Ian Morris. (2011) *Why The West Rules For Now*, Profile Books.
54. For Lichens see: https://britishlichensociety.org.uk/learning/what-is-a-lichen
55. CF Delwiche and ED Cooper (2015) The evolutionary origin of terrestrial plants. Current Biology 25. 899-910
56. E.g. https://en.wikipedia.org/wiki/Gut_microbiota
57. Marc Imhoff and Lahouari Bounouna (2006) *Exploring global patterns of net primary production carbon supply and demand using satellite observations and statistical data*. J Geophys. Res.: Atmospheres. https://doi.org/10.1029/2006JD007377
58. Yinon M Bar-On, Rob Phillips and Ron Milo (2018) The biomass distribution on Earth. PNAS,USA. 201711842
59. Daniel Everett (2017) *How Language Began. The Story of Humanity's Greatest Invention*. Profile books
60. Yuval N. Hariri. (2011) *Sapiens: A Brief History of Humankind*, Vintage Books.
61. Toby Ord (2021) *The Precipice: Existential Risk and the Future of Humanity*, Bloomsbury
62. Martin Rees (2018) *On the Future: Prospects for Humanity*. Princetown University Press
63. James Lovelock (2019) *Novacene: The Coming Age of Hyper Intelligence*, Penguin
64. A.J. Lotka (1922a) 'Contribution to the energetics of evolution' [PDF]. Proc Natl Acad Sci, 8: pp. 147
65. A. J. Lotka (1922b) 'Natural selection as a physical principle' [PDF]. Proc Natl Acad Sci, 8, pp 151–4.
66. Howard T. Odum, *Environment, Power and Society* (1971 New York: Wiley-Interscience. p. 43.
67. M.W. Gilliland ed. (1978) *Energy Analysis: A New Public Policy Tool*, AAA Selected Symposia Series, Westview Press, Boulder, Colorado
68. Steven Lukes (2005) *Power: A Radical View Second Edition*, Palgrave Macmillan
69. Michel Foucault. https://en.wikipedia.org/wiki/Power-knowledge#References
70. Robert U Ayres and Benjamin Warr. *The Economic Growth Machine: How Energy and Work drives Material Prosperity*. Edward Elgar Pubs. Cheltenham.
72. Nicholas Georgescu-Roegen. (1971) The Entropy Law and the Economic Process. Havard University Press. Boston
73. Ernst Schumacher (1973) *Small is Beautiful: A Study of Economics as if People mattered*. Vintage Press, London.

74. Robert Costanza, Herman Daly et al. (1997) *An Introduction to Ecological Economics*. St. Lucie Press, Florida
75. Kate Raworth. (2018) *Doughnut Economics; Seven Ways to Think Like a 21st Century Economist*. rh Business Books
76. Tim Jackson. (2011) *Prosperity without Growth : Economics for a finite planet*. Earthscan.
77. Peter Victor. (2023) https://newsociety.com/books/e/escape-from-overshoot?variant=43651293348080
78. Andrew Jarvis (2018) Energy returns and the Long-run Growth of Global Industrial Society. Ecological Economics. 146. 723-729
79. Gailliot, M. T.; Baumeister, R. F.; Dewall, C. N.; Maner, J. K.; Plant, E. A.; Tice, D. M.; Brewer, B. J.; *Schmeichel, Brandon J. (2007). "Self-control relies on glucose as a limited energy source: Willpower is more than a metaphor"*. Journal of Personality and Social Psychology. 92 (2): 325–336. CiteSeerX 10.1.1.337.3766. doi:10.1037/0022-3514.92.2.325. PMID 17279852. S2CID 7496171/.
80. Vohs, Kathleen D.; et al. (2021). "A Multisite Preregistered Paradigmatic Test of the Ego-Depletion Effect". Psychological Science. 32 (10): 1566–1581. doi:10.1177/0956797621989733. hdl:10072/408369. PMID 34520296. S2CID 236708287.
81. Shai Danziger; Jonathan Levav; Liora Avnaim-Pesso (2011), "*Extraneous factors in judicial decisions*", Proceedings of the National Academy of Sciences, 108 (17): 6889–6892, Bibcode:2011PNAS..108.6889D, doi:10.1073/pnas.1018033108, PMC 3084045, PMID 21482790
82. Konrad Lorenz in *Behind the Mirror* (1977) quoted in Chaisson [27]
83. https://www.nrc.gov/reading-rm/basic-ref/students/history-101/too-cheap-to-meter.html
84. https://www.eea.europa.eu/publications/environmental_issue_report_2001_22
85. https://www.goodreads.com/author/quotes/89275.Fr_d_ric_Bastiat
86. Peter Frankopan (2023) *The Earth Transformed: An Untold Story*, Bloomsbury
87. https://en.wikipedia.org/wiki/List_of_countries_by_carbon_dioxide_emissions_per_capita https://ourworldindata.org/co2-gdp-decoupling
88. https://www.imf.org/en/Publications/fandd/issues/2019/09/tackling-global-tax-havens-shaxon
89. Peter Haff. (2014) 'Humans and Technology in the Anthropocene: six rules,' *The Anthropocene Review*, 1(2), pp. 126-36. https://doi.org/10.1177/2053019614530575
90. https://www.greenhousethinktank.org/climate-and-justice/
91. Dieter Helm. (2023) Legacy: *How to build a sustainable economy*. CUP
92. Joseph Schumpeter. https://en.wikipedia.org/wiki/Capitalism,_Socialism_and_Democracy
93. George Akerlof and Robert Shiller (2015) *Phishing for Phools; The Economics of Manipulation and Deception*. Princetown University Press
94. Julian Baggini (2015) Freedom Regained; The Possibility of Free Will, Granta
95. Walter Schneidel. (2017) *The Great Leveler; Violence and the History of Inequality*. Princetown.
96. Leon Brillouin (1956) *Science and Information Theory*. Dover Publ.

Acknowledgements. I would like to thank a number of colleagues for their advice, criticisms and encouragement – Robin Grove White and his colleagues, Brian Wynne, Bronislaw Szerszynski and David Tyfield; Ron Pethig; James Intriligator and his colleagues at Tufts; Timm Hoffman; the late John Raven; Isabelle Winder; John Llywelyn Williams; Tony Rippin and Roger Leigh.

I am also indebted to Edward O. Wilson (**Consilience: The Unity of Knowledge**) whose belief in the unity of knowledge I have come to share.

These hypotheses were first publish in Welsh in Ynni, Gwaith a Chymhlethdod: Saith Chwyldro Hanesyddol' – Darlith Flynyddol Edward Lhuyd y Coleg Cymraeg Cenedlaethol a Chymdeithas Ddysgedig Cymru 2017 and in Ynni Gwaith a Chymhlethdod yn/in O'r Pedwar Gwynt Vol.7 Spring and Summer. 2018.

see file:///Users/garethwynjones/Documents/Darlith%20Edward%20Llwyd%202017%20Ynni,%20Gwaith%20a%20Chymhlethdod%20-%20Llyfrgell%20y%20Coleg%20-%20Powered%20by%20Planet%20eStream.html

6
Overshooting Limits

This chapter is based on a lecture first delivered in Carmarthen/Caerfyddin in 2012 at a conference jointly organised by the University and the IWA. It was then published by the IWA in a book Wales' Organising Principle. Subsequently I published a version in Welsh in Y Traethodydd.

Overshooting Limits
'Seeking a new Paradigm'[1]

This essay is dedicated to my four grandsons, Euros, Aled, Eirig and Owain hoping that, in 50 years time, they will be understanding of the follies of older generations and have lived through non-catastrophic change.

1 / Competing paradigms

Characteristically humans espouse many and various life styles, political and economic systems, religions and philosophies. However perhaps the most far-reaching division is between those convinced that there are tangible limits to humanity's ability to extract goods and services from the planet and those who do not. Broadly speaking the latter believe that the creative and destructive powers of capitalism and free market forces, combined with scientific progress, human ingenuity and technical innovations, can circumvent all supposed limits. A few even appeal to divine intervention.

The 'optimists' argue that, just as in Edwardian London concerns about being overwhelmed by horse manure were obviated by the combustion engine, and the Liverpool smog of my youth succumbed to the Clean Air Acts, humanity can and will harness other better technologies of limitless potential. Human, horse and waterpower have been superseded by coal-fired engines, then by diesel, gas and electric motors. Now we have miniaturized electronics and nano-technology. Our technical prowess is astounding. Mobile cell phones, helicopters and aircraft have transformed communications even in the remotest corners of this planet.

Even the nuclear holocaust so feared in the 1950s and 1960s has been avoided. Indeed, communism itself appears to have been vanquished. Fewer children now die in infancy but Malthus has not been vindicated.

A significant and increasing proportion of this planet's 7 billion human inhabitants are enjoying historically undreamt of prosperity. We are living longer despite HIV/Aids. In most areas food supply has comfortably kept pace with the rising population. The picture may not be 'panglossian' but is nevertheless impressive. It is claimed we can and will increase wealth, measured by global GDP, indefinitely. Indeed we may even colonise other planets. By this account, economic growth, as currently construed, is not only a good measure of success but also the vital policy objective. Those worrying about 'limits' are deemed misguided and willfully undermining the hopes and aspirations of the poor.

Given this scenario it is surprising that many, including eminent sober scientists, remain deeply pessimistic about mankind's future and concerned at the scale of humanity's impact upon planet Earth.[2] Knowledgeable and reputable doom-mongers abound. Is this simply sour grapes, anti-capitalist propaganda or an ignorant contrarian spirit? Since pessimism about humanities' future has been rife for millennia, we must ask whether it is really different this time. Indeed, as some claim, is humankind now standing at a critical crossroad having to make, or perhaps refusing to make, decisions of epic significance? If you conclude that the free market, apparently 'optimistic' analysis is a dangerous mixture of self-deception and self-aggrandisement, often promoted by those who have appropriated to themselves a disproportionately large faction of this planet's resources, then these may be very dark and challenging times indeed. Without analysing this fundamental issue, all talk of applying 'sustainable development' in a Welsh or international context is potentially misleading, even pernicious.

Let us recognize at the outset that most of the planet's seven billion human inhabitants are not engaged with these issues. If poor (still perhaps 80 per cent of us), they remain weighed down by immemorial problems. If relatively rich, they resent any challenge to their aspirations and achievements, especially the seductive American dream. Others are worried simply about retaining their jobs or paying off their debts. Some put their trust in an all-powerful, external God or Allah to provide providentially all the answers and resources to true believers. Any acceptance of limits, and, even more crucially, that mankind is fast approaching them, has far reaching implications. Such a conclusion must impact profoundly on our ethical, political, religious, economic and social thinking.

For centuries the older generation has perceived youth as feckless, irresponsible, and ill prepared to shoulder their responsibilities. In contrast, may not fecklessness and irresponsibility and a lack of foresight be defining traits of an older generation, my own generation? Are we handing to our children and our children's children a veritable Eden or a poisoned chalice? Many voices have been raised to warn us, others to reassure. Maybe we are instinctively aware of the dangers but are in denial. Are we too self-engrossed and too self-serving to react; too comfortable to act decisively and intelligently?

2 / The ambiguity of affluence

In a global free market there is variation on Gresham's law; bad capitalism tends to drive out good.
John Gray

The possibility of a tainted inheritance is powerfully illustrated in two books, both now over 40 years old. Superficially they appear to address very different topics. Certainly, they are formulated by authors from very different traditions, rooted in contrasting disciplines. Incidentally, however, we can thank Boston, Massachusetts for both. Nevertheless the central message in both is remarkably similar.

The first volume is *The Affluent Society* by the Scots-Canadian Harvard economist, John Kenneth Galbraith.[3] He was an academic patrician closely involved with the US Democratic Party and US policy for many decades, including as Kennedy's ambassador in India in the 1960s. First published in 1958 his book has enjoyed popular success and run into numerous editions over 40 years. It foresaw, with startling precision, the global financial disarray, which has gripped us since 2007-8. The second enjoys its 40th anniversary this year. *The Limits to Growth* was first published in 1972, describing work supported by the Club of Rome.[4] The initial volume by a group based at MIT led by Donella Meadows, Jorgen Randers and Dennis Meadows, has been followed by two other volumes, *Beyond the Limits* in 1992, and *Limits to Growth: the 30-year update* in 2004.[5] A further update will appear in late 2012.

Despite their popularity both books have been widely and wildly abused and misconstrued. Galbraith, as a humane political economist, welcomed the great leap in much of western capitalist society from abject poverty to affluence in a couple of centuries. All but a minuscule élite have moved from scraping a living on minimal resources which used to meet our most urgent needs – food, safe water, shelter, clothing – to widespread affluence.

By the late 1950s an impressive range of consumer goods had become available in the USA, in "you've never had it so good" Britain and elsewhere. Now vastly more have been added. Middle class housing has improved beyond recognition; central heating and at least one car is the norm. Every home should boast a huge flat-screen television, a clothes drier, dish washer and, of course, computers, the internet and a variety of mobile phones and iPads. This affluence has spread. As the goods and services on offer have multiplied many-fold, so some hundreds of millions of Chinese, Brazilians and Indians as well as Europeans and Americans have joined the comfortable middle class.

In Galbraith's analysis, when basic human needs are satisfied, continued affluence depends on an implicit, largely unspoken, economic 'bargain' between three main parties – the general public, industrialists and business, and politicians. Each of these parties has a rational self-interest in adhering to this bargain (see Figure 1). The public, individually and collectively, wishes to retain, and, if at all possible increase, its affluence and spending power. This in turn depends on the availability of preferably well paid jobs to maintain individuals and their families and protect their buying power and social status as well as work in the public services.

Jobs require flourishing businesses and a solvent government, being derived from the provision of goods and services. To maintain this dynamic, the range and quantity of these goods and services needs to be renewed and expanded continuously through private and public innovation and investment. These innovations can range from new IT gadgets or new cars, to providing care for an ageing population or new drugs. The new affluence itself has bred new wants, from holidays in the sun, to caravans in Abersoch, gourmet foods, fine wines and the latest fashions. More recently this has been eloquently described in "the story of stuff".[6]

Realising in Clinton's famous phrase that "it's the economy stupid", politicians must oil the wheels of this 'bargain' if they are to be re-elected by a relatively contented populace. Industrialists have a strong vested interest in ensuring the continual turning of the cycle so that their businesses may prosper. To help sustain the bargain, multi-billion industries such as advertising and product promotion, as well as novel design, have grown dramatically to convince the public of its urgent need for both traditional and innovative goods and services. It has also spawned brand loyalty, celebrity endorsement and a range of more subtle psychological strategies to stoke and stimulate our desires. If demand is waning or weak, it must be created!

Entirely reasonably and logically, business has also lubricated the wheel by devices such as hire purchase, easy credit, tempting special offers, all supplemented by ubiquitous credit and debit cards and loans. In many and subtle ways people are encouraged to take on debt, especially by buying and fitting out a home fulfilling the dream of a self-reliant citizen in a property-owning democracy. Companies must themselves borrow to compete and to invest in new exciting products, services or new outlets. In turn, government must borrow to play their part in keeping the economic wheels turning by investing in education, infrastructure and dampening the business cycles so as to reduce social discontent and seek to ensure its own re-election and political survival (sometimes propping up failures to do so). Credit and debt lubricate the whole system, giving huge power to financiers who emerge as the puppet masters.

Critically the cycle must continuously turn, recorded as an annual growth in economic activity measured by Gross Domestic Product (GDP) or its derivatives. The rate varies widely from country to country. In China growth at better than 7 per cent implies a doubling of activity

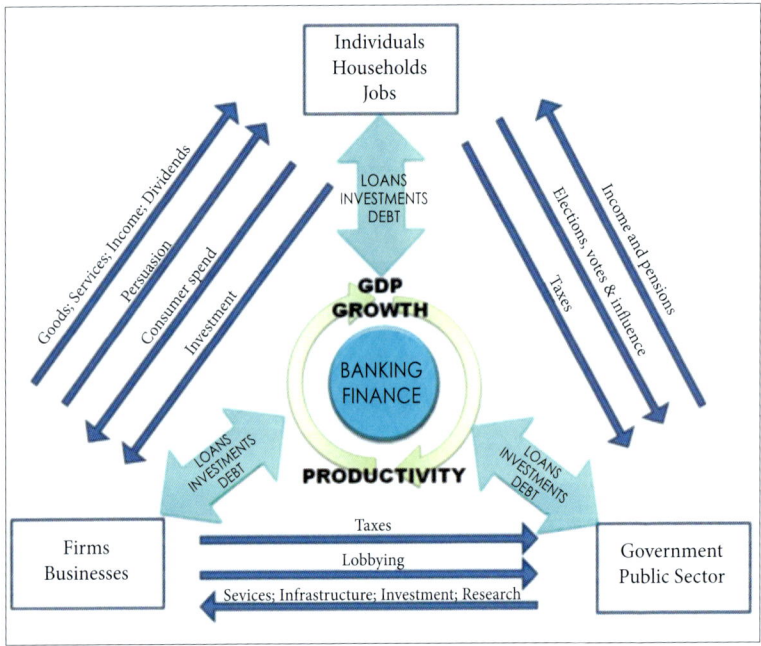

Figure 1: The Galbraithian bargain

every decade. The UK dreams of an annual three per cent growth rate. The model is predicated on continuous exponential growth in activity. It must spin like a top and grow like Topsy to retain stability. What seemed, in the first instance, a wonderful transformation for humanity (at least the affluent minority) has become a treadmill and a never-ending grind. As Galbraith foresaw, left to its own devices, the 'bargain' to which all the parties have enthusiastically subscribed, ensures vested interests turn a blind-eye to mounting debt. He saw the system as intrinsically unstable, leading inevitably to excessive, unsustainable debt, to **overshoot** and painfully bursting economic bubbles. He suggested that the natural tendency to overshoot was caused by over-optimism, aided and abetted by 'conventional wisdom'. Moreover, the human capacity for self-delusion delayed any appreciation of reality with, of course, an added injection of greed and self-interested deceit. The system required tight control.

According to what I refer to as the Galbraithian bargain, affluence requires a continually and indefinitely accelerating cycle of product innovation and consumer demand, supported by government and oiled by debt. Our economic success commits us to living on an ever-accelerating treadmill.

While emphasising the need for prudent regulation and careful economic management, Galbraith also recognized a tendency to public squalor despite private plenty. He suggested that public finances would inevitably come under pressure, as power and financial clout would accrue to a small élite. Affluence could reduce civic and collective pride and commitment.

In 2012 we can conclude unambiguously that our politicians and economists should have taken Galbraith's diagnosis much more seriously. However, central to his insight was the improbability of their doing so. Ironically, even countries such as Germany, reluctant since the war to promote excessive internal spending and debt, have been trapped in the maelstrom. Although the butt of much Anglo-American criticism for their caution, they are now caught up in the whirlpool as their exports depended on demand from debt-laden markets and their banks are in hock to the debts of other countries.

Galbraith recognised that, as intermediaries and the controllers of credit, bankers wielded enormous power. However, he did not anticipate several factors. First, was the invention of clever packages which would allow risky debts to be sold on as triple A-rated financial products only tangentially related to the real world. Not only would the money market encourage the poor to take on un-repayable debts, but these would be repackaged as deceitfully desirable, highly rewarding, and apparently safe investment products. This and an array of devices, such as credit fault swaps, derivatives and the like, led to increasing instability. Secondly, both in the USA and the UK, the cult of the infallibility of the unfettered market led to deregulation, first under Reagan and Thatcher, and then Bush, Blair and Brown. Both were contrary to Galbraith's recommendations.

In the UK the 'big bang' in the city, and in the USA the repeal of the Glass-Steagall Act by the Gram-Leach-Billey Act in 1999, helped ensure 'irrational exuberance' whose most influential cheerleader was, paradoxically, the Chairman of the US Federal Reserve, Allan Greenspan himself. A few statistics will be pertinent here. In 2010 global foreign currency transactions totalled $955 trillion. Off-exchange trading in financial derivatives was $601 trillion. Meanwhile, the traded volume of shares and bonds was $78 trillion, compared with the global gross domestic product of $63 trillion. This leverage and exploitation of vast sums of 'virtual' money has allowed the emergence of hyper-rich individuals to emerge in the real world.[7]

Thirdly, and certainly in the UK, government appeared to become reliant on the financial service sector for a significant part of their tax revenues and consequently on this sector to fund many public services. The sector became untouchable, holding, and continuing to try to hold the country to ransom. This was despite Adair Turner's recognition, as chairman of the Financial Services Authority in London, that much of their activity had no public benefit. Moreover, a recent report quotes the taxes paid by the financial sector from 2002 to 2008 as £193 billion but that from manufacturing in the period as £378 billion.[8] Following the crash, direct government support to the financial sector was put at £298 billion and with loans and underwriting at £1.7 trillion. If these figures are correct, then either the City's PR apparatus is quite exceptional or other darker factors are at play. The problematical situation has been compounded by leverage buy-outs converting capital into debt, by huge global trading imbalances, and by the ability of the rich to avoid paying taxes by recourse to tax havens and to use their enormous wealth to promote disinformation and protect their narrow self-interest.

Galbraith's fears have been realized, his diagnosis confirmed. The policy makers and politicians of successive governments in many countries – Ireland, Iceland, the UK and the USA to name only but four – have been seduced by their blind faith in the infallibility of the markets. Deregulation was the war cry of Tory and Labour alike, a process abetted by a number of prominent academic economists. Now we have to account for this folly. In the UK, the tension between the desire to lower the personal and national debt burden and the fear of stoking a downward spiral of activity and employment because of a lack of demand is being played out.

The lessons of Galbraith and also of Keynes seem drowned out as countries fear the verdicts of the rating agencies and subsequent reprisals by the same financial markets which were a major cause of the collapse. It is a bitter irony that the excesses of these same markets and their analysts who helped foul up in the first place, now need placating. Finance is now in short supply and economic growth to reduce unemployment is illusive. We are living the consequences of **overshoot**.

3 | Growth and overshoot

I don't think much of the science of the beastly scientists.
Victor Hugo
Science clears the field for technology.
Heisenberg

In large measure *The Limits to Growth* also deals with the danger of **overshoot**.[9] The initial slim volume in 1972 analysed a relatively simple model for the interdependence of population growth, industrial production, pollution, resource depletion and demand, including food production, land availability and capital demand. Central to their thinking were three factors. First, the nature of exponential growth; second, the time lags in the responses of most natural systems which lead inexorably to overshoot; and, third, the absence of a precise limit.

Meadows and her colleagues did not envisage our planetary system approaching a cliff edge, such as when the flat-earthers warned Columbus he would sail off the edge of the known world. Rather, the authors anticipated a decline, which could be gradual or rapid but would be greatly exacerbated by 'overshoot'. More and more human and natural resources would have to be directed to combating some limiting factor – be it reducing the negative impact of pollution, acquiring some scarce metal, or securing water or food supplies. In turn, this resource distortion would undermine living standards and create the potential for social disorder. I would add that this would also subvert the Galbraithian bargain, although Meadows and her colleagues did not refer to it. Critical to their analysis was exponential growth itself, as shown in Table 1.

Even a modest 2 per cent exponential growth in a factor such as population or GDP, or in demand for a valuable but rare metal or food, causes a doubling in only 35 years, about a generation. At an annual 2 per cent compound growth, our current global population of around 7 billion would theoretically, but of course improbably, reach 14 billion

Table 1: Relationship of exponential growth rate to doubling time – applicable to factors such as GDP, population, and resource use.

Growth rate (% per year)	Doubling time (years)
0.5	140
1.0	70
2.0	35
3.0	23
4.0	17.5
7.0	10
10.0	7

by 2046, and, even more improbably, 28 billion by 2080 compared with 3 billion in 1960. The equivalent UK population figures would be over 120 million in 2046 and near 250 million by 2080.

Politically an annual GDP growth of 1 per cent is regarded as lamentable, inadequate to maintain employment and rising living standards. Nevertheless, mathematically or in resource terms, it is significant. The conventional political aim is an annual GDP growth near to 3 per cent to ensure full employment, pay for pensions and public services for an aging population. But this implies a doubling every 23 years. Can this possibly be sustained? The fundamental conclusion of Meadows and colleagues was that it cannot and the current economic growth model is not sustainable.

Two major criticisms have been leveled at the *Limits to Growth*. One is that in the last 40 years we have not hit the brick wall. Indeed, on a global scale, economic growth has never been faster. More than 2 billion Indians and Chinese, or at least a skewed proportion of them, are reveling in an annual 7 to 10 per cent GDP growth. However, as noted, Meadows and her colleagues didn't foresee a precipitous sudden halt. Rather she anticipated the 'limits' gradually tightening their grip in the 21st not the 20th Century. Secondly, they were accused of underestimating human ingenuity and the dynamic power of modern technology and the market economy. This they deny.

Indeed, the latter argument has little substance. As discussed later, technological interventions are essential – the question is **which technologies**? Some low environmental-impact technologies have advanced rapidly, for example, greatly improved home insulation and waste recycling, CFC-free refrigerators, photovoltaic cells, more fuel-efficient cars. No doubt pollution of the UK's seas and fresh waters has declined. The 1987 Montreal Protocol to curb CFC emissions and allow the Polar ozone holes to recover, seems to be a success.[10] But switching technologies may bring their own resource constraints. Moving from fossil fuels to wind turbines may reduce demand for oil and gas, but increases the demand for rare earth metals for the electrical circuitry of the turbines. There are few, if any, free-meals.

If overshoot is a dreaded feature of our economic system, it is an intrinsic character of natural systems. Despite the Protocol, the ozone hole persists. Only by 2050 is it hoped that it will recover. If female fertility drops to the basic replenishment rate (2.2 children per woman), it will take decades for population growth to decline to zero because of a pyramidal population structure – that is the presence of many fertile women in their teens and early twenties. In relation to climate change, even if anthropogenic greenhouse gas emissions, especially carbon dioxide (CO_2), were to stabilize or even to fall to pre-industrial levels tomorrow, it would take hundreds, maybe thousands of years, for the climate, oceanic circulations and sea levels to re-equilibrate (see Figure 2). Some but not all over-used aquifers could refill in time but it would take many decades, not the few years during which they exploit them. The fish stocks of Newfoundland Grand Banks have not been replenished despite a moratorium imposed some 15 years ago.[11] If we continue to cause the extinction of species then, maybe, in time new evolutionary niches will arise for new species but on a millennial not a decadal time scale.

It is interesting to address some up-to-date examples of the 'limits' dilemma and their implications. I will consider, briefly, mineral, energy and food supplies because of their global ramifications. (Global water shortage is equally pertinent but has less resonance in Wales.) Between them they illustrate different facets of the global challenges with which we may not be sufficiently familiar.

A striking analysis of mineral supply is found in the 2011 Risk List published by the British Geological Survey, which assessed the supply risk of economically important chemical elements on a scale of 1 to 10.[12] The rating depended on:
- Scarcity, in terms of crystal abundance.
- Production concentrated in limited number of countries,
- Location and distribution of potentially exploitable base reserves.
- Political stability of producing countries.

On this basis, relatively little known elements such as antimony, the platinum group, mercury, and tungsten receive the highest risk rating of 8.5, followed by the rare earths and niobium. These elements have become widely used in high-tech materials such as in microelectronics, catalysts, special alloys, semi-conductors and fire and temperature resistance products. At the other extreme lie the abundant and familiar elements: titanium, aluminium, chromium, iron, and sulphur.

Our advanced technologies have created new resource demands, including for materials or elements that are rare and often toxic. As some of the high supply-risk elements are essential to modern devices, securing an uninterrupted reliable supply is now a major objective of powerful nations especially China. Similarly, maintaining its gas and oil supplies has long been an American obsession, even at the expense of others. So it is apparent that the 'risk' is not simply one of geological abundance or even mining technology but the interaction between geology and economic, political and even military might. So for example, in late 2011 *Rare Earth Investment News* reported that the price of Cerium, used in energy efficient LED lights, increased from $50 to $413 per kilogram between January and June 2011 threatening the industry and leading to flight of companies to China. The limitations of market and state capitalism in accessing and allocating resources are all too apparent and fundamental problems remain unresolved.

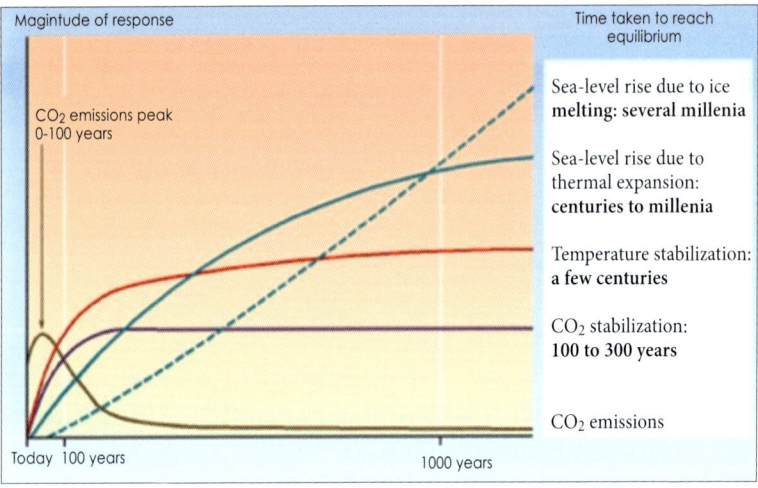

Figure 2: Lags in climate change and overshoots
Source: R.T. Watson and the Core Writing Team (eds.), *Climate Change 2001: Synthesis Report. A Contribution of Working Groups I, II, and III to the Third Assessment Report of the Intergovernmental Panel on Climate Change*, Cambridge University Press, Cambridge 2001.

Figures 3a and 3b illustrate the near exponential growth and price fluctuations in world copper production. These reflect the impact of global economic growth and futures speculation as well as possibly the feedback effects of demand and supply scarcity on the market price. In the wake of the 2008 crash the price fell back but has now been on rising trajectory for about 6 months, to ~$8,500 per tonnes. Again major companies and nations are hell-bent on securing a dominant market share of this resource.

Another even more challenging, but dauntingly complex, aspect of 'limits' is illustrated by the web of food supply, land availability, water for irrigation and climate change.[13] Despite the high levels of waste in western counties, the combination of more people, up to and maybe over 10 billion by the mid 21st Century, and an increasing standard of living equate to an increasing and changing food demand. This applies not only to the basic commodities, such as grains, potatoes, cassava, but especially to white and red meats and dairy products of all types.[14] Logically these demands can be met by either increasing yields from existing land or by expanding the land under cultivation or being grazed. Both options are replete with problems (see Figure 4).

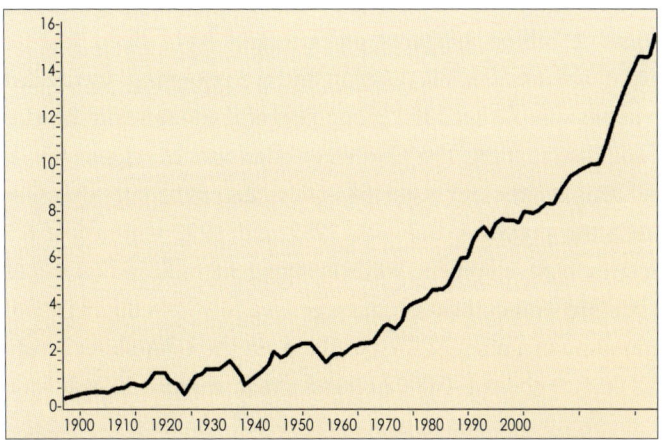

Figure 3a: Trends in world copper production 1900-2007.[15]

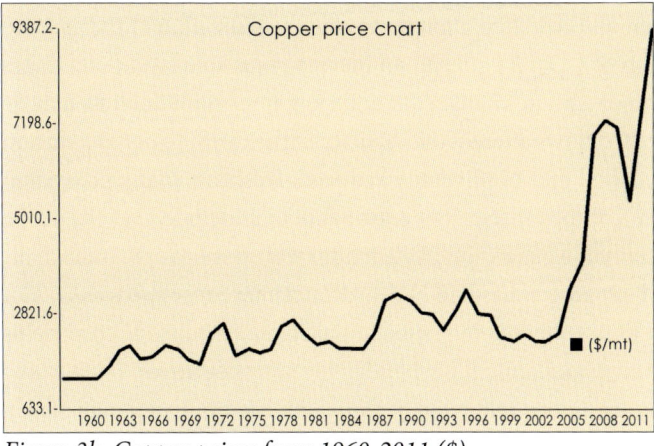

Figure 3b: Copper prices from 1960-2011 ($).

Unused arable land is in short supply and irrigation water even more so. In many regions even existing water supplies are threatened by the over-use of aquifers, rivers or reservoirs and competition from urban demands. Globally about 1.5 billion hectares of land are under cultivation annually, with estimates of the potential area in the range 2 to 4 billion hectares. However, such estimates are fraught as cultivation of inappropriate land can lead to serious difficulties.

Over the last millennium of urban growth and increasing agriculture activity, some 2 billion hectares of farmland have been lost to desertification, salinisation, alkalisation and waterlogging. In the last decade Syria has discouraged the ploughing of the semi-arid 'baadia' because of soil degradation. The 'Garden of Eden' is either desert or salt marsh depending on its exact historic location. Recent urban expansion has increased the pressure. Between 1989 and 1992 China lost 6.5 million hectares to development, while bringing 3.8 million hectares of forest and pasture into cultivation. A large area is lost to the march of urbanization annually, often of the best land. In the USA alone before the financial crunch about 170,000 hectares a year was lost.[16] Worldwide the area probably approaches or may even exceed 1 million hectares.

Worryingly, bringing grassland or forest into arable production is not environmentally cost-free. Ploughing causes oxidization of the soil and the emission of CO_2. As a result an increase in arable land exacerbates the problem of climate change.[17] In addition, important habitats may be lost, watershed flows altered, the capacity of the forest to act as a carbon sink diminished and biodiversity reduced. In short, many ecosystem services are compromised. It is reasonable to conclude that the scope for the sustainable expansion of agricultural land is limited. Indeed, ill-considered expansion has had highly deleterious consequences.

Without doubt there is very substantial scope to improve crop yields and better crop and animal husbandry in many countries. Usually on-farm yields are a poor reflection of those obtained in comparable local, well-managed trials. This applies to both irrigated and rain-fed systems, although historically most of the additional food required to meet the demands of population growth has come from the former. Unfortunately in a number of places, irrigated systems are threatened by diminishing water supplies and urban competition. Moreover, the poor yields have proved resistant to improvement. The problems arise from issues as diverse as land tenure, equitable water rights, farmer education, market access, spiraling debts, cost and availability of agro-chemical inputs, lack of appropriate, well-adapted modern seed and poor quality advice from many National Agricultural Research Systems. Another issue can be the diminished size of farm holdings as land is subdivided between growing family members. Broadly it appears that the easy gains in the most fertile areas have been achieved. Future progress is likely to be won more slowly and more expensively.

Many studies suggest that climate change will reduce water availability in much of the Middle East, Central Asia, Australia and parts of the US, both for rain-fed and irrigated crops.[18] River flow rates are projected to fall in the semi-arid areas, and soil moisture content to decline. Although higher CO_2 levels may stimulate vegetative growth, yields are generally projected to decrease because of greater climatic variability, especially periods of extreme drought, as occurred in Russia in 2010 and Australia in 2008-2011, and catastrophic flooding, as in Pakistan in 2010. Specific phenological issues such as terminal heat stress and poor germination, when rains are late or fail, and decreases in the length of the growing season are a threat. Poorer water availability and greater evapo-transpiration at higher temperatures will combine to limit yields.

Greater intensity and high yields require greater fertility. Less well known is that the application of fertilizer leads to the release of a gas, nitrous oxide (N_2O). Unfortunately this is a far more effective greenhouse gas than CO_2 itself (x 310 over 100 years). Every tonne of applied nitrogen, whether in the form of manure or inorganic fertilizer, leads to the emissions of about six tonnes of CO_2 equivalent (the total atmospheric forcing of all greenhouse gases is often expressed as CO_2

equivalents –CO_2e).

In addition, commercial fixation of abundant atmospheric nitrogen (N_2) gas is itself very energy demanding. The price of the nitrogen fertilizers, so produced, strongly correlates with international oil and gas prices. Since crop yields are near linearly related to the amount of nitrogen available up to several hundred kilograms of nitrogen per hectare, this is not a trivial issue. Paradoxically, the drive for increases in cropping intensity and yield could exacerbate the contribution of the food chain to climate change. Further, the global rate of increase in nitrogen application rivals the increase in atmospheric CO_2 and is ecologically damaging in its own right, leading to eutrophic anoxia in fresh and inshore waters. So intensification is not without cost, and made worse because much of the corn and wheat harvest is fed inefficiently to animals.

The climate change impacts of farm animals and animal products cause much angst in the UK as ruminants such as goats, sheep and cattle, sheep produce a large percentage of global methane (CH_4) emissions. Many of these animals in the US and elsewhere are fed on grain and other concentrates in feed-lots, wastefully diverting resources from direct human use as well as producing methane. On a 100-year basis, methane is about 24 times as effective a greenhouse gas as CO_2. There is no doubt that the world food problem and the global warming impact of the food chain would be reduced dramatically if our diets were to incorporate fewer animal products. We have calculated that, based on current trends in global food demand and population growth, feeding ourselves will **alone** perpetuate climate change and ensure that we breach the +2°C limit by 2040.[19]

This temperature value is usually considered as a threshold for triggering 'fast-forward' reactions such as the release of tundra methane or added solar radiation absorption from total loss of Arctic summer sea ice. It is important to emphasise that is not simply an issue of about 1.5 billion cattle and other ruminants but as much about the world's population of about 23 billion chickens. The latter have a lower individual greenhouse gas footprint, but collectively are hugely significant in global terms. Moreover, this is without factoring in any tendency to push arable production into unsuitable areas leading to erosion and loss of topsoil and organic matter, or the use of poor quality irrigation water inducing desertification and saline degradation. It is salutary to consider this stark reality even in the unlikely event of mankind being able to completely decarbonize our energy supply by mid century.

Perhaps we are nearing a threshold. Not perhaps an absolute limit but one based on current methodologies, priorities and practices. While staple food prices fell gradually for many decades, in the last years this trend has been reversed. In 2008 a spike in prices triggered food riots in a number of African countries and in the Philippines. Since then, following a small decline, the United Nations Food and Agriculture Organisation's food price index has increased further, reaching 225 in September 2011, compared with its previous peak of 213.5 in June 2008. It's worth recalling that serious unrest caused by food prices in Tunisia and Algeria in early 2011 was a prelude to, perhaps a harbinger of, the Arab Spring.

These examples dramatically illustrate a core message of *The Limits to Growth*. There is no single discrete problem nor is there a unique silver bullet to solve it. Rather we must wrestle with a complex web of interrelated factors – some physical, some social, some political-economic – which together will make the feeding the world's growing population more and more challenging and demand more and more attention and resources.

Paradoxically, if affluence expands then the demands, especially for 'high-end' meat and dairy products, will increase – worsening the climate change threat. Nevertheless the demand from the affluent for agricultural products, including meat, offers access to the market economy and therefore one route out of poverty for the rural poor.

Moreover, for many semi-arid, partially nomadic people, pastoral animal-dependent systems are their only option.

In the name of a climatic risk, largely caused by the rich, are we to ban the Masai from herding cattle on the Masai Mara, or the Basutho on the Maloti mountains, or indeed, the graziers of the Welsh hills? Would they and many others listen? How could such policy be imposed? Perhaps we should revert to a near-vegan diet in which case some issues would diminish. Yet the social resistance would be enormous, not just in the US or Australia but in much of Islamic world and Africa. Ethics suggests that we should differentiate between pasturalists who are depending on and using otherwise unused pasture resources, and the cattle and sheep barons heavily dependent on feeding concentrates that compete directly with human use. How could this possibly be achieved? If meat and animal products were to be somehow banned, denial and deceit are inevitable and overshoot a certainty. Moreover, in Galbraithian terms, if the demand for many products, not just red meat but pork, bacon, chicken, turkey, milk, cheese, eggs, yoghurt were to be curtailed, would not many consider that the bargain underpinning affluence had been broken, with incalculable consequences?

We are running on an accelerating treadmill, having to consume more and more time, effort and resources to stop ourselves falling off. Only increasing by 2 per cent annually, our speed doubles in roughly a generation. This growth carries wide environmental consequences, illustrated by the established relationship between GDP and energy use per capita (dominated, of course, by hydro-carbon burning). In general high GPD per capita nations as well as having higher per capita energy demand are often more profligate i.e. have lower energy efficiencies but countries such as Switzerland and Austria suggest that this is not inevitable.[20] Other data indicate that above $20,000 per capita wealth some energy decoupling is observed but this may be a distortion due to the importing of goods with high levels of embedded GHG from developing countries. Demand for goods, even if these are imports manufactured in China not at home, is a drain on raw global resources.

Undoubtedly, one factor accelerating the treadmill is population growth, but the other is the Galbraithian bargain which underpins our 'affluent' economy and, much more problematically, the prospects and aspirations of the poor and disadvantaged.[21]

As the demand for resources from a combination of population growth and higher living standards increases, the resilience of food supplies to climatic extremes decreases. At the same time our flexibility to make changes is strangled by the complexity of dealing with not just technical issues but social and politico-economic factors.

Much as free market zealots argued that the global financial system was largely self-regulating, so critics of *Limits to Growth*, and other similar propositions have declared that, through pricing and scarcity, the market mechanism will stimulate new innovations and circumvent problems. This ignores several factors. Firstly, many incipient problems such as anthropogenic climate change are economic externalities and so excluded from market calculations. This applies not only to atmospheric greenhouse gas emissions but also to the release of nitrogenous and phosphatic compounds into the environment. Secondly, global variation in regulation encourages capital and manufacture to move to the least regulated, usually most pollution-indifferent, low wage location. Thirdly, there is no logical reason to believe that alternatives to scarce resource will always be available. Certainly this applies to food and water. The behaviour of the Chinese government in buying up many of the global copper and other metal resources shows that they are not so complacent. Fourthly, the record shows that the speculative aspect of the free market system ensures that modest changes in supply will be massively magnified by such speculation.

4/ Reaching for sustainability

…truth comes out of error more readily than out of confusion.
Francis Bacon

Sustainability and sustainable development are much abused terms. We need to avoid obviously unsustainable actions such as an exponential increase in ocean fish catches. Such points are well rehearsed in the Meadows' books. But, if our current system is making unsustainable demands on the planet's resources and sinks, even though many are dirt poor and many communities in crisis, what might a more sustainable system look like?

Herman Daly has suggested three criteria for environmental sustainability, distinguishing renewable and non-renewable resource use and pollutant dispersal.[22] For the former he suggests that their rate of use should be no greater than the rate of regeneration or renewal of that source. So, for example, fish catches should be limited to the rate of growth of the fish population, water extraction from an aquifer should not exceed its recharge rate, and farming should not result in a net decline in soil organic matter or net increase in greenhouse gas emissions.

Non-renewable resource use should be no greater than the rate at which a renewable **resource** can be sustainably substituted for it. For instance, oil use could be considered to be sustainable if part of the profit were to be systematically invested in renewable energy sources so that when the non-renewable source is depleted an equivalent renewable source can come on-stream. In the case of a pollutant, the rate of emission should be no higher than the rate at which the pollutant can be recycled, absorbed or rendered harmless, in the way sewage can be degraded by micro-organisms without damaging aquatic ecosystems. One can, perhaps, add to the second Daly criterion that the use of a non-renewable resource should be no greater than the sum of substitution by a renewable resource, as above, **and** the enhanced capacity to recycle and recover the non-renewable resource from waste.

Applying Daly's logic to both nuclear fission power and hydrocarbon fuels is highly uncomfortable. In the case of nuclear power, uranium is a non-renewable resource but given a low supply-risk rating of 4 by the British Geological Survey. Availability at current usage rates appears a modest issue. Should nuclear power become the international electric generating system of choice, regardless of any security considerations, the situation would change dramatically. Supply would rapidly become a problem using current technologies.

However, dealing with the radioactive waste even from the present small number of plants for hundreds or even thousands of years is formidable enough to tax our managerial capacity. Since the 1960s successive UK governments have failed to solve the issue of long-term waste storage.[23] Only after many centuries when the radiation has decayed to a low level does the material become biologically harmless, so fulfilling the Daly's criterion. An international agreement on containment, at a European level at least, is essential. It should be noted here that few who oppose nuclear power, or campaign for 'nuclear free zones' also consider nuclear medicine and the increasing use of radio-nucleids and radio-pharmaceuticals.

In their original 1972 edition of *Limits to Growth*, Meadows and her colleagues were careful to discuss constraints arising both from sources and sinks. Interestingly, reflecting the concerns of the day, they made little of anthropogenic CO_2 emissions (that is, sink impacts) from the use of non-renewable gas, oil and coal resources compared with the future of those reserves themselves. By now it is starkly apparent that both sources and sinks pose problems. Applying Daly's criteria to today's sink data in relation to our hydrocarbon use is truly formidable. Global emissions of the main greenhouse gases would have to return to circa 1870 levels: about 270 ppm for carbon dioxide (from ~390 ppm currently), methane to ~700 ppb (from current ~1800 ppb), and nitrous oxide to 270bpm (from 320 ppb). There is, of course, no prospect of this happening. But the critical issues arise from the continued rapid increases in emissions, and the overshoots and time lags within the system.

Due to the global demand for cheap energy, carbon dioxide emissions illustrate graphically the 'limits' problem in relation to both geophysical and geopolitical reality. There is near-universal, international political agreement that we should strive to avoid a mean global temperature rise of more than 2°C. This roughly equates to atmospheric greenhouse levels of ~450ppm CO_2e – that is, total greenhouse levels expressed as carbon dioxide equivalents. As CO_2e levels are already ~425 ppm, the 450 ppm threshold is certain to be exceeded. The physical problem is that since carbon dioxide is retained in our atmosphere for many decades, it is the total **accumulated carbon dioxide load** that is critical.[24]

If we continue to emit at current rates then the scope for low non-damaging emissions in future decades is diminished. In 2010 global CO_2 emissions reached 35Gt. This represented an increase of record 6 per cent, despite the Kyoto Accord and many protestations of political commitment. International inaction arises from a mixture of short termism and self-interest and is partly compounded by some scientific uncertainty. That is to say, the equating of 450 ppm CO_2e to a ~2°C temperature rise depends on 'global climatic sensitivity'. This refers to how the mean global temperature would rise due a doubling of atmospheric greenhouse gases from pre-industrial levels. The possible figures vary from 4.5 to 1.5°C with a median of ~3°C. This alone gives special interests plenty of scope to sow doubt, even though the global trends are unambiguous. Also we humans have great difficulty distinguishing our local highly variable weather from global climatic trends, or in perceiving time scale.

Historically, 'western' emissions have dominated and our affluence is based on cheap hydrocarbon fuels. Thus it is unsurprising that the rest of the world should expect us to take the initiative and cut our emissions first as they rush to catch up. However, our political leaders are understandably reluctant to tamper with the 'affluence' bargain despite the risks.

Equally, the stance of some in the developing world may be both shortsighted and self-defeating. Countries such as India and China are in the front line for being severely affected by the projected changes. By 2010 global CO_2e emissions per head were about 6 to 7 tonnes.[25] In Wales we emit about 17 tonnes per head. But Chinese emissions have already reached the global mean and they can no longer be counted amongst the innocent. Amongst the innocent are the inhabitants of the low-lying Pacific coral islands who are low carbon emitters but most exposed to sea level rise. The best 'projections' suggest we need to reduce emissions globally to about 2 tonnes per head by 2050, assuming the UN population 'projections'. Few, therefore, can escape their share of responsibility.

Of course, future 'projections' are always difficult to address when 'real politics' is about the here and now. Financial markets deal in profit on a second by second basis. Businesses operate on an annual or multi-year horizon. How can such a system cope with 30 and 40-year 'projections'? How can they deal in scientific probabilities hedged by uncertainty? These problems are surely a recipe for lagged responses and overshoot.

In relation to supply we also face formidable problems. While the concept of 'peak oil' is hotly debated, there is every prospect that oil and gas prices will continue to rise from their current ~$110 per barrel of Brent crude oil, despite the near recession in North America and Europe. Our search for oil takes us deeper and deeper into the oceans and into more problematic territory – the Lula field off Brazil is seven kilometres down. Fracking of tar oils is not only environmentally suspect, but expensive. The commodities director for Barclays, Paul Horsnell, is quoted in *New Scientist* as forecasting oil at $137 per barrel by 2015 and $185 by 2020.[26] Of course, others differ.

A number of economists have related poor economic growth, in the traditional GDP sense, to high oil prices. It has been suggested that more than 4.5 per cent of GDP is spent on oil in the United States recession ensues. But that figure equates to a price of just $90 per barrel in the United States.[27] So it may be the good news that the price mechanism will force governments, companies and individuals to move rapidly away from oil and gas hydrocarbons. The bad news would be if this meant a rush to coal, a far less greenhouse gas efficient fuel per unit kW produced without carbon capture. This latter unproven technology could, of course, alter the scenario dramatically but would undoubtedly add to costs.

The other reality is that, although there is much talk of developing renewable energy sources and/or nuclear electricity, these are themselves replete with technical, environmental and political problems. The scale of human demand is staggering, of which electricity supply *per se* is only a modest element.

Given our dependence on hydrocarbons, critical questions remain unresolved. Will change occur by the pricing mechanism or by fiat and will it occur fast enough? In physical terms the importance of lags cannot be overstated. Desultory action, which has been made more likely by the outcomes of Copenhagen and Durban climate summits, will probably lead to global CO_2e levels exceeding 550ppm by 2050. This is both dangerous and hugely irresponsible. An unpredictable chain reaction would then be likely to occur, making catastrophic changes possible in a world of 9 to 10 billion people. We would be set on course for a 3 to 4°C mean temperature rise, a massive methane release from the tundra, and a six metre rise in sea level as a result of a major melt of the Greenland ice cap. What then of Bangladesh or the Nile delta or the Fens or indeed low-lying London and New York? Still worse would be any loss of the annual glacial melt which feeds the great Himalayan rivers that sustain billions in India, China and the rest of south and east Asia.

5 / On human frailty

Man is practised in disguise; he cheats the most discerning eye.
John Gay

It is human to be occasionally dishonest, somewhat hypocritical and capable of self-deceit and self-justification. Psychological experiments show that our honesty is limited and deceases if we feel we enjoy impunity and/or power or, at the very least, an opportunity to escape the consequences of our actions.[28] The least corrupt societies are those with open, well-informed and educated citizens, enjoying freedom of information such as in the liberal social-democratic relatively egalitarian societies of Scandinavia. The broken societies of Iraq, Afghanistan, Burma, and Somalia are worst, while the UK and US lie poised in the middle.

Bribery as a social phenomenon is ingrained in corrupt countries and individuals born into such society are most likely to indulge, even if removed to a new community. But given the close links between power, wealth and corruption, we should be most concerned about the behaviour of powerful, wealthy people. This is because it is the most affluent who might feel they have the most to lose personally from a more sustainable world. They are the ones most able to put out misleading information, buy political influence and invest in inappropriate technologies. Human nature being what it is, no doubt they can also convince themselves that they are acting with integrity and even in the public good.

On the other hand, there is ample psychological and sociological evidence that humans have a keen sense of fairness and justice.[29] In countries where these characteristics are encouraged people appear to enjoy better health and greater fulfillment.

Our current global regime is economically, socially, environmentally, and politically unstable and probably unsustainable. Seeking to maintain and project such a regime into the next 50 to 100 years will almost certainly grossly exceed the planetary limits. Indeed, we may have already done so but our impact is masked by the physical lags. It has been calculated that currently our demands are equivalent to the resources of about one and half Earths. In Wales we may require some three times of our surface area to support us.[30]

Even in this scenario we privileged a few who are deemed, not for very obvious reasons in the case of bankers and many financiers, to be wealth creators. We assume that a version of trickle-down development will improve the living standards of the majority – in the US the so-called 99 per cent. This policy has been singularly ineffective and has resulted in huge discrepancies in income and wealth. Despite attaining national 'affluence', conditions for the poor and many in the middle classes of the USA and UK, the primary cheerleaders for a unfettered free market capitalism, have improved little, if at all, in 40 years.

Meanwhile, Mrs Jones, Llanrug, living on her pension, her equivalent on remittances in the Maloti mountains of Lesotho, or Abdul in his Bangladeshi fishing village have low expectations and low carbon footprints. Yet, although largely blameless compared with the rich or even high-flying scientists, they are more susceptible to rising fuel and food costs and tides. Riches have accrued to a privileged élite who buy power and political influence, and too often peddle half-truths and misinformation about environmental and social issues, avoid paying taxes and even circumvent their legal responsibilities. How can this possibly continue?

It would be facile not to acknowledge the appeal of consumerism and of 'retail therapy', the convenience and appeal of the supermarket or to underestimate our dedication to a particular brand of 'affluence'. Holidays in the sun seem almost a human right. Certainly, they are understandably important to the citizens of the rotting old mill towns or mining villages. All the more so if they are surviving in boring poorly paid jobs in dismal northern European winter weather. As budget airlines will attest, any suggestion that airfares should reflect their full environment cost will be fiercely resisted – the 'bargain' would be jeopardized.

Maybe we are dissatisfied with our materialism and suffering from 'Affluenza'.[31] However, the condition is deeply ingrained and not easily eradicated. We quite enjoy our 'fragile' condition. We may be aware of socially disruptive forces at work in our society, of communities that are marginalised and despairing. But we are swept along on the flood tide of consumerism, taking our holiday while we still can. Can we imagine any other way?

6 Responses

The human crisis is always a crisis of understanding; what we genuinely understand we can do.
Raymond Williams

However bleak the situation, it is not axiomatic that the inhabitants of this planet must follow those of Easter Island or Palmyra or indeed many other civilizations to disaster.[32] Some, perhaps many, will retreat into a personal bubble hoping that they and their immediate family can escape or can buy their way out. Others will simply despair and argue that nothing can be achieved. The 'we're all doomed' scenario is corrosive but understandable.

Certainly, allowing the current trends to continue unabated will cause environmental disaster and massive social upheavals. No country or region will be immune. Although northwest Europe is projected to be spared the worst physical threats, our interconnected world makes it inconceivable that we and forthcoming generations will be left to enjoy a quite life unscathed by international social and economic turmoil. There is every probability that, despite the time lags, we are fast approaching a 'limits' crisis – certainly within the lifetimes of our grandchildren, if not earlier.

That being so, the Earth will impose its own logic on humans and human welfare. Although masters of self-deception, we cannot 'cheat' or deceive the bio and geo-chemical cycles that control life on our planet. No doubt many will emerge chastened to reignite human culture. In my judgment the overriding issue is not 'limits' *per se,* but whether we can navigate a transition to a sustainable global future in a humane and relatively orderly way. The alternative is that the transition will occur brutally and chaotically, as suggested by James Lovelock in his *The Vanishing Face of Gaia: The Final Warning* (2009).

We will have to work very hard to achieve a softish landing. If we stopped to consider it, humankind has an overwhelming vested interest in reducing its environmental impacts. However, as so often with the 'common good', it is questionable whether we are capable of rising above our narrow personal, sectoral and national interests. Are there ways we can be motivated to do so? Quoting Gramsci – we must harness 'the optimism of the will' despite 'the pessimism of the intellect'. Human environmental impact can be represented ([30]) as:

Impact = Population x Affluence x Technology.[33]
Each factor is important and deserves serious analysis.

Population

First, consider population and the implications for energy and greenhouse gas emissions as a specific exemplar. It is obvious, but in my experience never admitted by politicians, that the larger the human population then the lower are the individual emissions compatible with limiting global climate change to about +2°C (or indeed any other target). It will also be harder to achieve such a target. The same applies, equally obviously, to demand for resources such as food and water (see also Figures. 2 and 4).

As noted earlier, total but rapidly rising emissions of about 45 Gt CO_2e in 2010 equate to about 6-7 tonnes per head. But by 2050 total global emissions need to be reduced to about 15 to 18 tonnes. If, as anticipated, the population has reached 9 to 10 billion, this equates to 1.5 to 2 tonnes each. Currently in Wales we are individually responsible for about 17 CO_2e tonnes, so we must look to about a 90 per cent cut (assuming a constant population of 3 million). From this perspective, everyone has a huge interest in curbing population growth, locally, nationally, and globally. If at all possible, the global population should not exceed 8 to 9 billion, despite the teachings of the Pope, some Islamic leaders and true believers in the miraculous powers of unfettered capitalism.

Experience suggests that population control is best achieved by female education, empowerment and enfranchisement. Indeed, that is a win-win situation for all but the most reactionary. However, a growing population creates additional demands for electricity, housing and other resources and facilities and can lubricate GDP growth. Populous cities are often generators of economic growth, while rural depopulation brings a raft of problems. But this thrust can only diminish our chances of meeting even the official target of an 80 per cent greenhouse gas cut by 2050. It is nevertheless quite remarkable that a developed country such as the UK seems incapable of producing its own skilled professionals and senior industrialists. Instead, it relies on importing medical and technical expertise, nurses and carers, plumbers and information technologists, and even bankers and traders from other often much poorer counties.

Ceasing population growth has profound economic and social implications for individual nations and also globally, especially when linked to a longer life expectancy. The proportion of those of working age will decline and of the aged increase. Those in their fifties and sixties will be expected to work for significantly longer, possible blocking employment and advancement for the young and middle aged. The problem of pensions will be acute, especially if stock market values stall, as is likely if the Galbraithian bargain unravels. But perhaps the bleakest prospects arise from the failure of healthy active old age to keep pace with absolute longevity. Thus the number and the time span of those requiring support and care will increase more rapidly than the actual number of over seventies. In the UK the historic answer has been to import labour, especially young labour, to build, nurse and care. This is not readily compatible with sustainable development as it denudes other societies of the skills they need. On the other hand, it is an area where much employment could be generated and human welfare enhanced at low environmental cost if ways can be found to pay those involved a reasonable wage.

Technology

To achieve a more sustainable and stable regime, technological enterprise and initiatives are essential. Some hanker after a 'simple' solution, a reversion to small self-sufficient communities. This may have worked (partially) for a global population of a less than a billion prior to 1800 but for one an order of magnitude greater in 2050, it would be highly problematic. Few would wish to return to a period in history when the great majority lived in squalor. At that time our ambition was to escape from the horror of a subsistence hand to mouth existence.

Such scenarios may arise from major catastrophes, but not from choice. It is not even clear that localism in food chains or, were it feasible, a dependence on entirely local production will ensure the lowest carbon footprint. Should Wales or regions therein, seek to achieve self-sufficiency in grain, for example, the consequential ploughing of grassland would increase CO_2e emissions significantly for decades. Nevertheless there is indeed room for more local exchanges, better food chains and a 'reinvention' of seasonal food. At all times the priority must be a low greenhouse gas footprint and a sustainable economy, not romance.

We must privilege technological changes, inventions and companies that reduce human environmental impacts. Paradoxically, although modern free market capitalism is at the core of both our transient affluence and our environmental problems, we must seek to harness its dynamism to generate the tools to extract humanity from the hole into which it has dug itself. Given our capacity for self-deception and cheating this will not be easy and rigorous criteria must be applied and audited.

However, an appreciation of the logic of exponential growth is of value. In becoming the workshop of the world the Chinese economy has been growing at just below 10 per cent a year – that is, doubling in eight to nine years. Even if significant resource decoupling is achieved, demand for hydrocarbon energy or copper or other raw material can be expected to double in 12 to 18 years. Hence, their policy of buying as many raw material sources in Africa and South America as possible. Clearly China does not take the view that resource constraints are unimportant or adopt a pure free market perspective that human technological invention will always find alternatives or substitutes.

Energy is central. Meadows and colleagues recognised the importance of harnessing technology to produce renewable low greenhouse gas emitting energy. They acknowledged the need to move towards a new hydrogen and electricity economy, to reduce energy use and increase energy use efficiency. Let us recall in this sector we are talking about reducing mean **per capita** global emissions from all sectors of energy generation and use, including aircraft, ships, vehicles, space heating, industrial processes such as cement manufacture, and the whole of the food chain, from between 6-7 tonnes in 2010, to 1.5 to 2 tonnes by 2050. A new global Marshall plan will be required. Is it too naive to point out that we invest each year globally some $1.5 trillion dollars in armaments and the military (the USA alone nearly $700 billion) and that foreign exchange transactions (largely speculation) amount to $955 trillion? The world would be much safer and fairer if a sizable fraction were invested in technologies to increase environment sustainability as defined by Daly. Might such a shift not create much more valuable and rewarding work than currently obtains? Can we create an enviro-industrial complex as powerful as the military-industrial complex?

The overriding question is simply this: do we take 'limits' seriously? These should include all the Earth's renewable and non-renewable resources and the strength and availability of sinks and dispersal times. If we were to do so, then I would readily join the optimists and would be confident that science, technology and human ingenuity would achieve an orderly transition. But to do so requires a new paradigm, a complete rethink and overhaul of our political and financial systems, and unprecedented global cooperation. The new system would need to put bio and geo-chemical 'limits' as its key criterion against which all activities are assessed, not as an incidental add-on. Therein lies the rub. Can such a change be contemplated?

Affluence

Gross Domestic Product (GPD) has little to commend it as a measure of human welfare or even personal economic success, a view that its originator recognized.[34] GDP is equal to the sum of private consumption, gross investment, government spending, and exports minus imports) within a particular area. Its inadequacies are apparent. After all, if one were to move 30 miles further from work to a large drafty house with a huge heating bill then one would be increasing one's contribution to GDP while diminishing one's own living standard. Going to prison might make a further contribution, especially if it meant either the state or private enterprise building a new one. If forest clearance in an upper watershed leads to downstream flooding, expensive insurance claims and an investment in major flood control schemes, they would all contribute commensurately to GDP.

Nevertheless, despite its lack of discrimination, there appears to be a broad correlation between GDP and the ability of an economy to create jobs, generate tax income and sustain the business of government. However, it is no guide to sustainability. Aside from its indifference to human welfare, the exclusion of environmental externalities – be that air quality in Liverpool, loss of biodiversity, flood risk or climate change – GDP will only change when the purchase price of an item is altered or an investment made. It is almost designed to encourage overshoot.

Various efforts have been made to modify GDP to better reflect the real world such as the Index of Sustainable Economic Welfare (ISEW), the Genuine Progress Indicator (GPI) and the UN System of Integrated Environmental and Economic Account.[35] For example, in ISEW, personal consumption (as measured in GDP above) is modified by the addition of public non-defensive expenditures but reduced by private defensive expenditure, for example a need for private policing or to fit CCTV or burglar alarms. Capital formation is added as are services for domestic labour (ignored in GDP), while estimates of environmental degradation and the depreciation of 'natural capital' are subtracted. The treatment of 'natural capital' such as oil reserves, soil or biodiversity as income is one of the main criticisms highlighted by Schumacher in his famous book, *Small is Beautiful*.[36]

It is apparent from these paragraphs that, despite its obvious deficiencies, any move away from GDP poses a major political and intellectual challenge. Politicians, especially in democracies where shelf lives are short, are as hooked on GDP as our society is on cheap energy. Many of the additional measurements and data required are difficult to acquire and are potentially contentious. Would not different factions value wind-turbines very differently even if there was an agreed price for their electricity and CO_2 emission reductions? Politically, it is easy to imagine rival parties parading a raft of often-spurious arguments while the public, with its ingrained scepticism, would wonder if this was not a massive con. Perhaps our obsession with placing a monetary value on everything is a symptom of our malaise? Is pandering to it part of the problem? Nevertheless it can be asserted with confidence that sustainable development cannot be made a 'central organizing principle' if GDP is retained as the measure of economic development and its associated growth remains a vital policy objective.

Fairness is the critical issue. Affluence has entailed seeking to emulate an élite – keeping up with the neighbours, acquiring better cars, and more and longer holidays. In this pursuit our society becomes more unequal. Income disparities and working hours for those in work have soared. We have acquiesced in the expectation that it is the price to be paid for maintaining or even increasing affluence, and turning Galbraith's treadmill. The 'wealth-maker' rewards could hypothetically be justified, even welcomed, if their efforts allowed the rest to enjoy a rising standard of living: a version of the much criticized theory of international development known as 'trickle-down' development. However, these pretensions have been dissipated and the balloons pricked.

Regrettably it appears that the less affluent will be paying a disproportionate amount of the bill to pay off debt and bail out the financial services sector. I am not aware of any senior executive in the financial services sector who has been required to face the legal consequences of their greed and deceit. In the mean time, we must also expect rising energy and food costs to further disadvantage the disadvantaged. Galbraith also suggests that a highly unequal society finds it hard to maintain the bargain and keep the consumer wheel turning. He observed that the middle and lower income groups could be relied upon to spend their income on goods and services and not to hoard unproductively, whereas the elite super rich are under no such pressure.[37] Perhaps the 'limits' are creeping up on us in unanticipated ways.

Sustainability and equity have to be close partners, with all this implies for policy in Wales, within the UK, within Europe and globally. For example, it requires that greenhouse gas emissions must be addressed per head. Should people or nation be inclined after such a point to increase their population, would it not be equitable to cap their total emissions? If an individual has compelling reasons for a larger share, s/he will have to buy it from the spare capacity of others. Similarly fair taxes would have to be paid by all and tax havens abolished. Currently it appears that the poor need to be galvanized to work hard by their lowly status and by freezing their wages, while the rich need tax breaks to encourage them to work harder.

7 | Humanity

Hope is not the conviction that something will turn up, but the certainty that something makes sense regardless of how it turns out.
Vaclav Havel

At the heart of the problem lie the relationships between our political-economic regime, our human natures, our moral codes and our ethical and religious value systems, and the natural world. In recent decades greed has been declared good. The super-rich, be they bankers, sport stars or celebrities are lauded and seen as role models. In parts of the USA even Jesus is deemed to make believers rich and successful. He appears to partake of the American dream, consort with American exceptionalism and to be a guarantor of security. More prosaically in the UK, WAGs (wives and girlfriends) are synonymous with conspicuous consumption, bankers, rock stars and footballers with excess. All but the bankers are regarded as desirable role models.

In following their leads we have plunged into debt but gained little satisfaction. How very different from the values of Welsh society a few decades ago and the values espoused by virtually all humanities, great thinkers and sages including of course Jesus. Setting aside Galbraith's insights and recommendations, our partly mythical affluence has been built on debt, both as individuals and as a country (now totaling nearly 500 per cent of UK GDP) and on exploiting natural capital. Individual welfare is being rapidly and painfully eroded. Despite its heavy environmental and social cost, our recent 'affluence' was partly a mirage. The 'emperor' had few clothes. The huge social cost and the betrayal of post war aspirations by casino capitalism has inspired Stephane Hessel to 'Outrage' and the 'occupy Wall Street' movement.[38]

Much has been written about the 'tragedy of the commons', that is the relative ease with which common group assets are mismanaged and denuded.[39] This appears a persuasive analogy for the problem of atmospheric greenhouse gas pollution and consequential climate change, or maybe even the stripping out of global capital assets for the benefit of a few. But there are differences:

- It is not a commonly owed resource but an un-owned resource upon which, nevertheless, we all depend.
- The effects of the pollution will not be consistent throughout the Earth nor experienced equally by people.
- The atmosphere and climate, both local and global, as un-owned resources, encourage free loaders and individual countries and sectors to plead their special cases.

Worse we face intergenerational inequity with our generation of the 'affluent' acting as free loaders at the expense of their children. In the cases of common land and un-owned global resources, the danger of others benefiting whilst we 'do the right thing' is a real dilemma – potentially, making a mockery of any 'sacrifice'.

In moving forward we need to honestly inform citizens about our human predicament, as this is perhaps the only way to ensue a measure of compliance. It will not be easy. We have singularly failed to hold the city financiers and traders to account. Feted by Labour, new and old, Tories of all complexions and Lib Dems alike, the City élite has every reason to feel smug and untouchable. Consequently people are deeply sceptical and cynical. Although the old order must be changed, it is difficult to conceive of a scenario where a London government will act decisively. Paradoxically, our economic malaise gives bankers, financiers and industrialists more power. Individually and collectively we are desperate for our town or region or children to benefit from some windfall investment or new jobs. Similarly government is desperate to avoid major companies relocating and, at the very least, to maintain its tax take.

Short of a totalitarian response, perhaps our only hope lies in the democracy and openness of the web. The Animal Farm model of governance tempts all-powerful leaders, whether they start as communist, fascist, Welsh nationalist, Valley's Labour or old Etonian. Power not only corrupts but blinds. The crisis of 'limits' can only increase the chances of an authoritarian leadership. What price democracy in Greece or Italy when staring at the abyss of economic collapse? What price the rights of workers if the Government response to the crisis is to make it easier for companies to sack their staff?

Faced with these choices, our three Westminster parties and US Republicans and Democrats have all chosen to save the banks and bankers, fearing even worse disaster. Only 'people power' can hold government, power brokers and polluters to account. Only the people can achieve the paradigm shift required to attain even a measure of sustainability. For this they must be convinced that it is in their and their families' best interests and that comparative justice will be done. Can we avoid lurching into a neo-fascist state? Can we achieve an orderly transition or will it be disorderly and brutal, as predicted by James Lovelock amongst others? Can we harness human energy and creativity, so apparent in self-interested capitalism, not to exploit the planet's resources but to maximise long-term human welfare and satisfaction? In asking these questions we must recognize an acute dilemma, that the currently rich, powerful and influential will be the biggest losers in the short run.

8 For Wales see Cymru

"… to be truly radical is to make hope possible rather than despair convincing."
Raymond Williams

The 1998 and 2006 Government of Wales Acts stipulated the pursuit or promotion of 'sustainable development'. Elected in May 2011 the current Welsh Government not only intends to make sustainable development 'its central organizing principle' but to embed it in primary legislation. This is therefore a crucial opportunity to establish effective principles to ensue long-term progress. In addition to providing intergenerational equity, sustainable development is often interpreted as a triple bottom line of a 'viable successful economy' supporting 'vibrant communities' in ways, which are 'environmentally sustainable'. The objective is to create a culturally and economically successful country to be inherited by our children and children's children. This is, of course, greatly to be welcomed but the implications have not yet been explored nor, I would suggest, widely understood. This essay is intended as a contribution to this dialogue but we are many years from an economy which is truly sustainable in economic, social or environment terms.

Even without addressing many vital community and cultural issues, the discussion to date has demonstrated the size and complexity of the challenge. A serious attempt to grapple with the issues is found in Tim Jackson's book *Prosperity without Growth* which is based on the conclusions of the UK Sustainable Development Commission disbanded in 2011.[40] However, if we are approaching – and possibly have reached – global 'limits' internationally, these will tend to define spending priorities. Inevitably such a crisis will re-orientate activity towards mitigating the social and economic damage inflicted by the breached 'limits', and towards securing adequate work, energy, food and related goods and services and social support. It appears that the international debt crisis is already carrying us down such path. However, regardless of current preoccupations, we have only a few decades to find sustainable paths and solve the paramount problem of reducing our greenhouse gas emissions drastically. How, therefore, can a country relatively low in the affluence league in European terms, with limited fiscal and policy levers, move forward? Are there practical ways of approaching a sustainable form of development, and achieving a realistically based prosperity for Wales? Can we take a lead, much as we did during the industrial revolution of the 19th Century?

I suggest we need to distinguish short-term, urgent actions that can be pursued locally and immediately from a longer-term comprehensive strategy. Also, although some initiatives can be pursued within the National Assembly's current powers, others will depend on acquiring additional powers to allow Wales to increase its sustainability. Critically, and despite people's distrust of authority and jaundiced view of politicians, their trust must be won. In many contexts the 'expert' does not always know best; certainly this is the common feeling. Initiatives should be informed by but not led by so-called 'experts' or 'consultants'. In Wales, where there isn't a strong tradition of 'institutions', we need to find ways of motivating citizens to take action for themselves, trusting to their common sense, not nannying.

Wales has a few modest comparative advantages. Fortunately we are not in hock to the financial services sector and, hopefully, are less anti-European than our English neighbours. Hopefully, we can recognise that many facets of sustainable development can best be achieved within an active cooperative European framework and that we have much to learn from initiatives outside the UK. Only on such a broad canvas may we be able to create a sufficiently powerful economic zone of common laws, regulations and ambitions to promote fairness, protect individual rights and the environment and provide a mechanism to counterbalance the destructive powers of big business and high finance.

Certainly, the successful pursuit of enhanced sustainability demands action at all levels – international, EU, UK, Welsh and local. Some, such as regulation of emissions from air and sea travel, carbon trading and potentially carbon taxes can only be contemplated meaningfully at the EU level. Other issues around the management and reliability of electrical supply through the National Grid have a vital UK and cross channel component. Similarly, action at an UK or British and Irish level is required in matters such as inter-conurbation travel and regulating the City. Although from a Welsh perspective we have little direct influence, we must ensure that our voice is heard.

Fortunately, Wales still retains a sense of communal cohesion and is sufficiently small to sustain an internal dialogue and debate about these issues and achieve a measured response, which could be even more problematical in large heterogeneous country. In his *Small is Beautiful* Schumacher did not argue exclusively for 'smallness'. Rather, he observed that our allegiance to the gargantuan can be misplaced. He argued that the human scale was the most appropriate. Interestingly, Cymry means 'fellow members of a group in which negotiated outcomes are possible'. Welsh on the other hand is the English word for 'others'. We need to see ourselves as 'cymrodyr', 'y Cymry', and not defined by otherness. We have potentially a highly educated and motivated population with deep cultural and historic roots. An unappreciated facet of Welsh life is our closeness to the rural world. Even in Cardiff people are within easy reach of the countryside. Nowhere is but a few miles from a rural or coastal area or a National Park. If, as I would conclude, the emerging sustainable economy will seek inspiration and individuals find solace and enjoyment from clear ties with the natural world, then we in Wales are extremely fortunate.

We have good natural resources in terms of renewable energy, water, timber and food, especially pasture-derived animal products, as well as a long history of technological innovation and scientific research. However, to date moves to utilize these assets effectively have been uncertain and hampered by the division of responsibility between the

Welsh Government and Westminster.

Historically, Wales has lacked many essential economic and political levers, although our comparative impotence makes us representative of many other nations and regions. This was partially rectified by the successful referendum on legislative powers for the Assembly in March 2011. Additional powers will required in relation to our natural resources. At the same time, the new low impact technologies that will emerge cannot be contained in a region or single nation. Research findings are by their nature international but if well focused can provide local or national advantage. In the remainder of this essay I will concentrate on national and local actions, while recognising that supra-national action may lag, as exemplified both by the comparative failure of climate change negotiations at Copenhagen and Durban.

Given the depth of the global challenge and our small geographic, economic and political space, Wales needs to address not only **mitigation** – that is, how to play our part in limiting the emission of pollutants and the irresponsible use of resources – but also **adaptation** and **resilience**. The former addresses ways to lighten the impact of impending crises such as peak oil on our society, and the latter ways we can increase our national capacity to withstand sudden, unexpected storms.

Daly defines an environmentally sustainable system as one with a 'steady state economy'. This is one which retains a constant stock of physical capital, which does not erode its own natural resources or those of elsewhere by its imports, in a world that enjoys a rate of material through-put within its regenerative capacities and adsorptive capabilities. While one cannot fault this logic, we are light years from achieving it and to move precipitously to this condition would be invite danger and could be counter-productive. The Daly criteria have the advantage of being quantifiable but do not extend readily to economic, social and cultural issues.

As a working definition in a Welsh context, sustainable development can be considered a political-economic system which promotes human prosperity through a range of work opportunities and the provision of quality services in and through the private and public sectors, while:

- Progressively adhering to the Daly principles of environmental sustainability.
- Enhancing cultural and social activities, including specifically through the medium of the Welsh language.
- Conserving biological diversity and natural beauty.
- Stimulating vigorous communities of self-confident compassionate individuals.

If this can be accepted, then we need a broadly agreed strategy to progress to such sustainability over 20 years. This, I suggest, implies a sequence away from Anglo-American casino-capitalism, first to a Nordic-style social democratic, mixed economy. This should entail an enhanced and continuing decoupling of a redefined growth model away from energy use and production, from the emission of greenhouse gases and from the production of wasteful rubbish and other pollutants. Over several decades this would lead to a fully decoupled 'steady state' economy by 2030 to 2040. It must be emphasized that decoupling *per se* is not straightforward and can be dangerously misconstrued.

For example, technological innovations can increase output at lower financial and environmental cost. But any gains could be overwhelmed if demand consequently increases substantially. On a personal level, we might invest to improve the energy efficiency of our homes, so saving on heating costs. However, if these savings were then spent on extra flights to Spain or longer car journeys, all the good work would be undone. Consequently public understanding and consent, although currently largely absent, is essential.

There can be little disagreement that energy supply and use and the attendant mitigation of greenhouse gas emissions lie at the heart of the sustainability issue. Energy costs are embedded throughout all aspects of the economy. Virtually everything revolves around energy: from housing to transport to industrial production and the food chain. As early as the 1960s intensive agriculture was recognised to be essentially a method of converting oil into food.[41]

The public may be more open to arguments about **adapting** to the increasing cost and likely insecurity of electricity and fuel supply than the more esoteric prospect of **mitigating** our overshooting the capacity of the atmospheric carbon sink. Few will volunteer to do away with the 'convenience' of their own freewill, so interventions will need to be tailored to this fact. Self-interest can be aligned with 'sustainability' and government must apply a judicious use of both sticks and carrots to achieve this.

A further note of caution is needed. The scenarios discussed earlier seem to imply that the service economy should offer low greenhouse gas economic returns but even this can be misleading. Carbon foot-printing of Gwynedd found that, given its reliance on motor vehicles tourism made a very large contribution.[42] Sadly the Snowdonia Green Key project, which might have provided part of the solution, has floundered due to lack of imagination, communication skills and political will. The lesson is not that tourism should be discouraged but that every effort should be made in all sectors to create and promote attractive and credible options with low greenhouse gas footprints.

My objective is to broaden a practical and evidence-based debate that leads to setting some strategic initiatives for Wales, given our serious economic, social and environmental deficiencies. The concepts below are predicated on expanding the local and regional components of the economy, increasing the social and cooperative elements to provide greater resilience, and enhancing personal environmental responsibility. In pursing the objective of 'sustainable development' as our 'central organising principle' we should acknowledge that the private, public and social enterprise spheres will each have unique and complementary roles. Each will have to contribute to job creation but it is highly unlikely that we can deliver a pure, free enterprise, low tax, and low spend system. Nevertheless all must meet criteria for sound economic management. Isolationism is not an option. Continued trade will be vital. Enterprise must be encouraged but with a re-balancing to favour local multipliers. In the medium term the Welsh Government must:

1. Commit the country and government to adopt the principles of 'one planet' living which would embed the following ten features:
   - Zero Carbon
   - Zero Waste
   - Sustainable transport
   - Sustainable materials
   - Local and sustainable food
   - Sustainable water
   - Land use, wildlife and ecosystems
   - Language, culture and heritage
   - Equity and inclusion
   - Health and happiness

2. Commit to using a version of the Index of Sustainable Welfare, which measures sustainable growth and properly 'subtracts' negative actions that destroy natural, social and cultural capital. The move from casino capitalism to true sustainability cannot be navigated without better analytical tools, both to develop new innovative policies and technologies, and to allocate scarce resources. First and foremost, this requires the adoption of more discriminatory economic tools and a better 'real growth' measurement than GDP to guide investment and policy. Valuable work has been already been carried out on a Sustainable Economic Welfare index for Wales and measuring 'progress' at an European level.[43] These must be built on.

3. Acquire real power over energy, planning and transport. This would entail a statutory power over Ofgen in Wales and the regulation of energy distribution to facilitate public involvement in dispersed renewable energy production and energy saving. We will also need to develop planning tools to ensure better energy conservation, and

provide much firmer controls of car-based, out-of-town developments. We should give serious consideration to a 60mph speed limit on dual carriageways. In the medium term we should establish a network of electrical and later hydrogen charging stations for vehicles. To make good use of the Welsh Government's enhanced powers a larger elected membership is essential for rigorous policy analysis and scrutiny. In addition high quality, dedicated public servants are necessary. The Government should encourage civil servants to be more proactive and innovative. The current divided-loyalty between Cardiff and Whitehall must be resolved. Finally the Welsh Government must accept the challenge of responsibility for its own research budget.

4. Establish a 'regional' Bank on the lines of those in North Dakota or the German Lander to retain local savings and support sustainable local business, industry and commerce. This would recognise the twin issues of improving the local economy in the face of the centripetal and income-lowering forces of globalisation and mobile City investment, and of stimulating low environmental impact growth and employment. We should provide legal and commercial advice to companies on structures that could protect them from unwelcome external takeovers. The Glas Cymru mutual model that operates Welsh Water in one example. It is interesting, both the Bank of England and the World Bank have recently recognised that the impacts of the free flow of capital, unfettered company relocation and aggressive take-overs are not necessarily positive.

5. Reinvigorate the Welsh local economy by using the purchasing power of the Welsh Government, local authorities, and related bodies such universities, schools, colleges and hospitals to support local business and increase local multipliers. We should resolve once and for all the conflict between benefiting the local economy and a slavishly over-stringent interpretation of European procurement law.[44]

6. Promote public engagement, fairness, language, culture and justice. In general and in its current form, sustainable development has a bad

press. Many, perhaps the great majority, regard it as a threat to their lives and aspirations, and as an excuse for more interference by government and highhanded individuals. While the current crisis in 'conventional capitalism' may have soured their view of that system, few can conceive of any other. Hence the desperate demands for GDP growth. In current 'conventional thinking' it seems inconceivable that an environmentally responsible, truly sustainable economy could be fulfilling and offer exciting prospects to the ambitious and hard working. Nevertheless, the exploitative unjust core of the current unsustainable system is so obvious that we have David Cameron and Ed Miliband competing on how to curb its excesses.

Of course, social justice is not a uniquely Welsh concept. However, we have a proud legacy in the ideas of Robert Owen, Lloyd George and Aneurin Bevan as well as countless trade union and religious leaders, including St David's reputed commitment to doing small communal actions 'gwnewch y pethau bychain'. As Michael Sandel argues, justice does not eliminate value judgments but requires them.[45] Fairness and social justice between generations, nations and people, with a strong emphasis on social and individual responsibility, lie at the heart of sustainable development.

In some ways it is all very Welsh! Instead of competing for new toys and gadgets, we should express our human competitiveness through sports, culture, the arts and sciences. This is entirely compatible with the Welsh eisteddfod tradition. As well as being an important component of our heritage, the Welsh language should be a stimulus to individual and social enterprises and local entrepreneurs. In many ways, its future is a measure of our sustainability. The negative forces undermining the language are frequently those that reduce the sustainability of development. Only with vibrant, self confident communities can a minority language itself flourish. We should embrace bilingualism, not just for those who currently speak Welsh, but as part of all our heritages.

9 Epilogue

...there is wealth only in people and in their land and seas. Uses of wealth which abandon people are so profoundly contradictory that they become a social disaster, on a par with the physical disasters, which follow from reckless exploitation of land and seas.
Raymond Williams

This essay argues that our core problem arises from the highly successful, selectively trans-formative, economic system that has evolved initially in the West over the last few hundred years. It has created unprecedented affluence for a proportion of humanity, while centralising power and money in élites, encouraging irrational exuberance, and ensuring overshoot. The system commits society, both globally and regionally, to infinite exponential growth. Measured by GPD, this growth is insensitive to the erosion of natural, social and cultural capita and indifferent to the source of capital, be it debt finance, money printed by government or financial perfidy.

The system contains the seeds of its own destruction. It is seductive but unsustainable. Nowhere is this more apparent than in the energy-climate change crisis. The forces sucking the life out of our mining and quarrying communities, eroding the Welsh language and hurtling us globally toward damaging climate change are inextricably intertwined. In time the planetary bio-geo-physical realities will reassert themselves, undoubtedly with unanticipated consequences. Major technological feats, such a new low carbon energy sources may be developed but can only buy time for a more profound reappraisal and allow space for adjustment. Just as the robust certainties of Newtonian physics have been overtaken by the subtle strangeness of Einsteinian quantum physics, and pre-Darwin-Wallace-Mendel natural history transformed into modern evolutionary and molecular biology, a new social-economic paradigm must be sought.

Given the magnitude of the challenge and pressure of current economic and social problems, even in this country, still more in the developing world, the understandable and inevitable reaction of politicians is to plump for conventional growth – to seek economic space for readjustment. Sustainability is seen as a distant desirable objective – 'Oh God make me sustainable but not just yet'! This is a sure receipt for overshoot. Much more work is required to map out a 20-year pathway to a new paradigm. First, however, we must face the issues honestly. This is a process that cannot start too soon.

The new paradigm will not be anti-business or anti-science. Innumerable opportunities will occur for enterprise and new businesses, which lower human environmental impact and increase real individual prosperity and facilitate social interactions. Dispersed renewable energy generation could be the key as it would provide local additional income sources and personalise the attainment of sustainability, including energy efficiency.

At the same time it will underscore the scale of the problem. We will be very hard pressed to be energy self-sufficient and, will be unlikely to solve the issue of energy storage. We need action at various levels, both technically and politically. A sustainable future is not anti-technology, nor low on ingenuity. Indeed it is quite the opposite. But the economic incentives and disincentives must be consistent with lessening our global and local impacts. We need to work with the grain of bio-geo-physical constraints and not in defiance of them. Some of the essential levers to effect such a change will lie in Wales, others do not and will not.

Many brilliant minds are engaged in maintaining the Galbraithian bargain and in advertising and promoting our 'wants'. It is not inconceivable that these talents and our growing understanding of human psychology can be used to encourage a far less destructive system. It is also entirely probable that such systems will create a greater degree of real well-being and contentment. This essay started by contrasting the apparent optimism of the advocates of unfettered 'growth' with the pessimism of the doom-mongers. I suggest this is simplistic and misleading. Rather we should be dismayed by attempts to prop up a destructive system. Despite the huge challenges we should enthuse about the prospects for real human prosperity and fulfillment from making the change to the new paradigm of sustainable development.

Footnotes

1. The current Welsh administration is committed to legislating to make 'sustainable development' the 'central organising principle' of our National Assembly. I hope this analysis will help inform that debate and the issue internationally. I am grateful to Dr Havard Prosser, Professor Ross Mackay, Dr James Intriligator and Dr Dafydd Trystan, and especially Dr Einir Young for their comments and suggestions for this essay. The faults are, of course, mine alone.
2. See Rees, Martin, *Our Final Century: Will the human race survive the 21st century?* Heinemann, 2003; Lovelock, James, *The Vanishing Face of Gaia: The Final Warning*, Penguin, 2009; and Flannery, Tim, Now or Never: Why we need to act now for a sustainable future, Harper, 2009.
3. Galbraith, John Kenneth, *The Affluent Society*, Pelican, 1958.
4. Meadows, Donella. H. et al., *The Limits to Growth*. Earth Island, 1972.
5. Meadows, D.H. et al., *Beyond the Limits*, Chelsea Green, 1992; and Meadows, D.H. et al., *Limits to Growth: 30 year update*, Earthscan, 2005.
6. http://www.storyofstuff.org/movies-all/story-of-stuff/
7. Hawranek, Dieter et al., *Out of Control: The destructive power of financial markets*, Der Spiegel-on-line, 22 August 2011.
8. Chakraborthy, A., writing in *The Guardian*, quoting the Centre for Research on Socio-Cultural Change, University of Manchester, 13 December 2011.
9. Meadows, Donella. H. et al., *The Limits to Growth*. Earth Island, 1972
10. Montreal Protocol on Substances that Deplete the Ozone Layer, UNEP, Kenya, 2000.
11. Hogan, C.M., *Overfishing*, Encyclopedia of Earth. NCSE, Washington D.C., 2010.
12. British Geological Survey, *Risk List 2011*.
13. See papers on the International Food Policy Institute web site, especially those of Mark Rosegrant, 2010-11. See also the websites of the Climate, agriculture and food security organisation and the Consultative Group on International Agricultural Research, for instance. http://www.cgiar.org/pdf/CCAFS_Strategy_december2009.pdf
14. See the United Nations' Food and Agriculture Organisation's online datasets, available at http://faostat.fao.org/site/380/default.aspx 2010.
15. http://en.wikipedia.org/wiki/File:Copper_-_world_production_trend.svg
16. Meadows, D.H. et al., *Limits to Growth: 30 year update*, Earthscan, 2005.
17. R.K. Pachauri, A. Reisinger, (Eds.), Contribution of Working Groups I, II and III to the Fourth Assessment Report of the Intergovernmental Panel on Climate Change, 2007. Also IPCC (1996) Revised 1996 IPCC Guidelines for National Greenhouse Gas Inventories, Intergovernmental Panel on Climate Change, www.ipcc.ch, 1996; and Intergovernmental Panel on Climate Change Emissions Databases, 2010: available at http://www.ipcc-nggip.iges.or.jp/EFDB/main.php
18. *Ibid.*
19. Wyn Jones, R. Gareth et al. (2012) *Climatic mitigation, adaptation and dryland food production*, Proceedings of the International Dryland Development Commission 2010 Conference, Cairo, 2012; see also Taylor, Rachel. C. et al., *Before 2050 human food chain will drive climate change*. In preparation.
20. Organisation for Economic Co-operation and Development, T*owards Green Growth: monitoring progress*, Report 2011.
21. *Per capita* UK energy use is growing at about 1 per cent a year. Given the population continues to rise, total use has grown faster – despite outsourcing emissions to China

and elsewhere by our importing of manufactures and food containing 'embedded' greenhouse gases.
22. Daly, H., 'Institutions for a Steady State Economy', in *Steady State Economics*. Island Press, Washington D.C., 1991.
23. The Dutch have been more successful with their short-term (100-year) repository: a centrally stored aboveground facility of the Central Organization for Radioactive Waste. This is largely due to transparency. The costs associated with the disposal of nuclear waste are accounted for in today's Dutch consumer electricity prices.
24. Baer, Paul. (2008) *Exploring the 2020 global emissions mitigation gap* (Analysis for the Global Climate Network). Woods Institute for the Environment, Stanford University, Palo Alto, California, 2008; Meinshausen, M. et al., *Greenhouse-gas emission targets for limiting global warming to 2^0C*, Nature 458, 1158-1162, 2009.
Ranger, N., A. Bowen, J. Lowe, L. Gohar, *Mitigating climate change through reductions in greenhouse gas emissions: the science and economics of future paths for global annual emissions*, December 2009 available at http://www.cccep.ac.uk/Publications/Policy/Policy-docs/bowen-Ranger_MitigatingClimateChange_Dec09.pdf
25. Carbon Dioxide Information Analysis Center, Oak Ridge Labs US, October 2011.
26. Strahan, David, 'The Oil Maze', *New Scientist*, 3 December 2011.
27. *Ibid*.
28. Spinney, Laura, 'The Underhand Ape', *New Scientist*, p. 43, 5 November 2011.
29. *Ibid*.
30. Dawkins, E. et al., Wales' *Ecological Footprint – Scenarios to 2020*, Stockholm Environment Institute for the Welsh Government, 2008.
31. James, Oliver, *Affluenza: How to be Successful and Stay Sane*, Vermilion, 2007.
32. Diamond, Jared, *Collapse: How Societies choose to Fail or Survive*, Penguin, 2006.
33. See Ehrlich, Paul, *The Population Bomb*. Buccaneer, 1968; and Ehrlich. P. and Ehrlich A., *Population Resources and the Environment*, 1970.
34. Kuznets, Simon, *National Income 1929-1932*, 73rd US Congress 2nd Session, 1934. http://library.bea.gov/us
35. See Daly, H. and Cobb. J., *For the Common Good*. Beacon Boston, 1989: and Nordhaus, W. and Tobin, J. *Is Growth Obsolete*, Columbia University Press, 1972. Professor Peter Midmore, of Aberystwyth University has constructed an ISEW for Wales.
36. Schumacher, E.F., *Small is Beautiful: A Study of Economics as if People really Mattered*, Vintage, 2011 (original edition, 1973).
37. Galbraith, J.K., *The World Economy Since the Wars*, Sinclair-Stevenson, 1994.
38. Hessel, Stephane, *Time for Outrage/Indignez-vous*, Quartet Books, 2011.
39. Hardin, Garrett, 'The Tragedy of the Commons', *Science*, 162, 1243-1249, 1968.
40. Jackson, Tim, *Prosperity without Growth*, Earthscan, 2011.
41. MacKay, David J.C., *Sustainable Energy – without the hot air*, UIT Cambridge, 2008; available online at www.withouthotair.com
42. Farrar, John. F. et al., *Report on Carbon Footprint to Gwynedd County Council*, 2004.
43. Munday, M. et al. 'An Index of Sustainable Economic Welfare (ISEW) for Wales (1990-2005), BRASS, Cardiff University Jan. 2006: Jackson, T. and McBride, N., Measuring Progress? Report to European Environment Agency, 2005 ; http://stiglitz-sen-fitoussi.fr/en/index.htm
44. see Kevin Morgan, 'Values for money' in *the welsh agenda*, IWA, Spring 2012.
45. Sandel, Michael. J., *Justice: What's the right thing to do*, Farrar Strauss and Giroux, 2009.

7
Croesi Ffiniau

Cyhoeddwyd yr erthygl hon yn wreiddiol yn Y Traethodydd yn Ionawr ac Ebrill 2013. Er nad yw yn ymdrin yn uniongyrchol a fy namcaniaethau ynglŷn phwysigrwydd creiddiol ynni yn ein bywydau ac yn yriant yr economi, mae'n trafod y cysyniad o gynaliadwyedd yn fwy cyffredinol. Credaf felly ei fod yn berthnasol i, ac yn ychwanegiad at, swmp y gyfrol hon.

Croesi Ffiniau
'Neu Geisio Paradeim newydd'[1]

Cyflwynir yr ysgrif hon i'r pedwar ŵyr, Euros, Aled, Eirig ac Owain gan obeithio ymhen 50 mlynedd, y byddant yn maddau ffolinebau cenedlaethau hŷn ac wedi byw trwy newid, ond nid trychineb. Ymddangosoedd fersiwn Saesneg o'r ysgrif yn 'Wales' Central Organising Principle'. gol. A. Nicholl a John Osmond. IWA. Caerdydd. Mai 2012.

1 / Paradeimau Gwrthgyferbyniol

Mae'n nodweddiadol o'r ddynoliaeth ei bod yn cofleidio amrywiaeth mawr o ffyrdd o fyw, systemau gwleidyddol ac economaidd, crefyddau ac athroniaethau. Ond hwyrach fod yr hollt fwyaf pellgyrhaeddol i'w gweld rhwng y rhai hynny sydd yn argyhoeddedig bod cyfyngiadau i allu'r ddynolryw i fyny nwyddau a gwasanaethau o'r blaned hon h.y. bod terfyn pendant, er un anodd ei ddiffinio, ar allu ein planed i gynnal bywyd, a'r rhai sy'n credu'r gwrthwyneb. Mae'r garfan olaf yn credu, yn fras, bod pwerau creadigol a dinistriol cyfalafiaeth a grymoedd y farchnad, ar y cyd â chynnydd gwyddonol, dyfeisgarwch dynol ac arloesi technegol, â'r gallu i oresgyn pob terfyn ymddangosiadol. Mae ambell un yn apelio hefyd at ymyrraeth ddwyfol. Yn union fel y gwnaeth dyfeisio'r injan tanio fewnol yn y cyfnod Edwardaidd leddfu pryderon y byddai Llundain yn diflannu o dan dwmpathau o dail ceffyl, ac fel y chwalwyd mwrllwch trwchus Lerpwl fy ieuenctid gan y Deddfau Awyr Glân, maent yn dadlau y gall y ddynoliaeth harneisio technolegau eraill amgenach sydd â phosibiliadau dihysbydd ac y gwnaiff hynny. Mae nerth bôn braich, nerth ceffylau ac ynni dŵr wedi eu disodli gan yr injan lo, ac yna gan beiriannau disel, nwy a thrydan. Bellach mae gennym electroneg fechan a nanodechnoleg. Mae ein galluoedd technegol yn anhygoel. O ganlyniad i ffonau symudol, hofrenyddion ac awyrennau, trawsnewidiwyd cymunedau hyd yn oed yn y parthau mwyaf anghysbell o'r ddaear. Osgowyd y gyflafan niwclear oedd yn peri cymaint o ddychryn yn y 50au a'r 60au. Ymddengys yn wir fod comiwnyddiaeth wedi ei threchu. Bellach mae llai o fabanod yn marw, ond nid yw proffwydoliaethau tywyll Malthus wedi cael eu gwireddu. Mae cyfran sylweddol a chynyddol o'r 7 biliwn o bobl sy'n byw ar y blaned hon yn mwynhau hawddfyd na welwyd ei debyg mewn hanes. Rydym yn byw'n hirach er gwaethaf HIV/Aids. Yn y rhan fwyaf o ardaloedd mae'r cyflenwad bwyd wedi parhau i ddiwallu'r boblogaeth gynyddol yn lled ddiymdrech.

Wrth reswm, nid yw'n fel i gyd ar draws y byd ond mae'r ffyniant yn drawiadol serch hynny. Honnir y gallwn ac y byddwn yn ychwanegu at ein cyfoeth, a fesurir gan y Cynnyrch Mewnwladol Crynswth (CMC) byd-eang, yn ddi-ben-draw (sef GDP ar lafar gwlad). Efallai'n wir y byddwn yn ymgartrefu ar blanedau eraill. Yn ôl y gred hon mae twf economaidd, fel y deellir y term heddiw, nid yn unig yn brawf ar lwyddiant ond yn amcan polisi hanfodol bwysig. Credir bod y rhai sy'n poeni am 'derfynau' neu ffiniau, yn ddiddeall ac yn mynnu tanseilio gobeithion a dyheadau pawb, yn enwedig tlodion y blaned hon.

O ystyried y sefyllfa uchod mae'n syndod faint o bobl, gan gynnwys gwyddonwyr syber ac enwog, sy'n dal i fod yn eithriadol o besimistaidd ynghylch dyfodol y ddynoliaeth ac yn bryderus ynghylch ei heffaith aruthrol ar y ddaear[1 2 3]. Ceir lliaws o broffwydi gwae gwybodus. Ai chwerwder personol neu bropaganda gwrth-gyfalafol neu ysbryd gwrthnysig ac anwybodus sydd ar waith yma? Gan fod pesimistiaeth ynghylch dyfodol y ddynoliaeth wedi bod yn beth cyffredin ers milenia, rhaid i ni ofyn, a oes unrhyw wahaniaeth go iawn y tro hwn? Tybed ydy'r ddynolryw, fel mae rhai'n honni, yn sefyll ar groesffordd allweddol ac yn gorfod gwneud, neu efallai'n gwrthod gwneud, penderfyniadau o bwys aruthrol? Os deuwn i'r casgliad bod gwerthoedd a dadansoddiadau y farchnad rydd yn gymysgedd peryglus o hunan-dwyll a hunan-ymchwydd, sydd yn aml yn cael ei hyrwyddo gan y rhai hynny sydd wedi cipio iddynt eu hunain gyfran anghymesur o fawr o adnoddau'r blaned hon, yna rydym yn wynebu cyfnod tywyll ac anodd iawn yn wir. Heb ddadansoddi'r mater sylfaenol hwn, mae'r holl sôn am gymhwyso'r cysyniad dadleuol 'datblygiad cynaliadwy' i'r cyd-destun Cymreig neu ryngwladol yn hollol ddiystyr.

Dylid cofio nad yw'r mwyafrif o'r 7 biliwn o drigolion dynol y ddaear hon yn malio dim am y materion hyn. Os ydynt yn dlawd (efallai fod 80% o drigolion y blaned yn y categori hwnnw o hyd) maent yn dal i fyw dan orthrwm problemau oesol. Os ydynt yn gymharol gyfoethog, maent yn ffromi at unrhyw awgrym fydd yn cyfyngu ar eu gallu i wireddu eu dyheadau a'u llwyddiannau, yn enwedig swyn y freuddwyd Americanaidd. Cadw eu swyddi neu dalu eu dyledion yw'r unig bethau sy'n poeni eraill. Mae rhai'n rhoi eu ffydd mewn Duw/Allah hollalluog, allanol a fydd trwy ei ragluniaeth yn cynnig yr holl atebion ac adnoddau i'w ddilynwyr selog. Mae cydnabod bod terfynau i gael, ac yn bwysicach fyth fod y ddynolryw yn prysuro tuag atynt, yn esgor ar oblygiadau pellgyrhaeddol, ac mae'n rhwym o gael effaith ddwys ar ein syniadaeth foesegol, wleidyddol, grefyddol, economaidd a chymdeithasol.

Ers canrifoedd mae pob cenhedlaeth hŷn wedi synied am y genhedlaeth iau fel rhai diamcan, anghyfrifol, amharod i ysgwyddo eu cyfrifoldebau. Ond mae'n bosib bod y gwrthwyneb yn wir, sef, bod agwedd ddifeddwl, anghyfrifol sy'n gwrthod ystyried y tymor hir yn nodweddion sy'n diffinio'r genhedlaeth hŷn bresennol, fy nghenhedlaeth i fy hun? A ydym yn cyflwyno Eden i'n plant ynteu afal gwenwynig? Mae sawl llais wedi codi i'n rhybuddio, eraill i'n cysuro. Efallai ein bod yn reddfol ymwybodol o'r peryglon ond yn gwrthod eu cydnabod: yn rhy brysur yn pluo ein nythod ein hunain i ymateb, yn rhy gysurus i weithredu'n benderfynol ac yn ddeallus, neu yn rhy hunanol i gymryd y camau cyntaf?

2 | Amwysedd cyfoeth

Mae'r posiblrwydd o drosglwyddo etifeddiaeth bwdr yn cael ei ddarlunio'n rymus mewn dau lyfr, y ddau bellach dros 40 oed. Ymddengys eu bod yn trafod pynciau gwahanol iawn, ac yn sicr maent wedi eu llunio gan awduron o draddodiadau gwahanol sydd â'u gwreiddiau mewn disgyblaethau cyferbyniol, er bod y ddau lyfr, fel mae'n digwydd, yn gysylltiedig â Boston, Massachusetts. Serch hynny, mae negeseuon canolog y ddau lyfr yn rhyfeddol o debyg.

T*he Affluent Society*[4] yw'r gyfrol gyntaf, gan John Kenneth Galbraith, economegydd o dras Canadaidd-Albanaidd: bonheddwr ac ysgolhaig ym Mhrifysgol Harvard oedd yn chwarae rhan flaenllaw ym mhlaid Ddemocrataidd yr Unol Daleithiau a pholisi'r Unol Daleithiau dros sawl ddegawd, gan gynnwys bod yn gennad Kennedy yn India yn y 1960au. Cyhoeddwyd y llyfr gyntaf yn 1958 ac mae wedi mwynhau llwyddiant poblogaidd a'i ailargraffu sawl gwaith dros y degawdau ers hynny. Fe ragfynegodd ac fe'n rhagrybuddiodd, yn syfrdanol o gywir, am yr anhrefn ariannol fyd-eang sydd wedi ysgubo trwy'r byd ers 2007/8. Mae'r ail gyfrol yn dathlu ei phen-blwydd yn 40 oed eleni. Cyhoeddwyd *Limits to Growth*[5] am y tro cyntaf yn 1972. Gwaith grŵp yn MIT dan arweiniad Donella Meadows, Jorgen Randers a Dennis Meadows. Cyhoeddwyd dau dyweddariad ganddynt sef *Beyond the Limits* yn 1992[6] a *Limits to Growth: the 30-year update* yn 2004[7].

Er gwaethaf eu poblogrwydd camliwiwyd a chamddehonglwyd y ddau lyfr yn ddybryd gan lawer. Roedd Galbraith, fel economegydd gwleidyddol a dyngarol, yn croesawu'r naid fawr oedd wedi digwydd mewn y rhan helaeth o gymdeithas gyfalafol y gorllewin, o dlodi affwysol i hawddfyd mewn cwta dwy ganrif. Rydym ni (h.y. pawb ond elit bychan oedd eisoes yn dra chefnog) wedi symud o grafu byw ar adnoddau pitw, a'r rheini'n cael eu defnyddio i ddiwallu'n hanghenion mwyaf sylfaenol – bwyd, dŵr glân, lloches, dillad etc. i hawddfyd cyffredinol.

Erbyn diwedd y 50au roedd ystod ryfeddol o nwyddau defnyddwyr ar gael yn yr Unol Daleithiau, ym Mhrydain ('you've-never-had-it-so-good' oedd y slogan bachog) ac mewn mannau eraill. Erbyn hyn mae llawer mwy. Mae tai'r dosbarth canol wedi eu trawsnewid; gwres canolog ac o leiaf un car yw'r norm. Dylai pob cartref ymffrostio mewn set deledu anferth sgrin wastad, peiriant sychu dillad, peiriant golchi llestri, ac wrth gwrs cyfrifiaduron ac amrywiaeth o ffonau symudol, iPads etc. Mae mwyfwy'n mwynhau'r bywyd bras. Mater arall yw y pris emosiynol a chymdeithasol a delir. Wrth i'r nwyddau a'r gwasanaethau a gynigir amlhau, mae cannoedd o filiynau o Tsieiniaid, Brasiliaid ac Indiaid yn ogystal ag Ewropeaid ac Americanwyr wedi ymuno â'r dosbarth canol cyfforddus.

Yn ôl dadansoddiad Galbraith, mae parhad yr hawddfyd hwn, ar ôl bodloni anghenion dynol sylfaenol, yn dibynnu ar 'fargen' rhwng y cyhoedd yn gyffredinol, diwydianwyr a busnes a gwleidyddion (Ffigwr 1). Wrth reswm mae'r cyhoedd am gadw ei gyfoeth a'r arian sydd ganddo i wario; mae llawer yn dyheu am wella eu byd. Mae angen swyddi da er mwyn cynnal unigolion a'u teuluoedd a'u statws cymdeithasol. Felly rhaid i amrywiaeth a nifer y nwyddau a gwasanaethau sydd ar gael gynyddu'n barhaus trwy arloesi a buddsoddi preifat a chyhoeddus er mwyn creu'r galw sydd wedyn yn esgor ar swyddi a chyflogau da. Gall y dyfeisiadau newydd hyn gynnwys teclynnau TG newydd neu geir newydd, darparu cyffuriau newydd neu gofal i'r henoed niferus (y cynnydd yn nifer yr henoed a sut i ofalu amdanynt yw'r her fwyaf nesaf ar y gorwel). Yn ei dro mae'r cyfoeth newydd wedi creu anghenion newydd; gwyliau yn yr haul, carafanau yn Abersoch, bwydydd da, gwinoedd gwell a'r ffasiynau diweddaraf. Yn ddiweddar disgrifiwyd hyn yn fachog fel 'the story of stuff'. Rhaid i wleidyddion, gan atseinio dywediad enwog Clinton 'it's the economy stupid', iro olwynion y 'fargen' hon er mwyn cael eu h[ail]ethol; felly hefyd diwydianwyr a dynion a merched busnes os ydynt am redeg busnes llewyrchus. Felly gwelwyd twf aruthrol diwydiannau newydd gwerth biliynau o bunnoedd megis hysbysebu, hyrwyddo cynnyrch a dylunio dyfeisgar, er mwyn argyhoeddi'r cyhoedd bod arno ddirfawr angen nwyddau a gwasanaethau traddodiadol ac arloesol, sy'n arwain at ffyddlondeb brand, cefnogaeth y sêr ac amrywiaeth o strategaethau seicolegol mwy cynnil i ennyn ac ysgogi ein dyheadau. Os nad oes galw, rhaid ei greu!

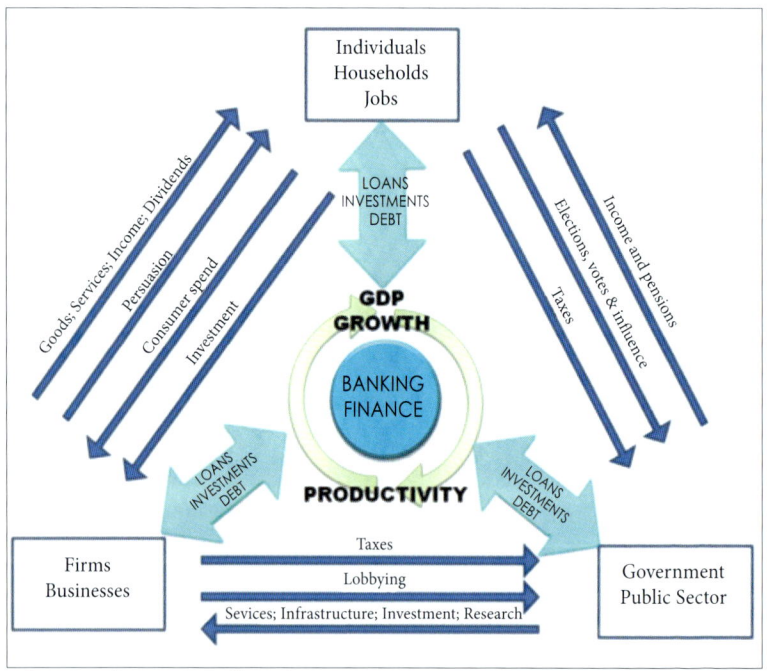

Ffigwr 1: 'Bargen' Galbraith

Mae'n dilyn yn rhesymegol bod busnes hefyd yn iro'r olwyn trwy ddyfeisiau megis hur-bwrcas, credyd rhwydd, cynigion arbennig apelgar, ac, at hynny, gardiau credyd a debyd a benthyciadau sydd ar gael ym mhob man. Mewn sawl dull a modd cynnil anogir pobl i fynd i ddyled; yn fwyaf eithafol er mwyn prynu a dodrefnu tŷ fel y gallant wireddu eu breuddwyd o fod yn ddinasyddion hunan-ddibynnol mewn democratiaeth sy'n rhoi pwys ar fod yn berchen ar eich tŷ eich hun. Rhaid i'r cwmnïau hwythau fenthyca arian er mwyn cystadlu a buddsoddi mewn nwyddau a gwasanaethau newydd a chyffrous neu mewn mannau gwerthu newydd. Rhaid i'r llywodraeth hithau fenthyca arian er mwyn chwarae ei rhan mewn cadw'r olwynion economaidd i droi trwy fuddsoddi mewn addysg, seilwaith a lliniaru ansefydlogrwydd cynhenid yr economi (gweler damcaniaethau Keynes) er mwyn lleihau anfodlonrwydd cymdeithasol ac er mwyn cael ei hailethol a goroesi'n wleidyddol (weithiau mae hynny'n cynnwys rhoi cymhorthdal i fusnesau sy'n methu). Un o ddamcaniaethau nodweddiadol Keyes yw pwysigrwydd creu galw trwy'r sector gyhoeddus i oresgyn dirwagiad os mae ddifyg galw preifat. Mae eraill yn dalau bod hyn yn arwain at chwyddiant, gor fenthyca ac yn mygu y sector preifat. Pa bynnag ddamcanaieth a dderbynir, mae credyd a dyled yn iro'r system gyfan. Eto mae'r hawddfyd hyfryd oedd yn addo cymaint yn gallu troi'n ddiflastod – yn lladdfa barhaus – a hynny sydd wedi digwydd. Fel y rhagwelodd Galbraith, heb ymyrraeth, mae'r 'fargen' y mae'r holl garfannau wedi ymrwymo iddi, yn sicrhau bod y rhai sydd yn elwa o'r system yn anwybyddu'r mynyddoedd cynyddol o ddyled. Credai fod y system yn ansefydlog yn ei hanfod, yn arwain o anghenraid at ddyled eithafol, anghynaliadwy, i anwybyddu ffiniau ac i greu swigod economaidd sy'n ffrwydro'n boenus. Awgrymodd fod tuedd naturiol mewn pobl at orymestyn a chroesi'r ffiniau, oherwydd ein bod wrth natur yn orobeithiol, yn rhoi gormod o ffydd yn 'doethinebu poblogaidd' (conventional wisdom). Sylweddolodd bod ein dychmygion hunanol yn ein rhwystro rhag amgyffred realiti, ac yn ychwanegol, yn cyfrannu ddogn go lew o drachwant a hunan-dwyll cyfleus. Roedd, yn ôl Galbraith, angen rheoli'r system yn dynn.

Er ei fod yn pwysleisio'r angen am reoleiddio doeth a rheoli economaidd gofalus, roedd Galbraith yn cydnabod bod yna duedd hefyd tuag at budreddi cyhoeddus er gwaethaf golud preifat ac awgrymodd ei bod yn anorfod y byddai'r pwrs cyhoeddus yn dod o dan bwysau oherwydd dyheadau a dymuniadau preifat a'r ffaith bod elit bychan yn graddol gronni grym.

Yn 2012 gallwn ddweud, heb os nac oni bai, y dylai ein gwleidyddion a'n heconomegwyr fod wedi cymryd diagnosis Galbraith yn llawer mwy o ddifrif. Ond rhan ganolog o'i weledigaeth oedd ei bod yn annhebygol iawn y byddent yn gwneud hynny. Beth sy'n eironig yw bod hyd yn oed wledydd fel yr Almaen, sy'n hwyrfrydig ers y rhyfel i hyrwyddo gorwariant a dyled, wedi cael eu cipio i ganol y corwynt. Er eu bod wedi dioddef llawer o feirniadaeth gan Brydain a'r Unol Daleithiau am eu hagwedd bwyllog, bellach maent wedi eu sugno i mewn i'r trobwll gan fod eu hallforion yn dibynnu ar alw gan farchnadoedd sydd mewn dyled fawr ac mae eu banciau'n gaeth i ddyledion pobl a gwledydd eraill. Er ymdrechion Llywodraeth Brydain i cynilo, ag ymyrraeth Banc Lloegr trwy chwystrellu £375 biliwn i'r economi [Quantitative Easing], mae'r dyled gwladol [Public Sector Net Borrowing] wedi codi i 45% i 65% o'n GPD blynyddol ers 2008.

Cydnabu Galbraith fod bancwyr fel cyfryngwyr a rheolwyr credyd yn meddu ar rym anferthol ond gwelir nifer o ffactorau nad oedd wedi eu rhagweld. Yn gyntaf dyfeisio pecynnau rhy glyfar a fyddai'n caniatáu i ddyledion peryglus gael eu gwerthu ymlaen fel cynhyrchion ariannol A-triphlyg nad oes a wnelont fawr ddim â'r byd go iawn. Felly nid yn unig gallai'r farchnad arian annog y tlodion i ysgwyddo dyledion na allent byth bythoedd eu had-dalu ond gellid ail-becynnu'r rhain yn gynhyrchion buddsoddi ymddangosiadol ddiogel, hynod o broffidiol a thwyllodrus o atyniadol. Trwy hyn a thrwy amryw o ddulliau eraill, cyfnewid credyd gwarantedig, gwarantau deilliadol [derivatives] a'u tebyg crëwyd sefyllfa fwyfwy ansefydlog. Yn ail, yn yr Unol Daleithiau a'r Deyrnas Unedig fel ei gilydd, roedd ffydd yn anffaeledigrwyddy farchnad ddilyffethair wedi arwain at ddadreoleiddio, o dan Reagan a Thatcher yn gyntaf ac yna o dan Bush a Blair a Brown – i gyd yn gwbl groes i argymhellion Galbraith. Yn y Deyrnas Unedig roedd y 'big bang' yn Llundain a disodli Deddf Glass-Steagall gan Ddeddf Gram-Leach-

Billey yn 1999 wedi helpu i sicrhau 'afiaith afresymol' ('irrational exuberance') a'r cefnogwr mwyaf dylanwadol, a hynny'n annisgwyl, oedd Cadeirydd yr US Federal Reserve, neb llai nag Allan Greenspan. Ychydig o ystadegau perthnasol – yn 2010 cafwyd cyfanswm o $955 triliwn o drafodion trwy'r byd mewn arian tramor; gwerth $ 601 triliwn o drafodion y tu allan i gyfnewidfeydd ar ffurf gwarantau deilliadol ariannol, er bod swm y rhandaliadau a'r bondiau a fasnachwyd yn $78 triliwn, o'i gymharu â'r Cynnyrch Mewnwladol Crynswth byd-eang sef $63 triliwn. Mae'r systemau hyn sy'n defnyddio arian 'rhithwir' wedi caniatáu i unigolion hel cyfoeth enfawr yn y byd go iawn[8]. Rwyn hoffi, er yn arswydo, dyfyniadau oddi wrth Niall Ferguson[9] – "Planet Finance is beginning to dwarf Planet Earth. And Planet Finance seems to spin faster". Hefyd "Money is not metal. It is trust inscribed. And it does not seem to matter much where it is inscribed – anything can serve as money. And it now seems in this electronic age nothing can serve as money to".

Yn drydydd, yn y Deyrnas Unedig o leiaf, mae'r llywodraeth fel pe bai wedi mynd i ddibynnu ar y sector gwasanaethau ariannol am gyfran helaeth o'i refeniw treth a thrwy hynny'n dibynnu arni i gyllido llawer o wasanaethau cyhoeddus. Nid oedd neb yn gallu cyffwrdd â'r sector a bu'n dal y wlad yn wystl, ac yn parhau i geisio gwneud hynny er bod Adair Turner, cadeirydd yr Awdurdod Gwasanaethau Cyllidol yn Llundain, wedi cydnabod nad oedd llawer o'i gweithgarwch o unrhyw fudd i'r cyhoedd. Ar ben hynny, mae adroddiad diweddar[10] yn datgan mai £193bn o drethi a dalwyd gan y sector ariannol rhwng 2002 a 2008 o'i gymharu â £378bn gan y sector gweithgynhyrchu yn yr un cyfnod. Yn dilyn y chwalfa ariannol amcangyfrifwyd bod y llywodraeth wedi rhoi £298bn o gymorth uniongyrchol i'r sector ariannol a gwerth £1.7 triliwn o fenthyciadau a gwarantau. Os yw'r ffigurau hyn yn gywir, yna naill ai mae mecanwaith cysylltiadau cyhoeddus y ddinas yn eithriadol neu mae ffactorau eraill tywyllach ar waith. Mae'r sefyllfa broblemus hon wedi gwaethygu oherwydd bod cwmnïau a busnesau wedi eu prynu â chymorth benthyciadau, sy'n trosi cyfalaf yn ddyled (megis ManU), anghydbwysedd masnachu enfawr ar draws y byd, a gallu'r cyfoethogion i osgoi talu trethi trwy ddianc i hafanau trethi a hefyd i ddefnyddio eu golud enfawr i gamarwain eraill ac i amddiffyn eu buddiannau hunanol. Rhyfeddol nodi bod rhwng 70 a 100 mil o unigolion gor-gyfaethog ar ein planed wedi 'cynilo' tua $21 triliwn [$2.1 x 1013] wedi ei guddio mewn hafanau di-dreth[11]. Swm iw gymharu a cyfaswn cyfoeth (CMC/GDP blynyddol) yr Unol Daliaethau o $15 trilwn a Siapan $5.8 triliwn!

Gwireddwyd ofnau Galbraith. Cadarnhawyd ei ddiagnosis. Ond mae gwneuthurwyr polisi a gwleidyddion sawl llywodraeth mewn sawl gwlad – Iwerddon, Gwlad yr Iâ, y Deyrnas Unedig a'r Unol Daleithiau i enwi pedair yn unig – wedi cael eu swyno gan eu ffydd gibddall yn anffaeledigrwydd y farchnad. Dadreoleiddio oedd cri y blaid Dorïaidd a'r blaid Lafur fel ei gilydd (a chlywir adlais unwaith eto??). Hybwyd y broses hon gan nifer o economegwyr academaidd blaenllaw. Yn awr rhaid i ni dalu am y ffolineb hwn. Ar hyn o bryd mae'r Deyrnas Unedig yn llwyfan i'r tyndra rhwng yr awydd i leihau baich dyledion personol a gwladol a'r ofn y bydd hyn yn ysgogi gostyngiad mewn gweithgarwch a chyflogaeth oherwydd diffyg galw. Ymddengys fod gwersi Keynes a Galbraith wedi mynd ar goll ymysg yr ofnau y bydd yr un marchnadoedd arian, a oedd yn un o brif achosion y chwalfa, yn ein cosbi ddwywaith. Dychwelwn at Keynes maes o law. Mae'n eironi chwerw mai trachwant y marchnadoedd hynny a'u dadansoddwyr a greodd y llanast yn y lle cyntaf. Nawr mae cyfalaf yn brin, twf mewn CMC/GDP yn sigledig a'n gallu i greu swyddi yn wantan. Rydym yn byw'r paradocs Galbraithaidd a chanlyniadau anwybyddu'r ffaith ein bod yn gorymestyn a **croesi'r ffiniau**.

3/Anwybyddu Ffiniau Twf

Prif fyrdwn "The Limits to Growth"[5] hefyd yw peryglon anwybyddu ein bod yn **croesi'r ffiniau**. Aeth y gyfrol denau a gyhoeddwyd yn 1972 ati i ddadansoddi model cymharol syml o'r rhyngddibyniaeth rhwng twf yn y boblogaeth, cynnyrch diwydiannol, llygredd, disbyddu adnoddau a'r galw amdanynt yn cynnwys cynhyrchu bwyd a'r tir oedd ar gael a galw cyfalaf. Roedd tri ffactor yn ganolog i'w syniadaeth. Yn gyntaf, natur twf esbonyddol, yn ail yr oediad cynhenid yn ymateb y rhan fwyaf o systemau naturiol sy'n arwain yn anorfod at **groesi'r ffiniau** ac yn drydydd absenoldeb terfyn diffiniedig syml. Nid oedd Meadows a'i chydweithwyr yn rhagweld y byddai system ein planed yn cyrraedd ochr dibyn, megis pan rybuddiwyd Columbus gan y rhai a gredai fod y byd yn wastad y byddai'n hwylio dros ymyl y byd. Yn hytrach roedd yr awduron yn rhagweld dirywiad, a allai fod yn raddol neu'n gyflym ond a fyddai'n cael ei waethygu wrth i ni '**groesi'r ffiniau**'. Byddai'n rhaid cyfeirio mwy a mwy o adnoddau dynol a naturiol tuag at wrthweithio unrhyw ffactor cyfyngol – boed yn lleihau effaith negyddol llygredd, cael gafael ar ryw fetel prin neu sicrhau cyflenwadau digonol o fwyd neu ddŵr. Byddai'r dargyfeirio hwn ar adnoddau, yn ei dro, yn tanseilio safonau byw ac yn arwain o bosib at anhrefn gymdeithasol. Byddwn yn ychwanegu (er na chyfeiriwyd ato yn y llyfr) y byddai hyn o ganlyniad yn tanseilio y fargen Galbraithaidd.

Rhan allweddol o'u dadansoddiad oedd twf esbonyddol (Tabl 1)

Mae hyd yn oed twf cymedrol o 2% yn y boblogaeth, dyweder, neu yn y GDP/CMC, neu yn y galw am fetel gwerthfawr ond prin neu am fwyd, yn achosi dyblu mewn 35 mlynedd yn unig – tua chenhedlaeth. Pe bai poblogaeth bresennol y byd, sef ~7 biliwn, yn tyfu ar raddfa gyfansawdd o 2%, gallai gyrraedd 14 biliwn erbyn 2036, er bod hynny'n annhebygol, ac yn fwy annhebygol fyth, 28 biliwn erbyn 2080 cf. 3 biliwn yn 1960. [Byddai'r ffigyrau cymharol am boblogaeth y Deyrnas Unedig dros 120 miliwn yn 2046 ac yn agos i 250 miliwn erbyn 2080]. O safbwynt

Tabl 1: y berthynas rhwng y gyfradd twf esbonyddol ac amser dyblu (gellir cymhwyso hyn i ffactorau amrywiol megis y Cynnyrch Mewnwladol Crynswth, poblogaeth, defnydd o adnoddau etc.)

Cyfradd twf (% y flwyddyn)	Amser Dyblu (blynyddoedd)
0.5	140
1.0	70
2.0	35
3.0	23
4.0	17.5
7.0	10
10.0	7

gwleidyddol ystyrir twf blynyddol o 1% yn y GDP/CMC yn beth i resynu ato, ac yn annigonol i gynnal swyddi a chostau byw cynyddol. Serch hynny mae'n arwyddocaol yn fathemategol – ac o safbwynt adnoddau. Y nod gwleidyddol confensiynol yw twf blynyddol o bron i 3% yn y GDP/CMC er mwyn sicrhau swyddi i bawb, talu pensiynau a darparu gwasanaethau cyhoeddus i boblogaeth sy'n heneiddio h.y. dyblu bob 23 blynedd. A oes modd cynnal hyn? Casgliad sylfaenol Meadows a'i gydweithwyr oedd nad yw'n bosib ac nad yw'r model twf economaidd presennol yn gynaliadwy.

Cafwyd dwy brif feirniadaeth ar 'Limits to Growth'. Y feirniadaeth gyntaf yw nad ydym wedi taro'r wal frics ar ôl 40 mlynedd. Fel arall yn wir, mae'r economi fyd-eang yn tyfu'n gyflymach nag erioed o'r blaen. Mae mwy na 2 biliwn o Indiaid a Tsieiniaid (neu gyfran ohonynt) yn mwynhau twf o 7 i 10% yn y GDP/CMC bob blwyddyn. Ond fel y dywedwyd, nid oedd Meadows a'i chydweithwyr yn awgrymu bod angen i'r economi sefyll yn ei hunfan. Yn hytrach rhagwelai y byddai'r 'terfynau' yn tynhau eu gafael yn raddol yn yr 21ain ganrif, nid yn yr 20fed ganrif. Yn ail fe'u cyhuddwyd o danbrisio dyfeisgarwch dyn a grym ac egni technoleg fodern ac economi'r farchnad. Maent yn gwadu hynny. Ychydig o sylwedd sydd i'r ddadl olaf hon: fel y gwelir yn nes ymlaen, rhaid wrth ymyriadau technolegol – y cwestiwn yw **pa dechnolegau?** Mae rhai agweddau ar dechnoleg amgen wedi datblygu'n gyflym e.e. celloedd ffotofoltaig llawer gwell, ceir sy'n defnyddio tanwydd yn fwy effeithlon. Nid oes dwywaith nad oes llai o lygredd bellach ym moroedd a dyfroedd croyw'r Deyrnas Unedig Llwyddiant hefyd fu Protocol Montreal (1987) i gyfyngu ar y defnydd o CFCs[12] a graddol ail-lenwi'r twll yn yr haen oson. Ond gall newid technolegau greu cyfyngiadau newydd ar adnoddau. Gall symud o danwyddau ffosil i dyrbinau gwynt leihau'r galw am olew a nwy, ond gall olygu mwy o alw am fetelau daear prin i adeiladu cylchedau trydanol y tyrbinau. Ond mae'r un mor hawdd rhestru methiannau. Mae cyfyngiadau cyfalafiaeth y farchnad wrth ddyrannu adnoddau bellach yn gwbl amlwg ac erys problemau sylfaenol heb eu datrys.

Hwyrach bod croesi'r ffiniau'n nodwedd ar ein system economaidd sy'n peri dychryn, ond mae'n nodwedd annatod ar systemau naturiol. Er gwaethaf y Protocol[12], mae'r twll yn yr oson yn parhau a dim ond erbyn 2050 y mae gobaith y bydd wedi diflannu'n llwyr. Os bydd ffrwythlondeb merched yn gostwng i'r isafswm i gynnal y boblogaeth [~2.2 o blant fesul merch], byddai'n cymryd degawdau i gyfradd twf y boblogaeth gyrraedd sero oherwydd ei strwythur pyramidaidd h.y. llawer o ferched ffrwythlon yn eu harddegau a'u hugeiniau cynnar. Os trown at newid yn yr hinsawdd, hyd yn oed pe bai allyriadau nwyon tŷ gwydr a achoswyd gan bobl, yn enwedig CO_2 yn sefydlogi neu hyd yn oed yn disgyn i lefelau cyn-ddiwydiannol yfory nesaf, byddai'n cymryd cannoedd, os nad miloedd o flynyddoedd i'r hinsawdd, cylchrediadau'r moroedd a lefel y môr ail gyd-bwysog (gweler Ffigwr 2). Gallai'r lefelau dŵr yn rhai o'r dyrfhaenau [aquifers] sydd wedi eu gorddefnyddio, godi eto ymhen amser, ond nid pob un, a byddai'n cymryd degawdau, o'i gyferbynnu â'r ychydig flynyddoedd a gymerodd i'w disbyddu. Nid yw stoc pysgod y Newfoundland Grand Banks wedi ei hadfer er gwaethaf gwaharddiad ar bysgota a osodwyd ryw 15 mlynedd yn ôl[13]. Os ydym yn parhau i achosi difodiant rhywogaethau yna, efallai, ymhen amser bydd cyfleoedd esblygol newydd yn codi ar gyfer rhywogaethau newydd, ond ymhen miloedd o flynyddoedd yn hytrach na degawdau.

Mae'n ddiddorol ystyried rhai enghreifftiau cyfredol o'r broblem 'terfynau' a'u goblygiadau.

Un enghraifft drawiadol yw adroddiad 2011 y British Geological Survey[14]. Ynddo ceir asesiad o'r perygl i gyflenwadau o rhai elfennau cemegol o bwys economaidd ar raddfa o 1 i 10. Mae'r sgôr yn dibynnu ar [a] digonedd yny gramen; dyfynnir rhai enghreifftiau isod [xx], [b] a ydynt yn cael eu cynhyrchu gan fwyaf mewn nifer fach o wledydd, [c] lleoliad a dosbarthiad cronfeydd y gellid eu defnyddio a [d] sefydlogrwydd gwleidyddol y gwledydd sy'n cynhyrchu. Ar y sail hon, mae elfennau cymharol anhysbys megis Antimoni (Sb) [0.2ppm] y grŵp platinwm (Pt) [(Pt) 0.0015ppm], mercwri (Hg) a twngsten (W) yn sgorio uchaf ar gyfer risg sef 8.5 ac ar eu hôl y prinfwynau a niobiwm. Bellach defnyddir yr elfennau hyn yn eang ym maes micro-electroneg, fel catalyddion, mewn aloiau arbennig, mewn lled-ddargludyddion ac mewn deunyddiau sy'n gwrthsefyll tân ac eithafion tymheredd h.y. defnyddiau technoleg uchel. Yn y pegwn arall ceir y metelau cyffredin a chyfarwydd, titaniwm, alwminiwm [84,149ppm], cromiwm, haearn [52,157ppm] a sylffwr [404ppm].

Mae'r datblygiadau mewn technoleg yn eu tro wedi creu galw newydd am adnoddau, gan gynnwys deunyddiau/elfennau prin ac/neu wenwynig. Gan fod cyflenwi rhai o'r elfennau sydd yn gwbl hanfodol i ddyfeisiau modern yn broses ansicr, mae sicrhau cyflenwad dibynadwy a di-dor bellach yn un brif amcanion cenhedloedd grymus yn enwedig Tsieina. (Yn yr un modd, yn hanesyddol ac hyd heddiw mae cynnal a chadw eu cyflenwadau nwy neu olew wedi bod yn obsesiwn gan yr Americanwyr a hynny ar draul eraill ar brydiau). Felly mae'n amlwg nad prinder daearegol neu hyd yn oed dechnoleg mwyngloddio yw'r 'risg' ond y berthynas rhwng y grymoedd daearegol, economaidd, gwleidyddol a hyd yn oed filwrol. (Adroddwyd yn Rare Earth

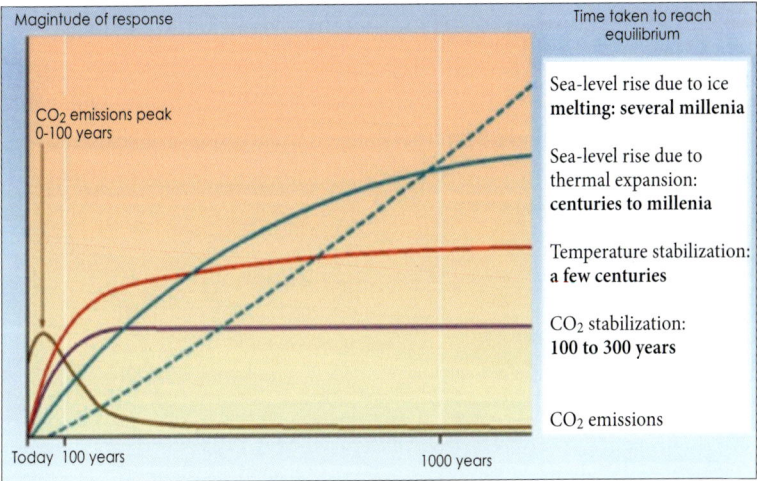

Figure 2: Lags in climate change and overshoots
Source: R.T. Watson and the Core Writing Team (eds.), *Climate Change 2001: Synthesis Report. A Contribution of Working Groups I, II, and III to the Third Assessment Report of the Intergovernmental Panel on Climate Change*, Cambridge University Press, Cambridge 2001.

Investment News yn niwedd 2011 fod pris Cerium, a ddefnyddir i wneud goleuadau LED ynni-effeithlon, wedi codi o $50 i $413 y kg rhwng Ionawr a Mehefin 2011 sydd yn bygwth y diwydiant ac yn golygu bod cwmnïau'n mudo i Tsieina.)

Mae Ffigyrau 3a a 3b yn darlunio'r twf sydd lled esbonyddol yng nghynnyrch copr y byd ynghyd â'r prisiau cyfnewidiol sydd yn dangos effaith twf yn yr economi fyd-eang a buddsoddi blaendrafodion [futures] ar y pris yn y farchnad. Ers y chwalfa economaidd yn 2008 mae'r prisiau wedi bod yn is ac yn gyson tua 6 mis ~$8,000 fesul tunnell fetrig.

Un agwedd arall ar 'derfynau' sydd yn gymhleth a hyd yn oed yn fwy heriol yw'r rhwydwaith o gysylltiadau rhwng y cyflenwad bwyd, y tir sydd ar gael, y dŵr i'w ddyfrhau a newid yn yr hinsawdd (gweler 18,19). Er gwaethaf yr holl fwyd a wastreffir yng ngwledydd y gorllewin, mae'r cyfuniad o fwy o bobl, hyd at 10 biliwn a mwy efallai erbyn canol y ganrif hon, a safon fyw uwch yn arwain at alw cynyddol am fwyd, a mathau gwahanol o fwyd hefyd. Mae hyn yn wir nid yn unig am y bwydydd sylfaenol sef grawn, tatws, casafa etc ond yn enwedig cigoedd gwyn a choch a chynhyrchion llaeth o bob math[18]. Yn rhesymegol, gellir diwallu'r galw trwy gynhyrchu mwy ar y tir sydd eisoes ar gael neu drwy ehangu'r tir âr neu'r tir pori. Mae'r ddau ddewis yn frith o broblemau.

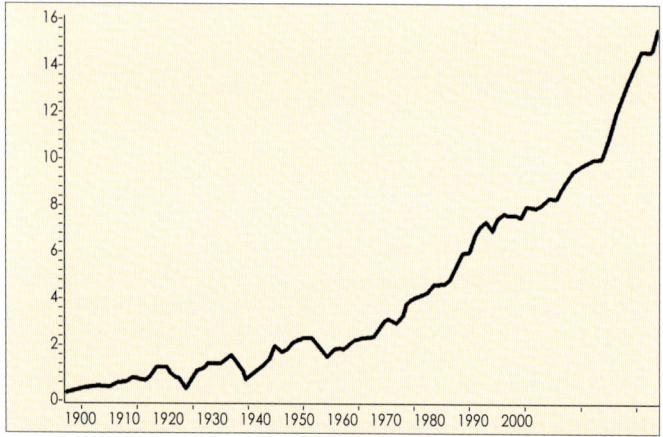

Ffigwr 3a: Tueddiadau yng nghynnyrch copr y byd 1900-2007[2]

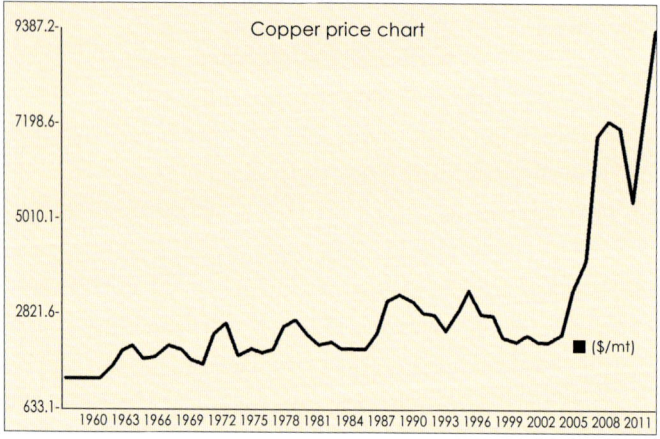

Ffigwr 3b: Prisiau copr (doleri'r Unol Daleithiau) o 1960-2011

Ychydig o dir âr sy'n segur ac mae dŵr i ddyfrhau'n brinnach fyth. Mewn sawl rhanbarth mae hyd yn oed y cyflenwadau dŵr presennol o dan fygythiad oherwydd gorddefnydd o acwifferau, afonydd neu gronfeydd dŵr ac oherwydd y gystadleuaeth o'r galw gan ardaloedd dinesig. Mae tua 1.5 biliwn hectar o dir âr yn cael ei amaethu bob blwyddyn trwy'r byd, ac amcangyfrifir y byddai'n bosib trin rhwng 2 a 4 biliwn hectar ond mae peryglon sylweddol ynghlwm wrth amcangyfrifon felly, oherwydd gallai amaethu tir anaddas greu anawsterau dybryd[7]. Yn ystod y mileniwm diwethaf, yn sgil twf ardaloedd trefol a chynnydd mewn amaethu, mae'n bosib bod rhyw ddau biliwn hectar o dir fferm wedi mynd i ddifancoll, wedi troi'n anialwch neu'n gors, y pridd wedi ei halltu neu ei alcaliaiddio. Yn y degawd diwethaf mae Syria wedi gwahardd pobl rhag aredig y tiroedd sych ('al badia') oherwydd diraddiad y priddoedd. Erbyn hyn mae Gardd Eden naill ai'n anialwch neu'n gors halen yn dibynnu ar ei union leoliad 'hanesyddol'. Mae'r ehangu sydd wedi bod ar ardaloedd trefol yn ddiweddar wedi dwysáu'r pwysau. Rhwng 1989 a 1992 collodd Tsieina 6.5 miliwn hectar i ddatblygiadau, ac ar yr un pryd trowyd 3.8 miliwn hectar o goedwigoedd a phorfeydd yn dir i dyfu cnydau. Bob blwyddyn mae ardal anferth, yn aml yn cynnwys y tir gorau, yn cael ei llyncu gan orymdaith y trefi. Yn yr Unol Daleithiau'n unig cyn y chwalfa ariannol collwyd tua 170,000 ha y flwyddyn[7]. Ar draws y byd rhaid bod yn agos i filiwn o hectarau neu fwy hyd yn oed wedi eu colli.

Mae'n destun pryder bod troi glaswelltir neu goedwigoedd yn dir âr yn gostus i'r amgylchedd gan fod CO_2 yn cael ei allyrru wrth i'r deunydd organig yn y pridd gael ei ocsideiddio, a hynny'n gwaethygu problem newid hinsawdd yn sylweddol[17]. Hefyd mae'n bosib bod cynefinoedd pwysig yn diflannu, llifeiriant dwr yn newid mewn dalgylch afon, gallu coedwigoedd i rwymo carbon yn lleihau a bioamrywiaeth yn crebachu. Mae'n rhesymol, felly, casglu mai ychydig o le sydd i ehangu tir amaethyddol mewn modd cynaliadwy ac y bydd ehangu difeddwl yn esgor ar ganlyniadau niweidiol iawn. Yn wir mae hynny wedi digwydd eisoes.

Nid oes amheuaeth nad oes cyfleon sylweddol mewn sawl gwlad i cynyddu cynnyrch cnydau a gwella hwsmonaeth anifeiliaid. Yn aml iawn mae'r cnydau ar ffermydd yn cymharu'n wael â'r cnydau a geir mewn treialon lleol tebyg sy'n cael eu rheoli'n dda. Mae hyn yn wir am systemau lle mae'r tir yn cael ei ddyfrhau ac am systemau sy'n dibynnu ar law, er yn hanesyddol mae'r rhan fwyaf o'r bwyd ychwanegol sydd ei angen i ddiwallu anghenion y boblogaeth gynyddol wedi ei dyfu o dan y drefn gyntaf. Mewn sawl lle, gwaetha'r modd, mae systemau dyfrhau yn cael eu bygwth gan ostyngiad yn y cyflenwadau dŵr a chan gystadleuaeth o du'r trefi. Nid gwaith hawdd chwaith yw ceisio gwella cnydau gwael. Mae'r problemau'n deillio o amrywiaeth eang o ffactorau, megis deiliadaeth tir, hawliau dŵr teg, addysg ffermwyr, mynediad at y farchnad, dyledion cynyddol, yr amaeth-gemegolion sydd ar gael ynghyd â'u cost, diffyg hadau modern priodol, sydd wedi eu haddasu'n dda. Hefyd yn amal cyngor gwael cier gan lawer o Systemau Ymchwil Amaethyddol Cenedlaethol. Mae daliadau tir sy'n crebachu wrth i'r tir gael ei rannu rhwng nifer cynyddol o aelodau teulu. Yn gyffredinol mae'n ymddangos bod y gwelliannau rhwydd yn yr ardaloedd mwyaf ffrwythlon eisoes wedi digwydd a bod cynnydd yn y dyfodol yn debygol o fod yn arafach ac yn ddrutach.

Mae sawl astudiaeth yn awgrymu yn sgil newid yn yr hinsawdd y bydd llai o ddŵr ar gael mewn ardaloedd allweddol o bwysig e.e. llawer o'r Dwyrain Canol, Canol Asia, Awstralia, rhannau o'r Unol Daleithiau, i gnydau sydd yn dibynnu ar law a rhai sydd yn cael eu dyfrhau[19]. Rhagfynegir y bydd cyfraddau llif afonydd yn disgyn yn yr ardaloedd lled-gras, ac y bydd llai o wlybaniaeth yn y pridd. Er y gall CO_2 uwch ysgogi llystyfiant, rhagfynegir y bydd cnydau ar y cyfan yn lleihau oherwydd y bydd mwy o eithafion tywydd, yn enwedig cyfnodau o sychder eithafol (*cf.* Rwsia 2010; Awstralia 2008-2011) a llifogydd trychinebus (Pacistan 2010). Daw bygythiad hefyd yn sgil problemau ffenolegol penodol megis straen gwres terfynol a diffyg egino pan fydd y glaw'n hwyr neu'n methu a lleihad yn hyd y tymor tyfu. Bydd cyfuniad o ddiffyg dŵr a chyfradd uwch o anwedd-drydarthu [evapotranspiration] oherwydd tymheredd uwch yn cyfyngu ar gynnyrch.

Ar gyfer amaethu dwys a chynnyrch da mae angen tir ffrwythlon. Ychydig sy'n ymwybodol bod defnyddio gwrtaith yn cynhyrchu N_2O. Gwaetha'r modd mae N_2O yn nwy tŷ gwydr llawer mwy effeithiol na CO_2 ei hun (x310 dros 100 mlynedd). Mae pob tunnell o nitrogen a ddefnyddir, boed ar ffurf tail neu cwrtaith anorganig [ar sail 0.0196 kg N_2O/kg N wedi ei luosi gan GWP 310] yn rhyddhau tua 6 tunnell o CO_2-gyfwerth[17]. Yn ogystal â hyn, mae rhwymo N_2 yn fasnachol yn defnyddio llawer o ynni. Gan fod cynnyrch cnydau'n codi mewn perthynas sydd bron yn llinol â'r cyfanswm N sydd ar gael hyd at rhai cannoedd o kg o N fesul ha, nid ystyriaeth ddibwys mo hon. Yn rhyfedd iawn, gallai'r amaethu dwysach a'r cynnyrch uwch olygu bod y gadwyn fwyd yn cyfrannu mwy at newid yn yr hinsawdd. Ar ben hynny, mae'r cynnydd yn y defnydd o nitrogen ar draws y byd yn debyg i'r cynnydd mewn CO_2 yn yr atmosffer ac mae hyn hefyd yn niweidiol i ecoleg ac yn achosi anocsia ewtroffig [tyfiant di oscigen] yn nyfrodd y glannau a dyfroedd croyw. Felly mae'r dulliau amaethu dwysach hyn yn costio'n ddrud; ac mae'r sefyllfa'n waeth oherwydd bod cyfran fawr o'r cynhaeaf ŷd a chorn yn cael ei bwydo'n aneffeithlon i anifeiliaid, neu, yn waeth, yn cael ei ddefnyddio i gynhyrchu bio-tanwydd[7].

Mae'r ffaith bod anifeiliaid fferm a chynnyrch anifeiliaid yn cyfrannu at newid yn yr hinsawdd yn destun pryder mawr yn y Deyrnas Unedig gan fod anifeiliaid sy'n cnoi cil (geifr, gwartheg, defaid, byffalo etc.) yn gyfrifol am ganran fawr o'r allyriadau methan (CH_4) ar draws y byd [17] [18]. Mae llawer o'r anifeiliaid hyn yn yr Unol Daleithiau ac mewn mannau eraill yn cael eu bwydo ar rawn ac ar borthiant dwys arall mewn llociau bwydo, sydd yn wastraffus ac yn cymryd adnoddau y gallai pobl eu defnyddio, a hefyd yn cynhyrchu CH_4. (Mae CH_4, dros gyfnod o 100 mlynedd, yn nwy tŷ gwydr sydd tua 24 gwaith mor effeithiol na CO_2.)

Nid oes amheuaeth na fyddai problem bwyd y byd a chyfraniad y gadwyn fwyd at gynhesu byd-eang yn lleihau'n aruthrol pe baem yn bwyta llai o gynnyrch anifeiliaid. Amcangyfrifwyd[18], cyn 2040, os bydd y tueddiadau presennol wrth ddiwallu anghenion bwyd y byd yn parhau, ar sail y cymysgedd bwyd a ragfynegir, y bydd hynny **ohono ei hun** yn parhau newid yn yr hinsawdd ac yn sicrhau ein bod yn croesi'r ffin +2C, sef y cynnydd yn y tymheredd ar draws y byd a ystyrir fel arfer yn drothwy ar gyfer ysgogi adweithiau cyflym e.e. rhyddhau methan o'r twndra neu gynnydd yn yr ymbelydredd solar a amsugnir oherwydd bod y rhew'n diflannu'n gyfan gwbl o fôr yr Arctig yn ystod yr haf[18]. Mae'n bwysig nodi nad ydym yn sôn yn unig yn y fan hon am ryw 1.5 biliwn o wartheg ac anifeiliaid eraill sy'n cnoi cil ond **hefyd** am y ~ 23 biliwn o gywion ieir, sydd er bod eu cyfraniad unigol at nwyon tŷ gwydr yn is, gyda'i gilydd yn creu effaith enfawr ac arwyddocaol. Nid yw hyn chwaith yn cymryd cyfrif o unrhyw duedd i fynnu trin y tir mewn ardaloedd anaddas sy'n arwain at erydu a cholli uwchbridd a deunydd organig neu ddefnyddio dŵr o ansawdd gwael i ddyfrhau sy'n gwneud y tir yn ddiffaith ac yn diraddio'r pridd trwy ei wneud yn fwy hallt. Mae'n werth cofio'r gwirionedd cignoeth hwn, os bydd y ddynoliaeth yn llwyddo i ddatgarboneiddio'r cyflenwad ynni'n llwyr ar draws y byd erbyn canol y ganrif hon, ac mae hynny'n annhebygol, fe all bwydo byd o 9 i 10 biliwn barhau i newid ein hinsawdd.

Efallai ein bod yn agosáu at drothwy, nid terfyn digamsyniol efallai, ond un a seilir ar fethodolegau, blaenoriaethau ac arferion heddiw. Er bod prisiau bwydydd hanfodol wedi disgyn yn raddol am sawl degawd, yn y blynyddoedd diwethaf gwelwyd gwrthdroi'r duedd hon. Yn 2008 gwelwyd cynnydd sydyn mewn prisiau a ysgogodd derfysgoedd bwyd mewn nifer o wledydd yn Affrica ac yn Ynysoedd y Philippines. Ers hynny mae mynegai prisiau bwyd yr FAO, ar ôl gostwng ychydig, wedi codi ymhellach, gan gyrraedd 225 ym mis Medi 2011, o'i gymharu â'r uchafbwynt blaenorol, sef 213.5 ym mis Mehefin 2008. Mae'n werth cofio bod y terfysg difrifol a achoswyd gan brisiau bwyd yn Tunisia ac

Algeria ddechrau 2011, efallai'n rhagargoel o'r Gwanwyn Arabaidd.

Mae'r enghreifftiau hyn yn adlewychu mewn modd pur drawiadol un o negeseuon y 'Limits to Growth'. Nid oes un broblem ac nid oes un ateb. Yn lle hynny rhaid i ni ymgodymu â gwe gymhleth o ffactorau cydgysylltiedig, rhai ffisegol, rhai cymdeithasol, rhai gwleidyddoleconomaidd, a fydd gyda'i gilydd yn ei gwneud yn fwyfwy o her i fwydo poblogaeth gynyddol y byd ac yn gofyn am fwy a mwy o sylw ac adnoddau.

Tristwch rhyfeddol nodi digwyddiau haf 2012; nid wrth gwrs y Gemau Olympaidd, ond y newid syfrdanol yn Cefnfor yr Arctig a gafodd nemor ddim sylw. Erbyn canol Medi roedd 80% o cyfaint [volume] yr Ia Morol Arctig wedi diflannu i cymharu a degawd yn ol. Un canlyniad tebygol yw i'r newidiadau gyflymu yn y blynyddoedd nesaf sef y sonywd eisoes gyda diflaniad posib yr ia erbyn hydref 2016 ac, yn arwyddocaol, i lefel y mor godi yn gynt na disgwyl. Ond, yn debygol, gwelwyd canlyniadau uniongyrchol y newidiadau yn barod. Tybir i'r "Llif Cyflym dros yr Iwerydd" a Hemisfer y Gogledd [jet stream] ddibynnu ar y gwahaniaethau tymheredd etc. rhwng yr Arctig a ardaloedd tymheraidd. Ond maent yn newid ac mae'r llif yn simsanu ac yn crwydro i'r de ac creu tonnau hirhoedlog (tonnau Rossby). O ganlyniad gawsom haf gwlyb, gwyntog ond nid oer. Ond yn llawer fwy pwysig dioddefodd a canol yr Unol Daliethiau a rhannau o Ddwyrain Europ haf hunllefys o boeth a sych. Cafwyd cnydau sal. Nawr gwelir prisiau bwyd yn codi i lefelau 2010/11: anffodus i ni, ond trychinebus i eraill. Er i erthyglau yn ein papurau newydd trafod y broblem plwyfol, bach iawn bu'r son am y gysylltiadau ehangach.

Rhai cydnabod cymhlethdodau eraill sydd yr un mor ddifrifol. Yn baradocsaidd, os oes cynnydd mewn cyfoeth yna bydd y galw, yn enwedig am gynhyrchion cig a llaeth o'r ansawdd gorau, yn cynyddu – fydd yn ei dro'n dwysau'r bygythiad o du newid hinsawdd. Serch hynny mae'r galw gan y rhai da eu byd am gynhyrchion amaethyddol, gan gynnwys cig, yn cynnig un ddihangfa o dlodi i'r tlodion gwledig ac yn agor drws iddynt i economi'r farchnad. Ar ben hynny, bugeilio anifeiliaid yw'r unig ddewis i lawer o bobl sy'n byw bywydau rhannol grwydrol mewn rhanbarthau lled-gras. A ydym ni, oherwydd perygl i'r hinsawdd, a achoswyd gan fwyaf gan bobl gyfoethog, i fod i wahardd y Masai ar y Masai Mara neu'r Basutho ar fynyddoedd y Maloti, neu yn wir, ffermwyr mynydd Cymru rhag bugeilio eu gwartheg? A fydden nhw a llawer eraill yn gwrando? Sut y gellid gorfodi polisi o'r fath? Efallai'n wir y dylem fabwysiadu diet sydd bron yn figan a fyddai'n lleihau rhai problemau ond byddai'r gwrthwynebiad cymdeithasol yn enfawr, nid yn unig yn yr Unol Daleithiau neu Awstralia ond mewn rhannau helaeth o'r byd Islamaidd ac Affrica. Mae moeseg yn awgrymu y dylem wahaniaethu rhwng pobl sy'n dibynnu ar borfeydd a fyddai fel arall yn segur a'r barwniaid da byw sy'n dibynnu'n drwm ar fwydo porthiant dwys sy'n cystadlu'n uniongyrchol ag anghenion pobl. Sut yn y byd y gellid gwneud hyn? Pe bai modd i wahardd cynhyrchion cig ac anifeiliaid, mae'n anorfod y byddai hynny'n esgor ar wadu a thwyll a byddem yn sicr o groesi'r ffin. Ar ben hynny, i fynd yn ôl at fargen Galbraith, pe bai tocio ar y galw am lawer o gynhyrchion, nid yn unig cig coch, ond porc, cig moch, cyw iâr, twrci, llaeth, caws, wyau, iogwrt, oni fyddai llawer un yn barnu bod y 'fargen' sy'n sail i gyfoeth wedi cael ei thorri, ac y byddai'r canlyniadau'n anfesuradwy?

Rydym yn rhedeg ar olwyn sy'n troi'n gyflymach o hyd, yn gorfod treulio mwy a mwy o amser, ymdrech ac adnoddau i'n rhwystro'n hunain rhag disgyn i ffwrdd. O gyflymu 2% yn unig bob blwyddyn, mae ein cyflymder yn dyblu mewn tua chenhedlaeth. Mae canlyniadau amgylcheddol eang i'r twf hwn a ddarlunnir gan y berthynas gadarn rhwng y GDP/CMC a'r defnydd o ynni[19] er ei bod yn llai os yw'r cyfoeth yn uwch na $20,000 per capita. Mae'r galw am nwyddau, hyd yn oed os cynhyrchir hwy yn Tsieina, ac nid gartref, yn disbyddu deunyddiau crai. Nid oes dwywaith nad yw'r twf yn y boblogaeth yn un ffactor sy'n cyflymu cylchdroi'r olwyn ond y llall yw 'bargen Galbraith' sy'n sail i'n heconomi 'gyfoethog' ac, yn llawer mwy problemus, yn sail hefyd i ragolygon a dyheadau'r tlawd a'r difreintiedig.

Wrth i'r galw am adnoddau gynyddu yn sgil cyfuniad o dwf yn y boblogaeth a safonau byw uwch, bydd gallu cyflenwadau bwyd i wrthsefyll eithafion tywydd yn lleihau, a bydd ein gallu i addasu a newid yn cael ei dagu gan gymhlethdod delio nid yn unig â phroblemau technegol ond â ffactorau cymdeithasol a gwleidyddol-economaidd hefyd.

Yn union fel roedd selogion y farchnad rydd yn dadlau bod y system ariannol fyd-eang yn hunanreoleiddiol i raddau helaeth, felly hefyd mae beirniaid y cysyniad o 'Derfynau', yn honni y bydd mecanwaith y farchnad trwy roi pris ar brinder yn ysgogi datblygiadau newydd ac yn osgoi problemau. Mae hyn yn anwybyddu nifer o ffactorau. Yn gyntaf mae nifer o broblemau megis newid yn yr hinsawdd yn deillio o ffactorau, na chânt eu hystyried gan economeg glasurol; fe'u gelwir yn 'allanolion' ac sydd felly ddim yn cael eu cynnwys ym mhris y farchnad. Mae hyn yn wir nid yn unig am allyriadau CO_2e yn yr atmosffer ond am ryddhau N a P i'r tir a'r môr. Yn ail, mae rheoliadau cynhyrchu yn amrywio ar draws y byd ac mae hyn yn annog cyfalaf a phrosesau gweithgynhyrchu i symud i'r lleoliadau lle mae'r lefel isaf o reoleiddio, a'r pryder lleiaf am lygredd a lle ceir y cyflogau isaf. Yn drydydd, nid oes unrhyw sail resymegol dros gredu y bydd dewisiadau eraill ar gael bob tro pan fydd adnoddau'n prinhau. Mae hyn yn sicr yn wir am fwyd a dŵr. Mae'r ffaith bod llywodraeth Tsieina wedi mynd ati i brynu llawer o'r adnoddau e.e. copr trwy'r byd ynghyd ag adnoddau metel eraill yn dangos nad ydynt mor ddibryder. Yn bedwerydd, go brin bod hanes economi'r farchnad rydd hyd yn oed o fewn ei libart ei hun yn cyfiawnhau ffydd diysgog ynddi ac mae bron yn siŵr y bydd newidiadau bach yn y cyflenwad yn cael effaith enfawr oherwydd hapfuddsoddi.

4 Ymgyrraedd at Gynaliadwyedd

Camddefnyddir y termau cynaliadwyedd neu ddatblygiad cynaliadwy yn ddybryd. Wrth gwrs rhaid i ni osgoi gweithredu mewn ffyrdd sydd yn amlwg yn anghynaliadwy e.e. cynnydd esbonyddol yn y dalfeydd pysgod môr. Mae llyfrau Meadows yn tanlinellu'r pwyntiau hyn yn dda. Ond os yw gofynion ein system bresennol ar adnoddau a suddfannau'r blaned yn anghynaliadwy, er bod llawer o bobl yn enbyd o dlawd a sawl cymuned yn wynebu argyfwng, sut olwg fyddai a'r system fwy cynaliadwy?

Mae Herman Daly[20] wedi awgrymu tri maen prawf ar gyfer cynaliadwyedd amgylcheddol, gan wahaniaethu rhwng y defnydd o anodau adnewyddadwy ac anadnewyddadwy a'r gallu i wasgaru llygredd. Yn achos adnoddau adnewyddadwy awgryma na ddylid defnyddio'r adnodd yn gyflymach nag y gellir ei adnewyddu neu ei atgynhyrchu e.e. dylai dalfeydd pysgod gael eu cyfyngu i gyfradd twf y boblogaeth bysgod ac ni ddylid tynnu dŵr o acwiffer yn gynt nag y bydd yn ail-lenwi, ni ddylai ffermio neu ddefnyddio tir arwain at leihad net yn y deunydd organig yn y pridd neu gynnydd net mewn allyriadau CO_2e. Ni ddylid defnyddio adnoddau anadnewyddadwy yn gyflymach na'r cyfnod a gymerir i greu adnodd adnewyddadwy, cynaliadwy yn eu lle e.e. gellid cyfrif y defnydd o olew yn ddefnydd cynaliadwy pe bai rhan o'r elw'n cael ei fuddsoddi mewn ffynonellau ynni adnewyddadwy fel pan fydd yr olew anadnewyddadwy wedi ei ddisbyddu y bydd ffynhonnell ynni adnewyddadwy cyfatebol yn barod i'w defnyddio. Yn achos llygrydd unigol ni ddylai'r allyriadau fod yn fwy na'r cyfanswm y gellir ei ailgylchu, ei amsugno neu ei wneud yn ddiniwed yn yr un cyfnod e.e. carthion sy'n cael eu diraddio gan ficro-organebau heb niweidio ecosystemau dyfrol. Gellir ymhelaethu ar ail faen prawf Daly trwy ddatgan, fel y dywedir uchod, na ddylai'r defnydd o adnodd anadnewyddadwy fod yn fwy na swm yr adnodd adnewyddadwy y gellir ei gynhyrchu yn ei le, **a hefyd** y gallu i ailgylchu ac adennill yr adnodd anadnewyddadwy o wastraff.

Mae cymhwyso rhesymeg Daly i ynni niwclear a thanwyddau hydrocarbon yn broses anghyfforddus iawn. Yn achos ynni niwclear, mae wraniwm yn adnodd anadnewyddadwy ond un a gafodd sgôr isel o ran y risg i'r cyflenwad, sef 4, gan y BGS[14]. O ystyried y defnydd presennol ohono ymddengys nad oes raid poeni am faint o wraniwm sydd ar gael. Ond pe bai ynni niwclear yn ennill poblogrwydd fel y brif system cynhyrchu trydan yn rhyngwladol, heb sôn am ystyriaethau diogelwch, byddai'r sefyllfa hon yn newid yn aruthrol ac yn fuan iawn byddai problemau'n codi o ran y cyflenwad crai o ddefnyddio'r technolegau sydd ar gael ar hyn o bryd. Ond mae delio â'r gwastraff ymbelydrol o'r nifer fechan o orsafoedd sydd gennym ar hyn o bryd am gannoedd neu hyd yn oed filoedd o flynyddoedd yn ddigon o dasg i drethu ein gallu rheoli. Mae pob llywodraeth yn y Deyrnas Unedig ers y 1960au wedi methu â datrys problem storio gwastraff yn yr hir dymor; mae'r Iseldiroedd wedi bod yn fwy llwyddiannus gyda'u storfa tymor byr (100 mlynedd), sydd yn storfa ganolog uwchben y ddaear yn perthyn i'r Sefydliad Canolog Gwastraff Ymbelydrol (COVRA). Mae hyn yn bennaf oherwydd eu dull cyfrifo agored e.e. mae'r costau sy'n gysylltiedig â gwaredu gwastraff niwclear yn COVRA yn cael eu cynnwys ym mhrisiau trydan defnyddwyr yr Iseldiroedd heddiw. Dim ond ar ôl sawl canrif pan fydd yr ymbelydredd wedi diraddio i lefel isel y bydd y deunydd yn ddiniwed yn fiolegol a thrwy hynny'n ateb maen prawf Daly – a rhaid wrth gytundeb rhyngwladol ar waredu gwastraff niwclear, o fewn Ewrop o leiaf. Dylid nodi hefyd mai ychydig o'r rhai sy'n gwrthwynebu ynni niwclear, neu'n ymgyrchu dros 'barthau di-niwclear' sy'n cofio am feddygaeth niwclear a'r defnydd cynyddol o deunydd ymbelydrol mewn meddygaeth.

Yn y gwaith gwreiddiol yn 1972[5] roedd Meadows a'i chydweithwyr yn gofalu trafod cyfyngiadau sy'n codi yn sgil ffynonellau a suddfannau. Yr hyn sy'n ddiddorol yw nad oeddent wedi rhoi fawr ddim sylw i allyriadau CO_2 a achosir gan ddynoliaeth (h.y. effaith suddfannau) oedd yn deillio o'r defnydd o adnoddau adnewyddadwy, nwy, olew a glo, o'u cymharu â dyfodol yr adnoddau hynny. Ond erbyn hyn mae'n greulon o amlwg bod y ffynonellau a'r suddfannau ill dwy yn broblem. Pan fyddwn yn ystyried hydrocarbonau a chymhwyso data heddiw i feini prawf Daly mae'r sefyllfa'n arswydus. Byddai'n rhaid i allyriadau'r prif nwyon tŷ gwydr drwy'r byd ddychwelyd i lefelau 1860-70: tua 270 ppm CO_2 [o ~390 ppm ar hyn o bryd], CH_4 i ~700 ppb [o ~1800 ppb ar hyn o bryd] a N_2O i 270bpm [o 320 ppb]. Nid oes unrhyw argoel y bydd hyn yn digwydd. Mae'r problemau dyrys hyn yn deillio o'r cynnydd cyflym a chyson mewn allyriadau a'r croesi ffiniau a'r oediadau amser yn y system.

Mae'r allyriadau CO_2 sy'n ganlyniad i'r galw am ynni rhad trwy'r byd yn ddarlun byw o broblem 'terfynau' boed y rheini'n rhai geoffisegol neu rai geowleidyddol. Mae gwleidyddion bron pob un wlad yn y byd yn gytûn y dylem ymdrechu i osgoi cynnydd yn nhymheredd ein planed o fwy na 2°C ar draws y byd; mae hyn yn cyfateb yn fras i lefelau nwyon tŷ gwydr yn yr atmosffer o ~450ppm CO_2e[17]. Gan fod lefelau CO_2e eisoes wedi cyrraedd ~425 ppm (fel y nodwyd, rhyw 390 ppm ar ffurf CO_2, 1800 ppb CH_4 a 320 ppm N_2O, y mae'n rhaid eu cywiro'n unol â photensialau niweidiol [forcing potential] dros gyfnod o 100 mlynedd), rydym yn siŵr o fynd dros y trothwy 450 ppm. Y broblem ffisegol yw, gan fod CO_2 yn aros yn yr atmosffer am sawl degawd, y llwyth CO_2 cronedig sy'n allweddol[21][22][23]. Os ydym yn parhau i allyrru nwyon tŷ gwydr ar y cyfraddau presennol, yna bydd llai o le i allyriadau isel nad ydynt yn niweidiol mewn degawdau i ddod. Yn 2010 cyrhaeddodd allyriadau CO_2 drwy'r byd uchafbwynt o 35Gt; cynnydd o 6 % mewn blwyddyn, mwy nag erioed o'r blaen (er gwaethaf Cytundeb Kyoto a'r holl leisiau a godwyd i ddatgan eu hymrwymiad gwleidyddol i leihau allyriadau). Hunanoldeb a diffyg gweledigaeth ac ofn dryllio 'bargen Galbraith' sy'n gyfrifol am amharodrwydd y gymuned ryngwladol i weithredu. At hynny mae rhywfaint o ansicrwydd ymysg gwyddonwyr e.e. mae dweud bod 450 ppm CO_2e yn cyfystyr â chynnydd o ~2°C yn y tymheredd yn dibynnu ar y 'sensitifrwydd hinsoddol byd-eang'[17] (hynny yw sut byddai'r tymheredd cymedrig trwy'r byd yn codi oherwydd bod nwyon tŷ gwydr yn yr atmosffer wedi dyblu o'r lefelau cyn-ddiwydiannol ac mae'r ffigyrau posib yn amrywio o 4.5 i 1.5°C gyda ~3°C yn ganolrif.) Mae hyn ynddo ei hun yn rhoi digon o gyfle i rai sy'n gwarchod eu buddiannau personol i hau amheuaeth, er bod y tueddiadau ar draws y byd yn ddigamsyniol. Hefyd rydym ni fel bodau dynol yn cael anhawster mawr i wahaniaethu rhwng ein tywydd lleol hynod gyfnewidiol a thueddiadau byd-eang yn yr hinsawdd, neu i ganfod tueddiadau dros gyfnod hir.

Yn hanesyddol y gorllewin fu'n gyfrifol am y rhan helaethaf o'r allyriadau ac mae ein bywyd bras ynghlwm wrth danwyddau hydrocarbon rhad. Felly nid yw'n syndod bod gweddill y byd yn disgwyl i ni arwain y ffordd a thorri ar ein hallyriadau gyntaf wrth iddynt ruthro i ddal i fyny gyda ni. Ond wrth reswm mae ein harweinwyr gwleidyddol yn hwyrfrydig i ymyrryd â'r fargen 'cyfoeth' er gwaetha'r peryglon. Ond mae diffygion hefyd yn safbwynt rhai gwledydd yn y byd sy'n datblygu a gallent fod yn ddi-weld ac yn gwneud niwed iddynt eu hunain. Mae gwledydd fel India a Tsieina ymysg y rhai fydd yn dioddef waethaf o ganlyniadau i'r newidiadau a ragwelir. Erbyn 2010 roedd allyriadau CO_2e ar draws y byd yn 6 i 7 tunnell *per capita*[24]; rydym ni yng Nghymru'n cynhyrchu tua 17 tunnell o allyriadau'r un. Ond mae Tsieina eisoes wedi cyrraedd y cymedr byd eang y pen ac ni ellir bellach ei chyfrif ymysg y dieuog. Ymysg y dieuog y mae trigolion ynysoedd cwrel isel y Môr Tawel sydd yn creu lefel isel o allyriadau carbon ond sydd yn debygol o ddioddef fwyaf pan fydd lefel y môr yn codi. Mae'r 'rhagamcaniadau' gorau yn awgrymu bod yn rhaid i ni dorri allyriadau ar draws y byd i ryw 1.5 i 2 dunnell y pen erbyn 2050[21][23] os derbyniwn 'ragamcaniadau' poblogaeth y Cenhedloedd Unedig. Ychydig ohonom, felly, a all ddianc rhag eu siâr o'r cyfrifoldeb.

O! Y gair 'rhagamcaniadau' – mae a wnelo gwleidyddiaeth go iawn â'r presennol. Mae marchnadoedd ariannol yn delio ag elw fesul eiliad; mae busnesau'n edrych ymlaen at y flwyddyn neu'r ychydig o flynyddoedd i ddod. Sut y gall system o'r fath ymdopi â 'rhagamcaniadau' 30 neu 40 blynedd neu ddelio â thebygolrwyddau gwyddonol lle erys ychydig o ansicrwydd o leiaf. Mae'n anorfod bod y cyfuniad hwn yn esgor ar oedi wrth ymateb a chroesi'r ffin.

Wynebwn broblemau dybryd hefyd mewn perthynas â chyflenwi. Er bod y cysyniad o 'olew brig' yn bwnc llosg, mae'r holl ragolygon yn awgrymu y bydd prisiau olew a nwy yn parhau i godi o'r ~$110 cyfredol fesul casgen o olew crai Brent, er gwaetha'r ffaith bod Gogledd America ac Ewrop ar ddibyn dirwasgiad. Wrth chwilio am olew rydym yn treiddio'n ddyfnach ac yn ddyfnach i'r cefnforoedd ac i diriogaeth fwy problemus: mae maes olew Lula oddi ar arfordir Brasil 7 km o dan wyneb y môr. Mae ffracio am olewau tar nid yn unig yn debygol o fod yn niweidiol i'r amgylchedd ond mae hefyd yn ddrud. Mae'r cylchgrawn y New Scientist yn dyfynnu Paul Horsnell, cyfarwyddwr nwyddau Barclays[25] a ragfynegodd y byddai olew yn $137 y gasgen erbyn 2015 a $185 erbyn 2020; mae eraill yn anghytuno wrth gwrs. Mae nifer o economegwyr wedi cysylltu twf economaidd siomedig, yn ystyr traddodiadol y GDP/CMC, â phrisiau olew uchel. Mae Kopits (gweler[25]) wedi awgrymu pan fydd mwy na 4.5% o GDP Gynnyrch Mewnwladol Crynswth yr Unol Daleithiau yn cael ei wario ar olew, bod dirwasgiad yn dilyn; yn yr Unol Daleithiau mae hynny'n cyfateb i $90 y gasgen! Hwyrach mai'r newyddion da fydd bod y mecanwaith prisio'n mynd i orfodi llywodraethau, cwmnïau ac unigolion i symud i ffwrdd yn gyflym o hydrocarbonau olew a nwy: Byddai'n newyddion drwg pe bai gwledydd oherwydd hyn yn rhuthro i ddefnyddio glo, tanwydd llawer llai effeithlon o safbwynt y nwyon tŷ gwydr a gynhyrchir fesul uned kW, heb rwymo carbon. Nid yw technoleg rhwymo carbon wedi ei phrofi eto a gallai newid y sefyllfa'n llwyr wrth gwrs ond nid oes dwywaith na fyddai'n ychwanegu at y costau!

Erys y cwestiwn heb ei ateb; a fydd newid trwy gyfrwng mecanwaith prisiau neu drwy orfodaeth yn digwydd yn ddigon cyflym? Mewn termau ffisegol ni ellir gorbwysleisio pwysigrwydd oediadau amser. Mae gweithredu mewn ffordd dila, sy'n fwy tebygol yn dilyn uwchgynadleddau Copenhagen a Durban, gyda lefelau CO_2e yn debygol o fod yn uwch na 550ppm erbyn 2050, yn hynod o anghyfrifol a pheryglus. Mae'n debygol y ceir adweithiau cyflym positif sy'n ei gwneud yn bosib y bydd newidiadau trychinebus mewn byd ansefydlog gyda phoblogaeth o 9 i 10 biliwn. Er y gall hyn ddigwydd ymhen degawdau byddem yn wynebu rhyddhau symiau aruthrol o CH_4 o'r twndra a chynnydd o ~6 medr yn lefel y môr o ganlyniad i ddadmer rhan helaeth o'r rhew sy'n gorchuddio'r Las Ynys a chynnydd cymedrig o 3 i 4°C yn y tymheredd. Beth ddaw wedyn o Bangladesh neu ddelta'r Nil neu ardal y Ffens neu yn wir, Llundain neu Efrog Newydd, ill dwy wedi eu codi ar dir isel? Gwaeth fyth fyddai colli unrhyw gyfran o'r rhewlifau sy'n toddi bob blwyddyn ac yn bwydo afonydd mawr yr Himalayas sy'n cynnal biliynau o bobl yn India, Tsieina, a gweddill de a dwyrain Asia.

5 / Gwendid y ddynoliaeth

Mae'n nodweddiadol o'r ddynolryw ein bod o bryd i'w gilydd yn anonest, braidd yn rhagrithiol ac yn gallu twyllo a chyfiawnhau ein hunain. Mae arbrofion seicolegol yn dangos bod cyfyngiadau ar ein honestrwydd a'i fod yn darfod yn llwyr os ydym yn credu nad oes modd ein cosbi ac/neu os ydym yn mwynhau grym, neu, o leiaf, cyfle i ddianc rhag canlyniadau ein gweithredoedd[26]. Y cymdeithasau lleiaf llwgr yw'r rheini lle mae'r dinasyddion yn agored, yn wybodus, wedi cael addysg, ac yn mwynhau rhyddid gwybodaeth megis yng nghymdeithasau cymharol egalitaraidd, sosioddemocrataidd a rhyddfrydol gwledydd Llychlyn. Y gwaethaf yw cymdeithasau toredig Irac, Afghanistan, Burma, Somalia, tra bod y Deyrnas Unedig a'r Unol Daleithiau rywle yn y canol. Mae llwgrwobrwyo'n ffenomen gymdeithasol sydd yn rhan annatod o wledydd llwgr ac unigolion sydd wedi eu geni mewn cymdeithas o'r fath sydd fwyaf tebyg o ymddwyn felly, hyd yn oed os symudant i fyd newydd. Mae'n dilyn y dylem boeni fwyaf am ymddygiad pobl bwerus a chyfoethog. Mae'n beth arswydus mai'r rhai cyfoethocaf, a allai gredu bod ganddynt fwy i'w golli'n bersonol o fyd mwy cynaliadwy, yw'r rhai sydd â'r gallu mwyaf i ledaenu gwybodaeth gamarweiniol, prynu dylanwad gwleidyddol a buddsoddi mewn technolegau amhriodol. Gan fod y natur ddynol fel ag y mae, mae'n siŵr y gallant hefyd argyhoeddi eu hunain eu bod yn gweithredu'n unplyg a hyd yn oed er budd y cyhoedd. Ar y llaw arall, mae llawer o dystiolaeth seicolegol a chymdeithasegol bod bodau dynol yn meddu ar ymdeimlad byw â thegwch a chyfiawnder[27]. Mewn gwledydd lle anogir y nodweddion hyn ymddengys fod pobl yn mwynhau gwell iechyd a mwy o foddhad.

Mae'n amlwg bod y gyfundrefn bresennol yn ein byd yn ansefydlog yn economaidd, yn gymdeithasol, yn amgylcheddol ac yn wleidyddol ac yn fwy na thebyg mae'n anghynaliadwy. Wrth geisio cynnal a pharhau cyfundrefn o'r fath am y 50-100 mlynedd nesaf mae bron yn sicr y byddwn yn mynd ymhell bell y tu hwnt i derfynau'r blaned hon. Efallai'n wir ein bod ni wedi croesi'r ffin eisoes ond bod yr oediad amser yn celu'r effaith. Cyfrifwyd[28] fod ein gofynion cyfredol yn cyfateb i adnoddau un Ddaear a hanner. Yng Nghymru byddai angen tua thair gwaith arwynebedd ein gwlad i'n cynnal[29].

Hyd yn oed yn y senario hwn rydym yn moesymgrymu i rai sy'n 'creu cyfoeth' (er nid yw yn achos bancwyr a llawer o arianwyr am amlwg paham). Rhagdybiwn y bydd rhyw fath o ddatblygiad 'diferu-i-lawr' yn gwella safonau byw'r mwyafrif (y 99% yw'r enw a roddir arnynt yn yr Unol Daleithiau!) Eto mae'r polisi hwn wedi bod yn neilltuol o aneffeithiol ac ae wedi arwain at anghysonderau mawr o ran incwm a chyfoeth. Er bod y Deyrnas Unedig a'r Unol Daleithiau'n mwynhau 'hawddfyd', nid yw amodau byw'r tlodion a llawer o'r dosbarth canol yn y gwledydd hynny, cefnogwyr selocaf cyfalafiaeth y farchnad rydd ddilyffethair, wedi gwella fawr ddim os o gwbl mewn 40 mlynedd. Nid yw disgwyliadau Mrs Jones, Llanrug, sy'n byw ar ei phensiwn a hen wragedd tebyg iddi sy'n byw ar gymorthdaliadau ym mynyddoedd Maloti Lesotho, neu Abdul yn ei bentref pysgota yn Bangladesh, yn uchel ac mae eu hôl troed carbon yn isel. Ond er eu bod yn gymharol ddieuog o'u cymharu â'r cyfoethogion neu hyd yn oed wyddonwyr enwog, maent yn fwy bregus yn wyneb costau cynyddol tanwydd a bwyd, a llanw uwch! Mae cyfoeth wedi llifo i gyfeiriad elit breintiedig sy'n prynu pŵer a dylanwad gwleidyddol, ac yn rhy aml yn pedlera hanner gwirioneddau a gwybodaeth gamarweiniol am faterion amgylcheddol a chymdeithasol, gan osgoi talu trethi ac yn anwybyddu hyd yn oed eu cyfrifoldebau cyfreithiol. Sut yn y byd y gall hyn barhau?

Byddai'n wirion peidio â chydnabod apêl y diwylliant prynu a 'therapi adwerthu', hwylustod ac apêl yr archfarchnad neu beidio â sylweddoli cryfder ein hymrwymiad i frand penodol o 'fyw'n fras'. Mae pobl yn meddwl am wyliau yn yr haul bron iawn fel pe baent yn hawl ddynol; hawdd gweld pam eu bod yn bwysig i drigolion hen drefi'r melinau neu'r pentrefi glofaol sydd bellach yn farwaidd. Mwy fyth os ydynt yn rhygnu byw ac yn gweithio am gyflogau isel yn nhywydd digysur gaeafol gogledd Ewrop. Fel y bydd y cwmnïau awyrennau rhad yn tystio, bydd gwrthwynebiad chwyrn i unrhyw awgrym y dylai prisiau tocynnau adlewyrchu'r gost amgylcheddol lawn – byddai hynny'n peryglu'r 'fargen'! Efallai ein bod yn anfodlon ar ein materoliaeth ac yn dioddef o 'Affluenza'[30] ond mae'r cyflwr yn rhan annatod o'n cymdeithas ac ni fyddai'n hawdd ei ddileu – rydym yn eithaf mwynhau ein cyflwr 'bregus'. Hwyrach ein bod yn ymwybodol o rymoedd anhrefn sydd ar waith yn ein cymdeithas, cymunedau sydd wedi cael eu gwthio i'r cyrion ac wedi colli gobaith, ond rydym wedi ein cipio gan lanw'r diwylliant prynu, yn mwynhau ein gwyliau tra medrwn. Allwn ni ddychmygu unrhyw ffordd arall o fyw?

Yn eu llyfr "How much is enough? – The love of money and the case for the good life"[31] mae'r Skidelsky's, gan ddilyn Keynes, yn cyffelybu cyfalafiaeth i fargen Faustaidd, un lle mae dynoliaeth wedi taro bargen gyda trachwant ac usuriaeth (yn gwbwl groes i Cristnogaeth cynnar a moesoldeb gwar) er mwyn sicrhau golyd. Disgwyliad Keynes oedd i glymau'r fargen llacio ac y byddai technoleg a thwf mewn cyfalaf erbyn y diwedd yr 20fed canrif yn ein galluogi i fwy yn fras ac y waraidd ac i ymatal ar fargen. Nid felly y bu. Bach iawn bu y lleihau mewn oriau gwaith, mawr iawn bu'r cynnydd yn ein chwantau am nwyddau. Mwy fyth dyfeisgarwch y dosbarth cyfalafol i sicrhau eu buddiannau ei hunain ar drael y gweddill. Nawr ceir son barhaus am 'gystadleuaeth' ac am 'gynnydd' – y rat race bondigrybwyll. Er nad yw y Skidelsky's yn trafod yn digonol y ffiniau amgylcheddol rwyf yn ei amlinellu, maen't yn dadlau dros y 'bywyd da' – 'y bywyd gwar' – ni ellir ei 'ennill' trwy cystadleuaeth cyfalafol di-diwedd.. Mae eu dadleuon yn rhai moesol ac athronyddol. Ceisiant ddefinio nodweddion y 'bwyd da' – fel iechyd, diogelwch, parch, personoliaeth/ human-rheolaeth, cyt-gord a natur, cyfeillgarwch a hamdden (nid yn yr ystys o segura ond amser a gofod i ddatglygu diddordebau a sgiliau er eu mwy eu hunain). Nid oes gofod yn wneud cyfiawnder a syniadaeth y Skidelsky's ond teg, yn fy marn, yw honi bod eu diffiniad yn cydweddu ag amcanion cynaliadwyedd ac i'r mae ddwy weledigaeth cydgerdded yn dwt.

6 Atebion

Ni waeth pa mor ddu yw'r rhagolygon, nid yw'n dilyn yn anorfod y bydd trigolion y blaned hon yn dilyn rhai Ynys y Pasg, Palmyra neu lawer o wareiddiadau eraill, i ddifodiant.[32] Bydd rhai, llawer efallai, yn encilio i swigen bersonol gan obeithio y gallan nhw neu eu teulu agosaf ddianc neu brynu eu ffordd allan. Bydd eraill yn colli gobaith ac yn dadlau na ellir gwneud dim. Mae'r agwedd 'mae ar ben arnom' yn peri loes ond yn ddealladwy. Yn sicr os caniateir i'r tueddiadau presennol barhau heb geisio eu hatal y canlyniad fydd trychineb amgylcheddol ac anhrefn enfawr mewn cymdeithas. Ni fydd yr un wlad neu rhanbarth yn dianc yn groeniach. Rhagwelir y bydd gogledd orllewin Ewrob yn osgoi'r bygythiadau ffisegol gwaethaf, ond serch hynny mae'n amhosib dychmygu yn ein byd rhyng-gysylltiedig y byddwn ni a'r cenedlaethau i ddod yn dianc yn ddianaf rhag anhrefn gymdeithasol ac economaidd i fwynhau bywyd ar ynys dawel a heddychlon. Mae'n debygol iawn ein bod yn agosáu, er gwaethaf yr oediadau amser, at argyfwng 'terfynau' – yn sicr o fewn oes ein hwyrion os nad cyn hynny. Os digwydd hynny, bydd y ddaear yn gorfodi ei rhesymeg ei hun ar y ddynolryw ac ar les y ddynolryw. Rydym yn feistri ar dwyllo ein hunan ond ni allwn 'dwyllo'r' cylchoedd biogeocemegol sy'n rheoli bywyd ar ein planed. Os bydd chwalfa yn dilyn ein hesgeulustod, mewn amser daw adfywiad. Ond yn fy marn i, nid 'terfynau' yw'r brif ystyriaeth ond p'un a allwn ganfod llwybr at ddyfodol cynaliadwy mewn ffordd ddyngarol, a (chymharol) drefnus neu a fydd hyn yn digwydd mewn anhrefn ddidostur (cf. proffwydoliaethau tywyll Lovelock,[2]).

Bydd yn rhaid i ni weithio'n galed iawn i ddod ohoni'n gymharol ddianaf. Pe bai'r ddynolryw yn oedi am ennyd i ystyried hyn oll gyda'n gilydd, byddem i gyd ar ein hennill o leihau ein heffaith ar yr amgylchedd. Ond rhaid i ni holi ein hunain a ydyw gosod y 'budd cyffredin' o flaen buddiannau cul ein sector neu ein gwlad o fewn ein gallu. Oes ffyrdd o'n cymell i ni wneud hyn?

I ddyfynnu Gramsci, rhaid i ni harneisio 'optimistiaeth yr ewyllys' er gwaethaf 'pesimistiaeth y deall'.

Mae'n hollbwysig ein bod yn gwneud hyn ond sut? Gellir cynrychioli effaith amgylcheddol pobl fel:

Effaith = Poblogaeth x Cyfoeth/Golud x Technoleg[33]

Mae pob un o'r ffactorau'n bwysig ac yn haeddu ei ddadansoddi o ddifrif.

Poblogaeth

Ystyrir y boblogaeth yn gyntaf gan ganolbwyntio ar enghraifft benodol allyriadau nwyon tŷ gwydr; mae hyn yn ddigon teg o gofio mor ganolog yw ynni i'n heconomi ac i'n bywydau personol. Mae'n gwbl amlwg, er nad yw gwleidyddion, hyd y gwela i, byth yn fodlon cyfaddef hynny, po fwyaf yw'r boblogaeth ddynol, mai isaf i gyd yw'r lefel allyriadau fesul pen sydd yn galluogi cyfyngu newid yn yr hinsawdd fyd eang i oddeutu +2°C (neu unrhyw darged arall yn wir). Hefyd bydd yn fwy anodd cyrraedd targed o'r fath. (Mae'r un peth yn wir, a'r un mor amlwg, am y galw am adnoddau megis bwyd a dŵr (gweler hefyd Ffigyrau 2 a 4)). Fel y nodwyd uchod, mae cyfanswm allyriadau 2010, sef tua 45 Gt CO_2e yn cyfateb i 6-7 tunnell y pen, ac mae'r cyfanswm yn codi'n gyflym. Ond erbyn 2050 bydd angen lleihau cyfanswm byd-eang allyriadau i oddeutu 15 i 18 G tunnell. Os bydd y boblogaeth wedi cyrraedd 9 i 10 biliwn, fel y rhagwelir, mae hyn yn hafal i 1.5 i 2 dunnell yr un. Ar hyn o bryd mae pob unigolyn yng Nghymru ar gyfartaledd yn gyfrifol am ryw 17 tunnell o CO_2e felly rhaid i ni edrych ar gwtogi ~90%, a bwrw bod y boblogaeth yn aros yn sefydlog ar ~3 miliwn. O'r safbwynt hwn, mae o fudd enfawr i bawb bod twf yn y boblogaeth yn cael ei atal; yn lleol, yn genedlaethol ac yn fyd-eang. Os yn bosib ni ddylai'r boblogaeth fyd-eang fynd dros 8 i 9 biliwn, er gwaethaf dysgeidiaeth y pab, arweinwyr Islamaidd a chredinwyr selog yng ngalluoedd gwyrthiol cyfalafiaeth ddilyffethair.

Mae profiad yn awgrymu mai'r ffordd orau o sicrhau hyn yw trwy addysgu, grymuso a rhyddhau merched er mwyn idynt gyfrannu o'i talentau a rheoil a lifeiriant/enedigaethau plant; sefyllfa sydd o fudd aruthrol i bawb ond y mwyaf adweithiol. Yng Nghymru neu'n rhyngwladol, mae poblogaeth gynyddol yn creu anghenion ychwanegol e.e. am drydan, tai ac adnoddau a chyfleusterau eraill a gall hwyluso twf yn y CMC/GDP. Yn aml iawn mae dinasoedd poblog yn ysgogi twf economaidd, tra bod diboblogi cefn gwlad yn creu lliaws o broblemau. Ond dim ond lleihau ein siawns o gyrraedd hyd yn oed y targed swyddogol o ostyngiad o 80% mewn nwyon tŷ gwydr erbyn 2050 wnaiff y twf hwn. Mae hyn y broblem gyfarwydd erbyn hyn! Serch hynny mae'n rhyfeddol bod gwlad ddatblygedig megis y Deyrnas Unedig neu Gymru i'w gweld yn analluog i gynhyrchu gweithwyr proffesiynol medrus a diwydianwyr profiadol ac yn dibynnu ar fewnforio arbenigedd meddygol a thechnegol a nyrsys a gofalwyr a phlymwyr a thechnolegwyr gwybodaeth a hyd yn oed fancwyr a masnachwyr o wledydd eraill sydd yn aml yn llawer tlotach.

Mae atal twf yn y boblogaeth yn creu goblygiadau economaidd a chymdeithasol dwys i genhedloedd unigol ac ar draws y byd yn enwedig os bydd pobl hefyd yn byw'n hirach. Bydd canran yr ifanc a'r rheini sydd mewn oed gweithio'n gostwng a chanran yr henoed yn codi. Bydd disgwyl i bobl yn eu 50au a'u 60au weithio am gyfnod tipyn hirach, a hynny o bosib yn atal yr ifanc a'r canol oed rhag cael swyddi. Bydd pensiynau'n broblem enfawr, yn enwedig os bydd gwerth y farchnad stoc yn aros yn ei unfan, fel sy'n debygol o ddigwydd os bydd bargen Galbraith yn dadfeilio. Ond efallai fod y rhagolygon mwyaf digysur yn gysylltiedig â'r disgwyliad oes hirach a fydd yn mynd y tu hwnt i henaint iach ac egnïol. Felly bydd nifer yr henoed fydd angen gofal a chefnogaeth yn cynyddu'n gynt na nifer y rhai dros 70 a'r cyfnod gofal ei hun yn ymestyn. Yr ateb hanesyddol yn y Deyrnas Unedig fu mewnforio llafur, yn enwedig llafur yr ifanc i adeiladu, nyrsio a gofalu. Nid yw hyn yn cyd-fynd yn hwylus â datblygu cynaliadwy oherwydd ei fod yn amddifadu cymdeithasau eraill o'u mewnbwn sgiliau. Ar y llaw arall, mae'n faes lle gellid creu llawer o swyddi a hybu lles pobl am gost amgylcheddol isel pe bai modd talu cyflog rhesymol i'r gweithwyr hyn.

Technoleg

I sicrhau cyfundrefn fwy cynaliadwy a sefydlog, rhaid wrth fenter a chynlluniau technolegol. Mae rhai yn hiraethu am ateb 'syml' – dychwelyd at gymunedau bach hunangynhaliol. Mae'n bosib bod hyn wedi gweithio (yn rhannol) cyn 1800 pan oedd y boblogaeth fyd-eang yn llai na biliwn ond pan fydd yn ddeng gwaith hynny yn 2050, bydd problemau enfawr yn codi. Ychydig o bobl fyddai am ddychwelyd i gyfnod mewn hanes pan oedd y mwyafrif mawr yn byw mewn budreddi. Ein huchelgais y pryd hynny oedd dianc rhag arswyd crafu byw; gallai trychineb mawr achosi sefyllfa o'r fath ond ni fyddai neb yn ei ddewis. Nid yw'n glir chwaith a fyddai cadw'r gadwyn fwyd yn lleol neu ddibynnu yn unig ar gynnyrch lleol, pe bai hyn yn ymarferol bosib, yn sicrhau'r ôl troed carbon isaf. Pe bai Cymru neu ei rhanbarthau'n ceisio bod yn hunangynhaliol mewn grawn, er enghraifft, byddai'n rhaid aredig glaswelltir a byddai hynny'n codi allyriadau CO_2e yn sylweddol am ddegawdau (gweler 34, 35 sy'n trafod hyn). Serch hynny, mae yna le i fwy o gyfnewid nwyddau'n lleol, gwella cadwynau bwyd ac ailgyflwyno'r syniad o fwyd tymhorol ond y maen prawf o reidrwydd fydd ôl troed carbon isel ac economi gynaliadwy, ac nid rhamant.

Rhaid i ni roi'r flaenoriaeth i newid a dyfeisio technolegol ac i gwmnïau sy'n lleihau'r effaith ar yr amgylcheddol. Yn baradocsaidd ddigon, er mai cyfalafiaeth y farchnad rydd sydd wrth wraidd ein cyfoeth byrhoedlog a'n problemau amgylcheddol, rhaid i ni geisio harneisio ei hegni i greu'r arfau i dynnu'r ddynoliaeth o'r twll a gloddiodd iddi ei hun. O gofio am y gallu sydd gennym i dwyllo ein hunain ac eraill nid yw hyn yn hawdd a rhaid gosod meini prawf llym, a'u harchwilio'n rheolaidd.

Ond mae'n werth deall rhesymeg twf esbonyddol. Mae economi Tsieina wedi bod yn tyfu ar gyfradd ychydig llai na 10% y flwyddyn h.y. dyblu mewn 8-9 mlynedd. Bellach dyma weithdy'r byd. Hyd yn oed os llwyddir i raddau sylweddol i ddatgysylltu twf ac adnoddau, gellir disgwyl i'r galw am ynni hydrocarbon neu gopr neu ddeunydd crai eraill ddyblu ymhen 12 i 18 mlynedd. Hyn sydd wrth wraidd eu polisi o brynu cynifer o ffynonellau deunyddiau crai yn Affrica a De America â phosib. Mae'n amlwg nad yw Tsieina yn diystyru cyfyngiadau ar adnoddau nac yn rhannu barn y farchnad rydd pur y bydd dyfeisgarwch technolegol y ddynolryw yn siwr o ddod o hyd i ddeunyddiau neu ddulliau amgen.

Mae ynni'n ganolog. Roedd Meadows a'i chydweithwyr yn llawn sylweddoli pwysigrwydd harneisio technoleg i gynhyrchu ynni sy'n isel o ran nwyon tŷ gwydr, i symud tuag at economi yn seiliedig ar hydrogen [H2] a thrydan di-garbon gan ddefnyddio llai o ynni a'i ddefnyddio'n fwy effeithlon. Gadewch i ni atgoffa'n hunain mae'r hyn sydd o dan sylw yn y sector hwn yw lleihau allyriadau cymedrig **fesul pen** drwy'r byd ym mhob is-sector sy'n cynhyrchu ac yn defnyddio ynni (awyrennau, llongau, cerbydau, gwresogi gofod, prosesau diwydiannol etc), cynhyrchu sment a'r gadwyn fwyd gyfan (yn cynnwys CH_4 a N_2O) o ~6-7 tunnell yn 2010 i 1.5 i 2 dunnell erbyn 2050. Mae angen Cynllun Marshall newydd i'r holl fyd. A ydym braidd yn ddiniwed yn tynnu sylw at y ffaith bod rhyw \$1.5 triliwn o ddoleri yn cael eu buddsoddi bob blwyddyn ar draws y byd mewn arfau a lluoedd milwrol [mae'r Unol Daleithiau'n unig yn gwario bron i \$700 biliwn?] a bod trafodion cyfnewid tramor (hapfuddsoddi gan fwyaf) yn dod i \$955 triliwn? Oni fyddai'r byd yn llawer mwy diogel a thecach pe bai cyfran sylweddol yn cael ei buddsoddi mewn technolegau i wella cynaliadwyedd amgylcheddol fel mae Daly yn ei ddiffinio? Oni fyddai'n creu gwaith llawer mwy gwerthfawr ac a fyddai'n rhoi llawer mwy o foddhad? A allwn greu cyfundrefn amgylcheddol-ddiwydiannol sydd yr un mor rymus â'r gyfundrefn filwrol-ddiwydiannol?

Dyma'r cwestiwn pwysicaf: A ydym yn cymryd 'terfynau' o ddifrif; gan gynnwys holl adnoddau adnewyddadwy ac anadnewyddadwy'r ddaear a pha mor gryf yw'r suddfannau a faint o amser sydd ei angen i wasgaru'r llygredd? Pe baem yn gwneud hynny, byddwn yn ddigon parod i ymuno â'r garfan obeithiol a byddwn yn ffyddiog y bydd gwyddoniaeth, technoleg a dyfeisgarwch y ddynolryw'n ein galluogi i oresgyn y cyfnod hwn mewn modd trefnus. Ond er mwyn gwneud hynny rhaid wrth baradeim newydd, ailfeddwl ac ailwampio ein systemau gwleidyddol ac ariannol yn llwyr a chydweithio ar draws y byd mewn modd nas gwelwyd ei debyg o'r blaen. Byddai angen i'r system newydd osod 'terfynau' biogeocemegol fel y maen prawf allweddol ar gyfer asesu pob un gweithgarwch, nid ystyriaeth wrth basio. Dyna'r maen tramgwydd!!! A ellid wynebu newid o'r fath?

Cyfoeth/Golud

Ychydig sydd o blaid defnyddio GDP/CMC[36] fel dull o fesur lles pobl neu hyd yn oed lwyddiant economaidd personol. Roedd ei grëwr yn cydnabod hyn[34]. Mae GDP/CMC yn hafal i swm defnydd preifat, buddsoddi gros, gwariant y llywodraeth ac allforion llai mewnforion, o fewn ardal benodol e.e. Y Deyrnas Unedig neu Gymru. Mae ei ddiffygion yn amlwg. Wedi'r cwbl pe bai rhywun yn symud 30 milltir yn bellach o'r gwaith i dŷ mawr drafftiog oedd yn costio'n ddrud i'w wresogi byddai'n cyfrannu mwy at y GDP/CMC ond ar yr un pryd yn gostwng ei safon byw ei hun. Gallai mynd i'r carchar wneud cyfraniad pellach, yn enwedig os oedd yn golygu bod y wladwriaeth neu fenter breifat yn adeiladu un newydd. Os yw clirio'r goedwig mewn gwahanfa ddŵr yn uwch i fyny'r afon yn creu llifogydd yn is i lawr, mae hawliadau yswiriant drud a buddsoddi mewn cynlluniau mawr i reoli llifogydd i gyd yn cyfrannu i raddau gwahanol at y GDP Cynnyrch Mewnwladol Crynswth. Er gwaethaf ei anallu i wahaniaethu, ymddengys fod cydberthynas fras rhwng y GDP/CMC a gallu'r economi i greu swyddi, cynhyrchu incwm o drethi a chynnal busnes y llywodraeth. Nid yw fodd bynnag yn unrhyw arwydd o gynaliadwyedd. Ar wahân i'r ffaith ei fod yn gwbl ddall i les pobl, ac yn diystyru allanolion amgylcheddol boed yn ansawdd aer yn Lerpwl, colli bioamrywiaeth, risg llifogydd neu newid yn yr hinsawdd ni fydd y GDP/CMC yn newid ond pan fydd pris prynu eitem yn newid neu pan fuddsoddir arian. Gellid honni ei fod wedi ei gynllunio i annog croesi'r ffin.

Gwnaed gwahanol ymdrechion i addasu'r GDP/CMC[36] i adlewyrchu'n well y byd go iawn megis y Mynegai Lles Economaidd Cynaliadwy (ISEW), y Dangosydd Cynnydd Go Iawn (GPI) a System Cyfrifo Amgylcheddol ac Economaidd y Cenhedloedd Unedig. Er enghraifft, yn yr ISEW Mynegai Lles Economaidd Cynaliadwy caiff defnydd personol (fel y mesurir ef gan y GDP/CMC uchod) ei addasu trwy ychwanegu gwariant cyhoeddus ac eithrio gwariant ar amddiffyn ond fe gaiff ei leihau gan wariant preifat ar amddiffyn e.e yr angen i dalu am blismona preifat neu osod systemau Teledu Cylch Cyfyng neu larymau lladron. Ychwanegir ffurfiant cyfalaf (capital creation) felly hefyd gwasanaethau llafur domestig (a anwybyddir yn y GDP/CMC) a thynnir o'r cyfanswm amcangyfrifon o ddiraddiad amgylcheddol a dibrisiant 'cyfalaf naturiol'. Mae'r ffordd y caiff 'cyfalaf naturiol' e.e. cronfeydd olew, pridd neu fioamrywiaeth eu trin fel incwm yn un o'r prif feirniadaethau a danlinellir gan Schumacher yn ei lyfr enwog, '*Small is Beautiful*'[39].

Mae'n amlwg o'r paragraffau hyn bod unrhyw symud i ffwrdd o'r GDP/CMC, er gwaethaf ei ddiffygion amlwg yn her wleidyddol a deallusol mawr. Mae gwleidyddion, yn enwedig mewn democratiaethau lle mae gyrfaoedd yn fyrhoedlog, yr un mor selog dros y GDP/CMC ag y mae ein cymdeithas dros ynni rhad! Mae llawer o'r mesuriadau a'r data ychwanegol sydd eu hangen yn anodd eu cael ac yn ddadleuol o bosib. Oni fyddai gwahanol garfannau'n rhoi gwerth gwahanol iawn ar dyrbinau gwynt hyd yn oed pe cytunid ar bris am y trydan ac unrhyw ostyngiadau mewn allyriadau CO_2? Mae'n hawdd dychmygu pleidiau gwleidyddol yn benben â'i gilydd ac yn cyflwyno gorymdaith o ddadleuon sydd yn aml yn ddi-sail a'r cyhoedd, sydd yn naturiol ddrwgdybus, yn meddwl tybed a yw'r holl beth yn dwyll llwyr. Efallai fod ein hobsesiwn efo rhoi gwerth ariannol ar bob dim yn symptom o'n hafiechyd? Ai rhan o'r broblem yw ein bod yn swcro'r agwedd hon? Serch hynny gellir datgan yn hyderus na ellir gwneud datblygu cynaliadwy yn 'brif egwyddor drefniadol'[40] os cedwir y GDP/CMC fel mesur datblygiad economaidd ac os bydd ei dwf yn parhau'n amcan polisi hanfodol.

Tegwch sy'n bwysig. Mae cyfoeth wedi ein gyrru i geisio efelychu elit. Cystadlu â'r cymdogion, prynu ceir gwell neu wyliau amlach a hirach. Wrth wneud hyn mae ein cymdeithas yn mynd yn fwy anghyfartal. Mae gwahaniaethau incwm ac oriau gwaith rhwng y rhai hynny sydd mewn gwaith wedi cynyddu'n ddirfawr. Rydym wedi cydsynio â'r disgwyliad mai dyma'r pris i'w dalu am gynnal neu hyd yn oed gynyddu cyfoeth (a throi olwyn Galbraith a cadw bargen Faust). Yn ddamcaniaethol byddai modd cyfiawnhau, hyd yn oed croesawu, elw'r gwneuthurwyr cyfoeth pe bai eu hymdrechion yn galluogi pawb arall i fwynhau safon byw well. Dyma un fersiwn o'r ddamcaniaeth ddatblygu ryngwladol a elwir yn ddatblygiad 'diferu i lawr' sydd wedi cael ei feirniadu'n hallt. Beth nesaf ar ôl chwalu'r credoau hyn a rhoi pin yn y swigen? Gwaetha'r modd ymddengys y bydd y llai cyfoethog yn talu cyfran anghymesur o'r bil er mwyn clirio'r ddyled ac achub y sector gwasanaethau ariannol. Nid wyf yn ymwybodol o unrhyw uwch reolwr yn y sector hwn y gofynnwyd iddynt wynebu canlyniadau cyfreithiol eu trachwant a'u twyll. Yn y cyfamser rhaid i ni hefyd ddisgwyl i gostau ynni a bwyd godi a fydd yn peri rhagor o loes i'r difreintiedig. Mae Galbraith[41] hefyd yn awgrymu bod cymdeithas anghyfartal iawn yn ei chael yn anodd cynnal y fargen a chadw'r olwyn defnyddwyr i droi. Sylwodd y gellir dibynnu ar y grwpiau incwm canolig ac isel i wario eu hincwm ar nwyddau a gwasanaethau ac i beidio â chasglu arian yn ddibwrpas, tra nad yw'r elit hynod gyfoethog o dan yr un fath o bwysau. Efallai fod y 'terfynau' yn dod ar ein gwarthaf mewn ffyrdd annisgwyl?

Rhaid i gynaliadwyedd a thegwch gerdded law yn llaw, gyda'r holl oblygiadau i bolisi yng Nghymru, yn y Deyrnas Unedig, yn Ewrop a thrwy'r byd i gyd. Mae'n golygu, er enghraifft fod yn rhaid delio ag allyriadau nwyon tŷ gwydr fesul pen. (Ar ôl hynny, pe bai pobl neu genedl yn dymuno cynyddu eu poblogaeth, oni fyddai'n deg capio cyfanswm eu hallyriadau?) Os oes gan unigolyn resymau cryf iawn dros fynnu cyfran fwy bydd yn rhaid iddo efo/hi ei phrynu o'r hyn sydd yn weddill gan eraill. Yn yr un modd byddai'n rhaid i bawb dalu trethi teg; a chael gwared â hafanau treth! Ymddengys ar hyn o bryd bod angen ysgogi'r tlodion i weithio'n galetach trwy eu statws iselradd a thrwy rewi cyflogau tra bo angen gostwng trethi'r cyfoethog i'w hannog i weithio'n galetach.

7 / Ddynoliaeth

Craidd y broblem yw'r cysylltiadau rhwng ein cyfundrefn wleidyddol-economaidd, ein natur ddynol, ein codau moesol a'n systemau gwerthoedd moesegol, hyd yn oed crefyddol a'r byd naturiol. Mewn degawdau diweddar datganwyd bod trachwant yn beth da. Mae'r tra gyfoethog, boed nhw'n fancwyr, yn sêr y byd adloniant neu'n bêl-droedwyr amlwg yn cael eu canmol a'u gweld yn esiamplau i'w dilyn. Mewn rhannau o'r Unol Daleithiau credir bod Iesu Grist hyd yn oed yn cynnig cyfoeth a llwyddiant i'w ddilynwyr. Ymddengys ei fod **yntau'n** cyfranogi yn y freuddwyd Americanaidd, yn credu yn natur unigryw'r Unol Daleithiau ac yn gwarantu eu diogelwch [gan gynnwys yn y byd a ddaw!] Yn y Deyrnas Unedig, yn agosach i'r ddaear, mae'r WAGs, sef, gwragedd a chariadon enwogion y byd chwaraeon yn symbol o wario gwyllt, bancwyr, sêr y byd roc a phêl-droedwyr yn symbol o fyw'n wyllt. Mae pawb ond y bancwyr yn cael eu cyfrif yn esiamplau i'w dilyn. Wrth ddilyn eu harweiniad rydym wedi mynd ar ein pennau i ddyled ond wedi cael fawr ddim boddhad [Affluenza[30]?]. Mae hyn yn wahanol iawn i werthoedd y gymdeithas Gymreig ychydig o ddegawdau'n ôl, a'r gwerthoedd a goleddir gan y meddylwyr mawr a doethion ein hil gan gynnwys Iesu Grist wrth gwrs. Diystyrwyd doethineb ac argymhellion Galbraith, ac mae ein rhith o gyfoeth wedi ei adeiladu ar ddyled, dyled unigol a dyled gwlad [bron 500% o Gynnyrch Mewnwladol Crynswth y Deyrnas Unedig] ac ar fanteisio ar gyfalaf naturiol. Bellach mae lles yr unigolyn yn cael ei erydu'n gyflym ac yn boenus. Rhith oedd ein 'cyfoeth' diweddar, er gwaethaf y gost amgylcheddol a chymdeithasol enfawr; mewn gwirionedd roedd yr 'ymerawdwr' yn hanner noeth. Mae'r gost gymdeithasol enfawr a bradychu dyheadau'r genhedlaeth ar ôl y rhyfel gan gyfalafiaeth gasino wedi ysbrydoli Stephane Hessel i ysgrifennu "Outrage"[42] ac wedi ysgogi'r mudiad 'Occupy Wall Street'.

Ysgrifennwyd llawer am 'drasiedi'r tir comin'[43], h.y. pa mor hawdd ydyw camreoli a blingo asedau cyffredin grwpiau, weithiau oherwydd diffyg pwrpas cyffredin neu ddiffyg rheoleiddio a thuedd unigolion i geisio elw personol tymor byr ar draul asedau cyffredin a budd cyffredin tymor hir. Mae hyn i weld yn ddarlun da o broblem llygru'r atmosffer gan nwyon tŷ gwydr a'r newid yn yr hinsawdd sy'n digwydd o ganlyniad neu hyd yn oed yn ddarlun o'r ffordd mae asedau cyfalaf byd-eang wedi cael eu cipio er budd yr ychydig. Rhaid cofio bod yna wahaniaethau [a] nid yw'r atmosffer yn adnodd sy'n eiddo i unrhywun . Er nad yw'n eiddo i neb serch hynny rydym i gyd yn dibynnu arno; [b] yn ail ni fydd effeithiau'r llygredd yn gyson ar draws y byd nac yn effeithio pawb yr un fath ac yn anffodus y tlawd fydd yn dioddef waethaf. [c] gan fod yr atmosffer a'r hinsawdd, yn lleol ac yn fyd-eang, yn adnoddau rhad ac am ddim nad ydynt yn eiddo i neb mae hyn yn annog unigolion, sectorau a gwledydd i'w hecsbloetio ac i ddadlau dros eu byddianau hunanol Yn waeth na hynny, mae yna annhegwch rhwng cenedlaethau gyda'n cenhedlaeth gyfoethog ni am y cyntaf i fanteisio ar yr adnoddau. Yn achos y tir comin – hynny yw, adnoddau nad ydynt yn eiddo i neb trwy'r byd, rydym yn wynebu cyfyng-gyngor mawr, sef y perygl y bydd eraill yn elwa os dewiswn ni wneud 'y peth iawn' – gallai hyn wneud ein 'haberth' yn gwbwl ddiwerth.

Er mwyn symud ymlaen rhaid cyd-drafod fel dinasyddion a lledaenu gwybodaeth amgenach am y trybini mae'r ddynolryw yn ei wynebu. Efallai mai dyma'r unig ffordd o sicrhau cydymffurfiad gwirfoddol. Ni fydd yn hawdd! Rydym wedi methu'n drychinebus â galw arianwyr a masnachwyr Dinas Llundain i gyfrif. Yn gyfeillion mynwesol i'r blaid Lafur, hen a newydd, i Doriaid o bob lliw a llun ac i Ddemocratiaid Rhyddfrydol yr un modd, mae gan elit y Ddinas fawr bob rheswm dros deimlo'n hunanfodlon ac yn ddiogel eu byd. Oherwydd hyn mae'r bobl gyffredin yn hynod ddrwgdybus a sinigaidd. Er bod yn rhaid newid yr hen drefn mae'n anodd dychmygu sefyllfa lle bydd llywodraeth Llundain yn gweithredu'n benderfynol. Yn rhyfedd iawn mae'r ffaith bod yr economi'n dihoeni'n rhoi mwy o bŵer i fancwyr, arianwyr a diwydianwyr. Yn unigol ac yn dorfol yr ydym yn hynod awyddus i'n tref neu ein rhanbarth neu ein plant elwa o fuddsoddiad annisgwyl neu swyddi newydd. Yn yr un modd mae'r llywodraeth yn awyddus iawn i atal cwmnïau mawr rhag symud o'r wlad, ac i sicrhau'r elw a ddaw iddi o drethi, o leiaf.

Heb ymateb totalitaraidd, ein unig obaith efallai yw llwyfan ddemocrataidd ac agored y we fyd-eang. Mae pob arweinydd grymus yn cael ei demtio gan fodel llywodraethu Animal Farm pa un ai comiwnydd oedd i ddechrau, yntau ffasgydd, cenedlaetholwr Cymreig, Llafurwr y cymoedd, neu gynnyrch ysgol fonedd. Mae grym nid yn unig yn llygru, mae'n dallu hefyd. Dim ond cynyddu'r siawns o gael arweinyddiaeth awdurdodaidd y mae'r argyfwng 'terfynau'. Beth yw gwerth democratiaeth yng ngwlad Groeg neu'r Eidal os ydym yn syllu dros y dibyn ar chwalfa economaidd? Beth yw gwerth hawliau gweithwyr os yw'r llywodraeth yn ymateb i'r argyfwng trwy ei gwneud yn haws i gwmnïau ddiswyddo eu staff. Yn wyneb y dewisiadau hynny, mae'r tair plaid yn San Steffan, ynghyd â'r Democratiaid a'r Gweriniaethwyr yn America fel ei gilydd, wedi dewis achub y banciau a'r bancwyr gan ofni trychineb gwaeth hyd yn oed. Dim ond 'grym y bobl' sy'n gallu galw llywodraethau, broceriaid grym a llygrwyr i gyfrif. Dim ond pobl sy'n gallu gwireddu'r newid paradeim sydd ei angen i ymgyrraedd at hyd yn oed ryw lun ar gynaliadwyedd. Er mwyn gwneud hyn rhaid eu hargyhoeddi mai dyma'r llwybr mwyaf buddiol iddynt hwythau a'u teuluoedd ac y bydd cyfiawnder cymharol yn cael ei wneud. A allwn ni osgoi baglu ar ein pennau i mewn i wladwriaeth neo-ffasgaidd? A allwn sicrhau bod ein byd yn newid mewn modd trefnus neu a fydd y newid yn un anhrefnus a didostur, yn unol â phroffwydoliaeth James Lovelock[2] ymysg eraill? A allwn harneisio ynni a chreadigedd pobl sydd mor amlwg mewn cyfalafiaeth sy'n hybu ei lles ei hun, nid er mwyn elwa o adnoddau'r blaned ond er mwyn hybu lles a boddhad y ddynolryw yn y tymor hir. Ond rhaid i ni gydnabod yr argyfwng enbyd sydd yn ein hwynebu – y rhai sydd yn awr yn gyfoethog, yn bwerus ac yn ddylanwadol fydd yn colli fwyaf yn y tymor byr.

8 | Am 'Wales' gweler Cymru

Datganodd Deddfau Llywodraeth Cymru 1998 a 2006 y rheidrwydd i hybu neu hyrwyddo 'datblygu cynaliadwy'. Mae Llywodraeth Cymru a etholwyd ym Mai 2011 nid yn unig yn bwriadu gwneud datblygu cynaliadwy yn 'brif egwyddor drefniadol' ond i'w ymgorffori mewn deddfwriaeth sylfaenol. Dyma felly gyfle allweddol i sefydlu egwyddorion effeithiol er mwyn sicrhau cynnydd tymor hir. Yn ychwanegol at ddiffiniad enwog Bruntland[44] a seiliwyd ar degwch rhwng cenedlaethau, dehonglir datblygu cynaliadwy'n aml fel datblygu sy'n seiliedig ar dri man cychwyn sef 'economi lwyddiannus a hyfyw', sy'n cefnogi 'cymunedau bywiog' mewn ffyrdd sydd yn 'amgylcheddol gynaliadwy'; gwlad lwyddiannus o safbwynt diwylliannol ac economaidd i'w hetifeddu gan ein plant a phlant ein plant. Mae hyn i'w groesawu'n fawr wrth gwrs ond hyd yma nid yw'r goblygiadau wedi eu harchwilio eto na'u deall yn iawn fe dybiwn i ac maent yn anodd eu mynegi ar ffurf feintiol. Bwriad yr ysgrif hon yw cyfrannu at y deialog hwn ond aiff blynyddoedd lawer heibio cyn bod economi sy'n gynaliadwy yn economaidd, yn gymdeithasol neu'n amgylcheddol yn bosib.

Mae'r drafodaeth hyd yn hyn, er nad yw wedi llwyddo i fynd i'r afael â llawer o faterion cymunedol a diwylliannol hollbwysig, wedi dangos maint a chymhlethdod yr her. Ceir ymdrech o ddifrif i ymgodymu â'r materion dan sylw yn llyfr Tim Jackson 'Prosperity without Growth' sydd yn seiliedig ar gasgliadau Comisiwn Datblygu Cynaliadwy'r Deyrnas Unedig a ddaeth i ben yn 2011[45]. Ond os ydym yn agosáu at 'derfynau' rhyngwladol – ac efallai wedi eu cyrraedd, bydd y rhain yn diffinio blaenoriaethau gwario. Mae'n anorfod bod argyfwng o'r fath yn ailgyfeirio gweithgarwch tuag at gyfyngu ar y difrod cymdeithasol ac economaidd a wnaed trwy fynd y tu hwnt i'r 'terfynau' a thuag at sicrhau digon o waith, ynni, bwyd a nwyddau a gwasanaethau cysylltiedig ynghyd â chefnogaeth gymdeithasol. Ymddengys fod yr argyfwng dyled rhyngwladol eisoes yn ein gyrru i lawr y llwybr hwnnw. Beth bynnag yw ein hofnau ar hyn o bryd dim ond ychydig ddegawdau sydd gennym i ddod o hyd i lwybrau cynaliadwy a datrys y brif broblem sef cwtogi'n aruthrol ar ein hallyriadau nwyon tŷ gwydr. Sut, felly y gall gwlad sydd yn gymharol dlawd o'i chymharu â gwledydd Ewropeaidd eraill ac ychydig iawn o ddylanwad o ran arian a pholisi symud ymlaen? Oes yna ffyrdd ymarferol o ymgyrraedd at ddull datblygu cynaliadwy ac o greu Cymru lewyrchus sy'n seiliedig ar realiti, gan ddangos y ffordd (neu un ffordd) fel ag a wnaethom yn ystod chwyldro diwydiannol y 19eg ganrif?

Awgrymaf fod angen i ni wahaniaethu rhwng camau gweithredu tymor byr y gellir eu cyflawni'n awr yn lleol a strategaeth gynhwysfawr tymor hir. Hefyd er bod yna rai cynlluniau y gellir eu datblygu o fewn pwerau cyfredol y Cynulliad Cenedlaethol y mae eraill sy'n ddibynnol ar ennill pwerau ychwanegol i alluogi Cymru i symud ymhellach tuag at gynaliadwyedd. O gofio diffyg ffydd pobl mewn awdurdod a'u dadrithiad gyda gwleidyddion rhaid i ni ennill ymddiriedaeth y bobl. Mewn sawl cyd-destun nid yr 'arbenigwr' sy'n gwybod orau bob tro. Yng Nghymru lle nad oes traddodiad cryf o 'sefydliadau' rhaid i ni ddod o hyd i ffyrdd o symbylu dinasyddion i weithredu ar eu liwt eu hunain, gan ymddiried yn eu synnwyr cyffredin, gan ddefnyddio gwybodaeth 'arbenigwyr' neu 'ymgynghorwyr', beth bynnag mae hynny'n feddwl, ond heb adael i'r rheini eu harwain neu eu trin fel plant bach.

Mae gan Gymru ychydig o fanteision bach o'u cymharu â Lloegr. Diolch byth nad ydym mewn dyled i'r sector gwasanaethau ariannol a gobeithio ein bod yn llai gwrthwynebus i Ewrop na'n cymdogion dros y ffin. Siawns na allwn gydnabod mai'r ffordd orau, ac efallai'r unig ffordd o wireddu sawl agwedd ar ddatblygu cynaliadwy yw o fewn fframwaith Ewropeaidd lle ceir cydweithredu arch-wladwriaethol Dim ond ar gynfas mor eang y gallwn efallai greu parth economaidd digon pwerus fydd yn cynnwys cyfreithiau, rheoliadau ac uchelgeisiau cyffredin er mwyn hyrwyddo tegwch, gwarchod hawliau'r unigolyn a'r amgylchedd a chynnig dull o wrthweithio pwerau dinistriol busnesau mawr ac uchelfannau'r byd arian. Yng Nghymru gallem wneud hyn yn glir. Er mwyn llwyddo i ddatblygu'n gynaliadwy mae gofyn i ni weithredu ar sawl lefel; yn rhyngwladol, yn yr Undeb Ewropeaidd, y Deyrnas Unedig, Cymru ac yn lleol. Byddai rhai o'r camau hyn, megis rheoleiddio allyriadau o awyrennau a llongau, masnachu carbon a threthi carbon o bosib yn ymarferol bosib dim ond ar lefel yr Undeb Ewropeaidd. Er mwyn cyflawni camau eraill megis rheolaeth a dibynadwyedd y cyflenwad trydan trwy'r Grid Cenedlaethol mae'n amlwg y byddai angen gweithredu ar lefel y Deyrnas Unedig a Ffrainc. Yn yr un modd mae'n hanfodol bwysig bod yna weithredu ar lefel y Deyrnas Unedig, neu rhwng Prydain ac Iwerddon, mewn perthynas â theithio rhwng trefi a rheoleiddio'r ddinas. Dyma ffactorau nad oes gennym fawr o ddylanwad uniongyrchol arnynt. Serch hynny rhaid i'n llais fod yn hyglyw.

Yn ffodus mae ymdeimlad cymunedol yn dal i fodoli yng Nghymru ac mae'n wlad ddigon bach i gynnal dialog a dadl fewnol am y materion hyn a chael ymateb synhwyrol, a allai fod yn fwy o broblem mewn gwlad fawr amrywiol ei natur. Nid oedd Schumacher yn '*Small is Beautiful*'[39] yn dadlau'n unig o blaid y 'bach'; dywedodd yn hytrach y gall ein teyrngarwch i'r anferthol fod yn gamgymeriad. Rhaid i ni weithredu ar raddfa briodol ac mae'r raddfa ddynol, y fro a'r genedl, yn hollbwysig. Yn ddiddorol iawn, ystyr Cymry yw 'cyd-aelodau o'r un grŵp lle mae'n bosib trafod a phenderfynu'. Mae'r gair Saesneg 'Welsh' ar y llaw arall yn golygu 'eraill'. Rhaid i ni synied amdanom ein hunain fel 'cymrodyr', 'Cymry' – yn hytrach na chael ein diffinio fel 'eraill'.

Hefyd mae gennym adnoddau naturiol da o ran ynni adnewyddadwy, dŵr, coed, gwynt a haul, yn ogystal â hanes hir o ddyfeisgarwch technolegol ac ymchwil wyddonol. Hyd yma fodd bynnag dim ond camau petrus a gymerwyd tuag at ddefnyddio'r asedau hynny a bu'r rhaniad cyfrifoldeb rhwng y Cynulliad Cenedlaethol a San Steffan yn rhwystr. Mae gennym boblogaeth sydd wedi cael addysg ac sy'n llawn brwdfrydedd a chanddi wreiddiau diwylliannol a hanesyddol dwfn. Un agwedd ar fywyd yng Nghymru nad yw wedi ei werthfawrogi'n llawn yw ein bod mor agos i'r byd naturiol. Hyd yn oed yng Nghaerdydd mae pobl o fewn tafliad carreg i'r wlad a does yna'r unlle sydd yn fwy nag ychydig filltiroedd i fwrdd o ardal wledig neu arfordirol neu Barc Cenedlaethol neu Ardal o Harddwch Naturiol Eithriadol. Credaf, os bydd yr economi gynaliadwy newydd yn ceisio ysbrydoliaeth o gyswllt agos â'r byd [lled] naturiol ac os bydd unigolion yn cael cysur a phleser ohono yna rydym ni yng Nghymru'n ffodus iawn.

Yn hanesyddol mae Cymru wedi bod yn brin o lawer o'r elfennau sy'n hanfodol i ddylanwad economaidd a gwleidyddol, er bod ein gwendid cymharol yn nodweddiadol o lawer o genhedloedd a rhanbarthau eraill. Ond mae'r sefyllfa hon yn cael ei hunioni'n rhannol yn dilyn refferendwm Mawrth 2011. Bydd angen pwerau ychwanegol dros ein hadnoddau naturiol ac mae angen mentrau ar bob lefel. Ond ni fydd yn bosib cyfyngu'r technolegau newydd effaith isel a ddatblygir i un rhanbarth neu wlad. Mae canfyddiadau ymchwil yn rhyngwladol yn eu hanfod, ond os cânt eu defnyddio'n bwrpasol gallant gynnig manteision yn lleol neu'n genedlaethol. Bydd gweddill yr ysgrif hon yn canolbwyntio ar gamau gweithredu cenedlaethol a lleol, tra'n cydnabod y gall camau gweithredu rhwng gwledydd fod ar ei hôl hi fel y dengys methiant cymharol y trafodaethau ar newid yn yr hinsawdd yn Copenhagen a Durban.

O gofio maint yr her fyd-eang a'n gofod daearyddol, economaidd a gwleidyddol bychan, rhaid i Gymru fynd i'r afael nid yn unig â **lliniaru** h.y. sut i chwarae ein rhan mewn cyfyngu ar allyriad llygryddion a'r defnydd anghyfrifol o adnoddau ond hefyd **addasu** a **gwytnwch**. Mae lliniaru'n golygu mynd i'r afael â ffyrdd o ysgafnhau effaith argyfyngau i ddod e.e. olew brig, ar ein cymdeithas, ac o gofio'r ansicrwydd gwyddonol mawr, mae addasu a gwytnwch yn cyfeirio at gynyddu ein gallu cenedlaethol i wrthsefyll stormydd sydyn, annisgwyl.

Mae Daly yn diffinio system amgylcheddol gynaliadwy fel un ag 'economi cyflwr sefydlog' sy'n cadw stoc gyson o gyfalaf ffisegol (h.y. nid yw'n erydu ei chyfalaf ffisegol/naturiol na chyfalaf mannau eraill trwy fewnforion) mewn byd sy'n defnyddio cyfran isel o adnoddau a dim mwy nag y gall eu hadnewyddu a'u hailgylchu. Er bod y rhesymeg hon yn ddifai rydym yn bell iawn o gyrraedd y nod a byddai symud yn rhy gyflym tuag at y cyflwr hwn yn ein gwneud yn agored i berygl a gallai'r canlyniadau fod yn andwyol. Mae i feini prawf Daly'r fantais ei bod yn bosib eu meintioli ond nid ydynt yn rhoi ystyriaeth i faterion economaidd, cymdeithasol a diwylliannol.

Yng nghyd-destun Cymru, gellir cynnig diffiniad dros dro o ddatblygu cynaliadwy fel "system wleidyddol-economaidd sy'n hyrwyddo ffyniant dynol trwy ystod o gyfleoedd am swyddi a darpariaeth gwasanaethau o ansawdd da yn y sectorau preifat a chyhoeddus a thrwyddynt, ac ar yr un pryd [a] yn glynu wrth egwyddorion cynaliadwyedd amgylcheddol Daly [b] yn hybu gweithgareddau diwylliannol a chymdeithasol, gan gynnwys rhai yn y Gymraeg yn benodol, [c] yn gwarchod amrywiaeth biolegol a harddwch naturiol a [d] yn ysgogi cymunedau egnïol o unigolion hyderus a charedig".

Os derbynnir hyn yna mae angen i ni gytuno'n fras ar strategaeth ar gyfer symud ymlaen at gynaliadwyedd o'r fath dros gyfnod o 20 mlynedd. Awgrymaf y byddai hyn yn golygu symud i ffwrdd oddi wrth gyfalafiaeth gasino Anglo-Americanaidd tuag at economi gymysg, ddemocrataidd gymdeithasol ar lun a delw gwlad fel Norwy, ond gyda phwyslais parhaol a chynyddol ar ddatgysylltu'r 'diffiniad newydd o dwf' oddi wrth ddefnyddio a chynhyrchu ynni, allyrru nwyon tŷ gwydr a chynhyrchu sbwriel gwastraffus a llygryddion eraill. Byddai hyn yn arwain at economi 'cyflwr sefydlog' wedi ei datgysylltu'n llawn erbyn 2030. Rhaid pwysleisio nad mater syml yw datgysylltu a gellir ei gamddeall yn ddybryd (gweler[45]).

Er enghraifft gall dyfeisiadau technolegol gynyddu allbwn am gost ariannol ac amgylcheddol is. Ond pe bai'r galw am y nwydd yn codi i'r entrychion gallai hyn ddiddymu unrhyw arbedion. A siarad yn bersonol, os ydym yn buddsoddi er mwyn gwneud ein tai'n fwy ynni effeithlon ac felly'n arbed costau gwresogi etc ac yna'n gwario ein cynilion ar hedfan yn amlach i Sbaen neu ar deithiau car hirach, mae'r holl waith da'n cael ei ddadwneud. Felly rhaid wrth ddealltwriaeth a chydsyniad y cyhoedd, er nad oes dim llawer o hynny i'w weld ar hyn o bryd. Fel y nodwyd eisoes byddai'n anodd anghytuno mai'r cyflenwad ynni a'r defnydd o ynni a'r allyriadau nwyon tŷ gwydr sydd wrth wraidd problem cynaliadwyedd. Mae bron popeth yn troi o gwmpas ynni: o dai i drafnidiaeth i gynnyrch diwydiannol a'r gadwyn fwyd. Cydnabuwyd, hyd yn oed yn y 60au, fod amaethyddiaeth ddwys yn ei hanfod yn ffordd o drosi olew'n fwyd (gweler Albert Bartlett a ddyfynnwyd yn[46]). Felly mae costau ynni wedi eu hymgorffori ym mhob agwedd o'r economi. Gall y cyhoedd fod yn fwy parod i wrando ar ddadleuon o blaid ym**addasu** i gost a diogelwch y cyflenwadau trydan a thanwydd nag i'r syniad o **liniaru** effeithiau croesi ffin capasiti'r suddfan carbon yn yr atmosffer. Ychydig o bobl fyddai'n ymwrthod o'u gwirfodd â 'chyfleuster' eu hewyllys rhydd ei hunain felly bydd angen teilwra unrhyw atebion gan gadw hyn mewn cof. Mae cynaliadwyedd yn llesol i'r unigolyn a rhaid i'r llywodraeth geisio cyrraedd y nod drwy ddefnyddio cydbwysedd o gosbau a chymhellion yn ddoeth. I seinio rhybudd arall – mae'r senarios a drafodwyd ynghynt fel pe bacnt yn awgrymu fod yr economi gwasanaethau yn creu elw economaidd heb gynhyrchu llawer o nwyon tŷ gwydr ond nid yw hyn yn wir o reidrwydd. Wrth fesur ôl troed carbon Gwynedd[44] gwelwyd bod twristiaeth yn gwneud cyfraniad mawr iawn, oherwydd ei bod yn dibynnu cymaint ar gerbydau modur. Mae project Goriad Gwyrdd Eryri a allai fod wedi cynnig ateb wedi mynd i'r gwellt oherwydd diffyg dychymyg ac ewyllys gwleidyddol. Y wers yw nid na ddylid annog twristiaeth ond y dylid gwneud pob ymdrech ym mhob sector i greu a hyrwyddo dewisiadau deniadol a chredadwy sydd ag ôl troed nwyon tŷ gwydr isel.

I gloi hoffwn agor trafodaeth ymarferol ar sail tystiolaeth sy'n cynnig rhai mentrau strategol i roi Cymru ar lwybr mwy cynaliadwy, o ystyried y diffygion economaidd, cymdeithasol ac amgylcheddol difrifol a'n hastudiaeth ar ryngddibyniaeth y materion hyn. Mae'r cysyniadau isod yn dibynnu ar geisio ehangu elfen leol/ranbarthol yr economi, gan gynyddu'r elfennau cymdeithasol a chydweithredol er mwyn creu mwy o wytnwch a hybu cyfrifoldeb amgylcheddol. Wrth geisio gwireddu amcan datblygu cynaliadwy fel 'prif egwyddor drefniadol Cymru' dylem gyfaddef y bydd swyddogaethau unigryw i fentrau preifat, cyhoeddus a chymdeithasol. Dylai pob un gyfrannu at greu gwaith a swyddi ond mae'n hynod annhebygol y bydd yn system menter rydd bur, gyda threthi isel a gwariant isel. Serch hynny rhaid i'r system gwrdd â meini prawf rheolaeth economaidd gadarn. Nid yw ynysu'n hunain yn opsiwn: rhaid parhau i fasnachu, rhaid annog menter ond newid y cydbwysedd er mwyn ffafrio busnesau lleol sy'n gwneud elw.

Yn y tymor canolig rhaid i Lywodraeth Cymru wneud y canlynol:

1. mynnu bod y wlad a'r llywodraeth yn mabwysiadu deg egwyddor byw 'un blaned'[48]:
   - Sero carbon
   - Sero gwastraff
   - Trafnidiaeth gynaliadwy
   - Deunyddiau cynaliadwy
   - Bwyd lleol a chynaliadwy
   - Dŵr cynaliadwy
   - Defnyddio tir, bywyd gwyllt ac ecosystemau
   - Iaith, diwylliant a threftadaeth
   - Tegwch a chynhwysiant
   - Iechyd a hapusrwydd

   Mewn ysgrif i ddod byddaf yn ceisio nodi awgrymiadau penodol ym mhob un o'r categorïau hyn y gellid eu cyfrif yn berthnasol i fywyd pob dydd ac a allai apelio'n gyffredinol.

2. Ymrwymo i ddatblygu a defnyddio mesur e.e. fersiwn o ISEW, sy'n mesur datblygu cynaliadwy a 'dileu gweithredoedd negyddol sy'n dinistrio cyfalaf naturiol, cymdeithasol a diwylliannol'. Ni ellir canfod y llwybr o gyfalafiaeth gasino i gynaliadwyedd gwirioneddol heb ddulliau dadansoddol gwell i ddatblygu polisi a thechnolegau newydd

ac arloesol; y peth cyntaf a phwysicaf yw mabwysiadu dull mwy gwahaniaethol o fesur twf economaidd a thwf gwirioneddol na'r GDP/CMC i gyfeirio buddsoddi a pholisi.

3. Ennill grym go iawn dros ynni, cynllunio a thrafnidiaeth. Ar gyfer y cyntaf mae angen cysylltiad statudol ag Ofgem a rheoleiddio dosbarthiad ynni er mwyn helpu'r cyhoedd i gyfrannu at gynhyrchu ynni adnewyddadwy gwasgaredig (gweler hefyd [E]). O ran cynllunio mae angen mynd ati i ail-lansio Cynlluniau Gofodol er mwyn gwella profiadau byw gydag ôl troed carbon bach, unwaith eto gyda gwell systemau cadwraeth ynni, rheolaeth gadarnach ar ddatblygiadau tu allan i drefi lle mae angen defnyddio car etc. Yn drydydd, ystyried o ddifrif gosod terfyn cyflymder o 60 milltir yr awr ar ffyrdd deuol a sefydlu rhwydwaith o orsafoedd gwefru trydanol ac yn nes ymlaen, rhai H2. Er mwyn wneud defnydd da o bwerau uwch Llywodraeth Cymru awgrymaf, er na fydd yn sicr yn awgrym poblogaidd, fod angen mwy o aelodau etholedig ar y Senedd er mwyn dadansoddi a chraffu ar bolisïau. Hefyd mae angen mwy o weision sifil o safon uchel. Rhaid i'r llywodraeth gydnabod bod yn rhaid annog gweision sifil i weithredu mewn modd mwy dychmygus ac arloesol a rhaid datrys problem y teyrngarwch rhanedig rhwng Caerdydd a Whitehall. Yn olaf rhaid i Lywodraeth Cymru dderbyn her cyfrifoldeb am ei chyllideb ymchwil ei hun.

4. Sefydlu banc 'rhanbarthol' ar lun a delw'r banciau yn North Dakota neu Landers yr Almaen i gadw cynilon lleol a chefnogi busnesau, diwydiant a masnach lleol cynaliadwy, gan gydnabod bod yn rhaid cyflawni dau beth, sef gwella'r economi leol yn wyneb grymoedd globaleiddio a buddsoddi symudol yn Llundain sydd yn denu arian ac yn gostwng incwm a gwneud hyn mewn ffyrdd sydd yn effeithio cyn lleied â phosib ar yr amgylchedd. Rhoi cyngor cyfreithiol a masnachol i gwmnïau a allai eu diogelu rhag cwmnïau mawr o'r tu allan sydd am eu llyncu cf. Glas Cymru. (Mae'n ddiddorol bod Banc Lloegr a Banc y Byd wedi sylweddoli'n ddiweddar nad yw llif rhydd

cyfalaf, adleoli cwmnïau heb gyfyngiadau a llyncu cwmnïau bach yn bethau cadarnhaol o anghenraid)

5. Adfywio economi Cymru trwy ddefnyddio pŵer prynu Llywodraeth Cymru, awdurdodau lleol, cyrff cyhoeddus anadrannol a chyrff cysylltiedig megis prifysgolion, ysgolion, colegau ac ysbytai i gefnogi busnesau lleol a chynyddu nifer y busnesau lleol sy'n gwneud elw. Datrys unwaith ac am byth y gwrthdaro rhwng hybu'r economi leol a dehongliad rhy gaeth o gyfraith gaffael Ewrop.

6. Cyfranogiad y cyhoedd, tegwch, iaith, diwylliant a chyfiawnder. Mae datblygu cynaliadwy ar ei ffurf gyfredol wedi cael cyhoeddusrwydd gwael ar y cyfan. Mae llawer, efallai'r mwyafrif mawr, yn ei ystyried yn fygythiad i'w bywydau a'u dyheadau – esgus dros ragor o ymyrraeth gan y llywodraeth, a hynny weithiau gan unigolion trahaus. Efallai fod argyfwng presennol 'cyfalafiaeth gonfensiynol' wedi suro eu hagwedd at y system honno, ond ychydig ohonynt sy'n gallu dychmygu unrhyw system arall. Felly'r galw taer am dwf yn y GDP/CMC. Yn ôl y syniadaeth bresennol mae'n amhosib dychmygu y gallai economi amgylcheddol-gyfrifol, gwirioneddol gynaliadwy roi boddhad a chynnig cyfleoedd cyffrous i bobl uchelgeisiol a gweithgar. Serch hynny mae craidd anghyfiawn ac ecsbloetiol y system anghynaliadwy bresennol mor amlwg nes bod David Cameron ac Ed Milliband am y gorau yn ceisio ffrwyno'r rhannau gwaethaf ohoni. Nid yw cyfiawnder cymdeithasol yn gysyniad Cymreig unigryw wrth reswm ond mae gennym hanes y gallwn ymfalchïo ynddo sy'n cynnwys Robert Owen, Lloyd George ac Aneurin Bevan yn ogystal ag arweinwyr undebau llafur ac arweinwyr crefyddol di-ri, hyd yn oed Dewi Sant a ymrwymodd yn ôl y sôn i 'wneud y pethau bychain' hynny sydd yn fuddiol i gymunedau. Fel Michael Sandel ar Sidelsky's yn dadlau[27 31], nid yw cyfiawnder yn dileu'r angen am farn ar werthoedd gwar; yn wir mae gostod safonau yn ofynnol. Wrth wraidd datblygu cynaliadwy y mae tegwch a chyfiawnder cymdeithasol rhwng cenedlaethau, cenhedloedd a phobl, gyda phwyslais cryf ar gyfrifoldeb cymdeithasol ac unigol. Mewn rhai

ffyrdd mae hyn i gyd yn Gymreig iawn! Efallai yn lle cystadlu am deganau a dyfeisiadau newydd, y gallwn fynegi ein hawydd dynol i gystadlu trwy chwaraeon, diwylliant, y celfyddydau a'r gwyddorau. Mae hyn yn gwbl gydnaws â'r traddodiad Cymreig, eisteddfodau ag ati. Mae'r Gymraeg yn rhan bwysig o'n hetifeddiaeth ond fe allai ac fe ddylai hefyd fod yn ysgogi mentrau unigol a chymdeithasol ac entrepreneuriaid lleol. Mae dyfodol yr iaith, mewn sawl ffordd, yn fesur o'n cynaliadwyedd; gan fod y grymoedd negyddol sy'n tanseilio'r iaith yn aml iawn yr un rhai ag sydd yn rhwystro datblygu cynaliadwy. Yn ail, dim ond mewn cymuned fywiog y gall iaith leiafrifol ffynnu. O ganlyniad rhaid i'r darpar fesur roi sylw i ddwyieithrwydd, nid yn unig i'r rhai sydd yn siarad Cymraeg ond fel rhan o dreftadaeth pob un ohonom. Mae'n eironig bod yr adroddiad llawn gan PWC o dan y teitl 'Effectiveness review of the Sustainable Development Scheme Report to Welsh Government'[49] ar gael yn Saesneg yn unig!

9 Diweddglo

O gofio mor enfawr yw'r her tymor hir sydd o'n blaen a phwysau problemau economaidd a chymdeithasol presennol, hyd yn oed yn y wlad hon, ac yn fwy fyth yn y byd sy'n datblygu, mae'n ddealladwy ond yn anorfod bod gwleidyddion yn glynu wrth y ddancaniaeth economegol confensiynol. Gwelir cynaliadwyedd fel amcan dymunol ond pell i ffwrdd. 'O Dduw, gwna fi'n gynaliadwy ond dim eto!' Mae'n anorfod y bydd hyn yn peri i ni orymestyn a chroesi'r ffiniau. Mae angen llawer mwy o waith i lunio llwybr 20 mlynedd at baradeim newydd ond rhaid i ni wynebu problemau'n onest a gorau pa gyntaf i ni ddechrau

Bydd lliaws o gyfleoedd yn codi i fentro ac i greu busnesau newydd sy'n lleihau effaith dyn ar yr amgylchedd ac yn gwneud unigolion a chymdeithas yn fwy llewyrchus. Yr allwedd i hyn, o bosib, fyddai cynhyrchu ynni adnewyddadwy gwasgaredig oherwydd byddai'n darparu ffynonellau incwm ychwanegol lleol ac yn hoelio sylw y bobl a'r materion dan sylw, gan cynnwys effeithlonrwydd ynni. Ond mae'n tanlinellu maint y broblem hefyd (gweler 45) gan fod y ffynonellau ynni hyn yn annhebygol o fod yn ddigonol – ac nid ydynt eto wedi datrys y broblem storio trydan[45]. Felly mae angen gweithredu ar wahanol raddfeydd, yn lleol, yn dechnegol ac yn wleidyddol[39]. Nid yw dyfodol cynaliadwy yn wrth-dechnolegol, nac yn brin o ddyfeisgarwch – i'r gwrthwyneb. Ond rhaid i'r cymhellion a'r gwrth-gymhellion economaidd gyd-fynd â lleihau ein heffaith ar draws y byd ac yn lleol, gan weithio'n unol â chyfyngiadau biogeoffisegol yn lle yn eu herbyn. Bydd rhai o'r elfennau sy'n hanfodol i wireddu newid o'r fath i'w cael yng Nghymru, eraill ddim.

Mae llawer o feddylwyr disglair wedi ymroi i gynnal bargen Galbraith ac i hysbysebu a hyrwyddo ein 'hanghenion'. Nid yw y tu hwnt i amgyffred y gellid defnyddio'r doniau hyn ynghyd â'n dealltwriaeth gynyddol o seicoleg pobl i annog system lawer llai dinistriol. Mae hefyd yn gwbl debygol y bydd systemau o'r fath yn esgor ar ymdeimlad cryfach o fodlonrwydd a dedwyddwch ond bydd perffeithrwydd bob amser y tu hwnt i'n heneidiau aflonydd. Man cychwyn yr ysgrif hon oedd cyferbyniad rhwng optimistiaeth y rhai hynny sydd o blaid 'twf' dilyffethair a phesimistiaeth y proffwydi gwae. Credaf fod hyn yn rhy syml. Yn lle hynny dylem resynu at yr ymdrechion i gynnal damcaniaeth ddinistriol, ac er gwaethaf yr her enfawr, dylem fod yn frwdfrydig am y rhagolygon am ffyniant a boddhad gwirioneddol.

Diolchiadau

Rydw i'n ddiolchgar iawn i Dr Havard Prosser, i'r Athro Ross Mackay ac yn enwedig i'r Dr Einir Young am ddarllen trwy'r ysgrif hon, am eu sylwadau adeiladol a'u hawgrymiadau. Carwn hefyd ddiolch yn cywir i Sylvia Prys Jones, Gareth Roberts ag Einir Young am eu cymmorth arbennig yn datglybgu'r ysgrif Gymraeg. Myfi'n unig sy'n gyfrifol am unrhyw wendidau a erys.

Cyfeiriadau

1. Rees, Martin. (2003) *Our Final Century: Will the human race survive the 21st century?* Heinemann
2. Lovelock, James. (2009) *The Vanishing Face of Gaia: The Final Warning*. Penguin
3. Flannery,Tim. (2009) *Now or Never:Why we need to act now for a sustainable future*. Harper
4. Galbraith, John Kenneth. (1958;1971) *The Affluent Society*. Pelican
5. Meadows, Donella. H. et al. (1972) *The Limits to Growth*. Earth Island
6. Meadows, D.H. et al. (1992) *Beyond the Limits*. Chelsea Green
7. Meadows, D.H. et al. (2005) *Limits to Growth: 30 year update*. Earthscan
8. Hawranek, Dieter et al. (2011) *Out of Control: The destructive power of financial markets* Der Spiegel. 22 Awst o Der Spiegel-on-line
9. Ferguson, Niall. (2009) *The Ascent of Money*
10. Chakraborthy, A. (2011) *Guardian* 13fed Rhagfyr: yn dyfynnu Centre for Research on Socio-Cultural Change, Prifysgol Manceinion
11. *Observer* (2012) 22nd June.
12. Montreal Protocol on Substances that Deplete the Ozone Layer (2000) UNEP, Kenya
13. Hogan, C.M. (2010) *Overfishing*, Encyclopedia of Earth. NCSE, Washington D.C.
14. Risk List – 2011, British Geological Survey, wedi ei lawrlwytho o wefan BGS
15. International Food Policy Institute (2010/11) gweler y papurau ar y wefan yn enwedig papurau Mark Rosegrant. Also CCAFS (2009) Climate, agriculture and food security. A CGIAR Challenge Program. The Alliance of the CGIAR Centers ans ESSP. Ar gael ohttp://www.cgiar.org/pdf/CCAFS_Strategy_december2009.pdf (cyrchwyd 13/02/2012)
16. Setiau data FAO (2010) ar gael ar lein ar http://faostat.fao.org/site/380/default.aspx (cyrchwyd 13/02/2012)
17. R.K. Pachauri, A. Reisinger, (Gol.). IPCC Synthesis Report 2007: Contribution of Working Groups I, II and III to the Fourth Assessment Report of the Intergovernmental Panel on Climate Change. Hefyd IPCC (1996) Revised 1996 IPCC Guidelines for National Greenhouse Gas Inventories, Intergovernmental Panel on Climate Change, www.ipcc.ch a chronfeydd data allyriadau IPCC (2010) ar gael ar http://www.ipcc-nggip.iges.or.jp/EFDB/main.php (cyrchwyd 13/02/2012)
18. Wyn Jones, R. Gareth et al. (2012) Climatic mitigation, adaptation and dryland food production Proceedings 2010 IDDC Conference, Cairo; hefyd Taylor, Rachel. C. et al. Cyflwynwyd. Before 2050 Human Food Chain will drive Climate Change.
19. OECD, Towards Green Growth: monitoring progress. (2011) ESA/STAT/AC.238, UNCEEA/6/11
20. Daly, H. (1991) Institutions for a Steady State Economy, yn Steady State Economics. Island Press, Washington D.C. (gweler hefyd y trafodaethau yn nghyfeiriadau 7 and 40)
21. Baer, Paul. (2008) Exploring the 2020 global emissions mitigation gap (Analysis for the Global Climate Network). Woods Institute for the Environment, Stanford University, Palo Alto, California (gweler hefyd gyfeiriadau 16 a 43)
22. Meinshausen, M. et al. (2009) Greenhouse-gas emission targets for limiting global warming to 2 0C. Nature 458, 1158-1162
23. Ranger, N., A. Bowen, J. Lowe, L. Gohar (2009) Mitigating climate change through reductions in greenhouse gas emissions: the science and economics of future paths for global annual emissions. Cyfarwyddyd polisi 2009 ar gael ar http://www.cccep.ac.uk/Publications/Policy/Policy-docs/bowen-

Ranger_MitigatingClimateChange_Dec09.pdf (cyrchwyd 13/02/2012)
24. Carbon Dioxide Information Analysis Center (2011) Oct., Oak Ridge Labs US
25. Strahan, David. (2011) *The Oil Maze*. *New Scientist* Rhagfyr 3.
26. Spinney, Laura. (2011) *The Underhand Ape*, *New Scientist* Tachwedd 5 t43.
27. Sandel, Michael. J. (2009) *Justice: What's the right thing to do*. Farrar Strauss a Giroux
28. Gweler y cyfeiriadau yn 7
29. Dawkins, E. et al. (2008) *Wales' Ecological Footprint – Scenarios to 2020*, LlCC [Stockholm Environment Institute]. Caerdydd
30. James, Oliver. (2007) *Affuenza: How to be Successful and Stay Sane*. Vermilion
31. Diamond, Jared. (2006) *Collapse: How Societies choose to Fail or Survive.* Penguin.
32. Skidelsky, R a Skidelsky, E. (2012) *How Much is Enough*. Allen Lane
33. Ehrlich, Paul. (1968) *The Population Bomb*. Buccaneer, NY; Ehrlich. P. ac Ehrlich A., Population Resources and Environment (1970) W.H Freeman, SF a thrafodaeth yng nghyfeiriad [41]
34. Wyn Jones, R.G. a Prosser, H. (2010) *Land Use and Climate Change*, adroddiad i Lywodraeth Cynulliad Cymru. http://wales.gov.uk/docs/drah/publications/100310landuseclimatechangereport.pdf (cyrchwyd 13/02/2012)
35. *Foresight: The Future of Food and Farming* (2011) Adroddiad olaf y project. Swyddfa Gwyddoniaeth y Llywodraeth, Llundain.
36. Gweler y drafodaeth ar y Cynnyrch Mewnwladol Crynswth ar Wikipedia; y drafodaeth o dan [42] a gwerslyfrau economeg safonol.
37. Kuznets, Simon.(1934) *National Income 1929-1932*, 73rd US Congress 2nd Session. http://library.bea.gov/us (cyrchwyd 13/02/2012)
38. Daly, H a Cobb. J (1989) *For the Common Good*. Beacon Boston: Nordhaus, W a Tobin, J. Is Growth Obsolete (1972) Columbia University Press – mae'r Athro Peter Midmore wedi llunio ISEW i Gymru hefyd.
39. Schumacher, E.F. (2011) *Small is Beautiful: A Study of Economics as if People really Mattered*. Vintage [Cyhoeddwyd yn wreiddiol 1973]
40. Wales' Central Organising Principle: Legislating for Sustainable Development. IWA Conference; Caerfyrddin Ionawr 2011
41. Galbraith, J.K. (1994) *The World Economy Since the Wars*. Sinclair-Stevenson.London
42. Hessel, Stephane, (2011) *Time for Outrage/Indignez-vous*, Quartet Books
43. Hardin, Garrett (1968) *The Tragedy of the Commons*, Science. 162, 1243-1249
44. Bruntland Report: *Our Common Future*. Report to World Commission on Environment and Development. Oxford University Press, hefyd atodiad i ddogfen Cynulliad Cyffredinol y Cenhedloedd Unedig A/42/ 427
45. Jackson, Tim (2011) *Prosperity without Growth*. Earthscan. London-Washington
46. MacKay. David J.C. (2008) *Sustainable Energy – without the hot air*. UIT Cambridge, 2008. ISBN 978-0-9544529-3-3. Ar gael ar-lein o www.withouthotair.com (accessed 13/02/2012)
47. Farrar, John.F et al., *Report on Carbon Footprint* to Cyngor Gwynedd circa 2004/5
48. *The Ten Principles of One Planet Living* http://www.oneplanetliving.org/index.html (cyrchwyd 13/12/2012)
49. Effectiveness review of the Sustainable Development Scheme Report to Welsh Government http://wales.gov.uk/topics/sustainabledevelopment/publications/effectivenessreview/?skip=1&lang=cy (cyrchwyd 13/02/2012)

8
Ynni, gwaith a chymhlethdod

Cefais wahoddiad yn 2017 i draddodi Darlith Edward Lhuyd o dan adain y Coleg Cymraeg Cenedlaethol. Arweiniodd hwn at ddwy erthygl, 'Ynni, Gwaith a Chymhlethdod', yn O'r Pedwar Gwynt yng Ngaeaf ac Haf 2018 a atgynhyrchir yna.

Yn ogystal trefnodd Sioned Rowlands i mi gael fy holi gan Cynig Dafis am y syniadaeth mewn rhifyn arall o

O'r Pedwar Gwynt. Atgynhyrchir y cyfweliad hefyd a diolchaf i Sioned ac i Cynog am eu diddordeb cyson a'u hysbrydoliaeth.

Gwerth nodi bod yr erthyglau yn pwysleisio'r elfennau bywydegol, esblygiadol ac anthropolegol o syniadaeth. Ers y cyfnod hwn rwyf wedi dysgu llawer mwy am yr agweddau ffisegol ac yn benodol am y perthynas clos rhwng ynni a gwybodaeth (data). Ni thrafodir hyn yn yr erthyglau yn O4G.

Ynni, gwaith a chymhlethdod

'Iâ yn ddŵr', Yr Ynys Las, 2014

Fy mwriad yn yr erthygl hon yw cyflwyno dehongliad o hanes bywyd ar ein planed dros 4.5 biliwn blwyddyn ei bodolaeth, gan dynnu sylw arbennig at rai cyfnewidiadau sylfaenol yn yr atmosffer a'r lithosffer. Gwnaf hyn yn nhermau'r berthynas sydd rhwng ynni a'r gallu a ddaw yn ei sgil i gyflawni gwaith. Craidd fy nehongliad yw bod modd adnabod chwe chwyldro ynni ffurfiannol yn hanes ein planed, ac ynddynt, amlygir patrymau cyson. Yn benodol, drwy ffrwyno ffynhonnell newydd o ynni, cyfyd y potensial nid yn unig i gyflawni gwaith ychwanegol; gall y gwaith hwnnw arwain at greu cymhlethdod materol, ac, yn ein byd presennol ni, gall arwain at gynhyrchu cymhlethdod cymdeithasol cynyddol. Ceisiaf ddadansoddi yn y fan hon y patrymau arwyddocaol a chyflwyno eu goblygiadau yn fanylach, gan eu hystyried mewn perthynas â'r seithfed chwyldro, sef y chwyldro sydd ar ein gwarthaf heddiw.

Ymdriniaf â chynfas eithriadol o eang, sy'n pontio sawl maes academaidd – ymgais fentrus, mi dybiaf, onid ffôl efallai. Does ond gobeithio, felly, y byddai'r athrylith mawr ei hun, Edward Lhuyd, yn cymeradwyo f'ymdrechion ac yn cytuno, efallai, â'm casgliadau!

Dymunaf bwysleisio fy mod yn adeiladu ac yn ymhelaethu ar waith sawl rhagflaenydd. Yn 1912 ysgrifennodd Wilhelm Ostwald, enillydd Gwobr Nobel mewn cemeg, y canlynol: 'ynni rhydd yw'r cyfalaf a ddefnyddir gan greaduriaid o bob math, a thrwy ei drosglwyddiad mae pob peth yn digwydd' (*Der Energetische imperativ*, 1912). Datblygwyd syniadaeth Ostwald gan nifer o awduron diweddarach, gan gynnwys Howard Odum, yr arloeswr ym myd ecoleg, a gyflwynodd fap o lif ynni a mater yng nghyfundrefn byd natur (*Environment, Power and Society*, 1971). Mae Václav Smil, yn ei gyfrol *Energy and Civilization: A History* (2017), hefyd yn trafod dylanwad ynni fel sylfaen holl brosesau natur a holl weithredoedd dynoliaeth, ac mae'r ymwybyddiaeth o bwysigrwydd llif ynni yn greiddiol yng ngwaith James Lovelock, tad damcaniaeth Gaia.

Yn y rhan gyntaf hon o'r erthygl amlinellaf brif nodweddion y chwe chwyldro hanesyddol ynghyd â goblygiadau'r cysyniad hanfodol, er llai cyfarwydd, o 'homeostasis', cysyniad a gaiff ei egluro'n fanylach isod. Ystyriaf fod y chwe chwyldro yn gyfraniadau ffurfiannol i ddatblygiad ein bydysawd a'n byd presennol, yn ogystal â chydnabod bod y gwyddorau biolegol, daearegol, cymdeithasegol a'r deallusol hefyd wedi cyfrannu'n allweddol i hanes ein cread. Yn ail ran yr erthygl, i'w chyhoeddi yn rhifyn Haf 2018 *O'r Pedwar Gwynt*, byddaf yn trafod y casgliadau a'r patrymau a amlygir yn y chwe chwyldro, gan ystyried eu cyfraniad wrth i'r byd wynebu natur heriol y seithfed chwyldro.

Fel yr awgrymais uchod, ym mhob un o'r chwe chwyldro gwelir ynni yn cael ei drawsnewid i greu cymhlethdod materol. Yn hyn clywir adlais o hafaliad enwog Einstein [$e = mc^2$] sy'n diffinio'r berthynas ffisegol rhwng ynni a mater. Wrth gwrs, nid wyf am hawlio perthynas fesuradwy debyg. Er bod y berthynas rhwng ynni a datblygiad materol yn greiddiol, ac yn sylfaenol, nid yw'n esbonio nifer o ddatblygiadau hynod bwysig a ddaeth yn ei sgil megis, er enghraifft, esblygiad celloedd amlgellog a gwahaniaethiad celloedd (*differentiation*) mewn anifeiliaid a phlanhigion.

Y Syniadaeth Wyddonol

Mewn ffiseg diffinnir ynni fel 'y gallu i wneud gwaith', er bod tarddiad yr ynni a'r gwaith a gyflawnir yn dra amrywiol. Defnyddiaf ddwy enghraifft i egluro hyn. Gellir ffrwyno'r ynni mewn dŵr sy'n llifo o ben y mynydd i'r dyffryn islaw – sef ynni a ddaw drwy rym disgyrchiant – i droi tyrbin i greu ynni mecanyddol; ac o gyplysu grymoedd electromagnetig i'r tyrbin, gellir cynhyrchu trydan, sef llif o electronau i lawr graddiant ynni. Gellir defnyddio'r trydan i gynhesu tân trydan (gwres ymbelydrol), neu i oleuo lamp (ymbelydredd gweledol), neu i bweru ein cyfrifiaduron a'n rhewgelloedd. Yr ail enghraifft berthnasol yw'r ynni cemegol sydd mewn sylweddau carbon fel petrol neu siwgr. Os yw'n sylwedd rhydwythiol (*reduced*), yn aml gyda sawl moleciwl o hydrogen ynddo, yna, o'i 'fudlosgi' (yn dechnegol, ei ocsideiddio), rhyddheir ynni. Yn achos petrol, mae'r ynni yn gyrru ein moduron (ynni mecanyddol a chinetig) ac, yn achos y siwgrau, mae'n cynnal ein cyrff. Yn ôl y cysyniad ffisegol, pŵer yw'r ynni a ddefnyddir i wneud gwaith dros amser penodol, heb wrth gwrs ystyried gwerth nac amcan y gwaith. Yn ganolog i'r erthygl hon mae'r syniad o ynni yn galluogi gwaith, a gwaith, yn ei dro, yn creu pŵer.

Yn ogystal â'r ddibyniaeth ar lif o ynni, y mae trosglwyddiad gwybodaeth yn hanfodol i alluogi bywyd i fodoli ac i genhedlu. Ymgorfforir gwybodaeth yn y cod genetig sydd yn rhan o linynnau DNA. Mae'n bwysig nodi mai dim ond mewn celloedd, sy'n berchen ar gyfansoddiad cemegol arbennig, ac sydd o dan reolaeth fanwl, y gellir darllen negeseuon DNA yn gywir. Hynny yw, mae celloedd yn gwbl hanfodol i fywyd. Y mae gan bob cell beirianwaith cywrain i sicrhau bod ganddynt gyfansoddiad biocemegol cywir a chyson sy'n eu galluogi i ymateb i newidiadau allanol er mwyn cynnal y cysondeb mewnol – proses a elwir yn homeostasis.

Y chwe Chwyldro Hanesyddol

Gwelir yn y tabl isod amlinelliad o'r chwe phrif chwyldro ynni hanesyddol, ynghyd â rhai o'u nodweddion ac amseriad y digwyddiadau.

Cyfeiriaf hefyd at y seithfed chwyldro yr ydym yn byw drwyddo ar hyn o bryd. Diddorol yw cymharu fy rhestr â rhai a luniwyd o safbwyntiau 'geneteg esblygiadol' (gw. Eörs Szathmáry, John Maynard Smith, 'The Major Evolutionary Transitions' yn *Nature*, 1995; 371) a 'systemau daearol' (Gaia) (gw. Tim Lenton, Andrew Watson, *Revolutions that Made the Earth*, 2011). Sut mae'r sgwarnogod hyn yn cyfrannu at y cyflwyniad?

Y Digwyddiadau: Y Saith Chwyldro

Chwyldroadau	Amseriad	Y Prif Elfennau
1. Egnïo Bywyd	~4 biliwn bldd.	Genedigaeth Celloedd Procariotig: Grym Graddiannau Protonau
2. Cynaeafu Ynni'r Haul	~2.7 biliwn bldd.	Esblygiad Ffotosynthesis: ehangu y gadwyn fwyd: ocsigeneiddio'r awyr a'r môr
3. Esblybiad Celloedd Eucariotaidd Egnïol.	~2 i 1.7 biliwn bldd.	Traflynciad bacteria i greu mitochondria i egnïo celloedd cymhleth. Sefydlu rhyw, detholiad naturiol a chystadleuaeth Ddarwinaidd.
4. Dyfodiad yr Hominid: Gallu nid Grym.	~2 miliwn bldd.	Rheoli tân i goginio bwyd gan ymchwanegu at yr ynni a'r maeth. Buddsoddiad yr ynni yn yr ymenydd a galluoedd meddyliol.
5. Y Chwyldro Amaethyddol.	~10 i 5,000 bldd.	Datblygiad amaeth gan fachu mwy o ynni fffotosynthetig i ddefnydd dyn. Sefydlu cymunedau poblog, sefydlog, yn meithrin galluoedd cymdeithasol.
6. Y Chwyldro Diwydiannol: Tanwydd Ffosil.	~250 bldd, yn ol.	Datblygu peiriannau stêm, trydan a phetrol. Twf technoleg, gwyddoniaeth, cyfalafiaeth a masnach fyd-eang.
7. Her yr Oes Anthropogenaidd.	Heddiw	Newid Hinsawdd a Chynhesu Byd Eang. Bygythiad i olud y rhai ffodus ond i fywydau y rhai tlawd.

Y chwyldro cyntaf – Egnïo bywyd drwy ffrwyno'r ynni mewn graddiant o brotonau ac electronau

Cyfeirir yn aml at enynnau DNA a'r helics dwbl, a'r dystiolaeth ryfeddol fod cod bywyd wedi ei argraffu mewn llythrennau (triawdau o fasau) mewn llinynnau DNA. Newidiodd darganfyddiadau James Watson a Francis Crick a'u cyfoedion, a arweiniodd at adnabod strwythur DNA yn 1953, nid yn unig ein byd-olwg ond hefyd ein gallu i arbrofi ac i ymyrryd ym meysydd peirianneg enetig a meddygaeth. Llwyddodd llyfrau fel *The Selfish Gene* (1976) i boblogeiddio'r maes ac i wneud yr awdur Richard Dawkins yn fyd-enwog – yn eilun i rai ond yn elyn i eraill. Ond mae DNA ynddo'i hun yn gemegolyn hynod sefydlog a dyna paham mae'n werthfawr mewn ymchwiliadau fforensig ac yn gymorth i olrhain esblygiad rhywogaethau, megis dyn.

Lawn mor bwysig i fywyd, er nad yw'n rhan o'r ymwybyddiaeth gyffredin, yw'r ffaith fod pob cell fywiog yn gwbl ddibynnol ar ddefnyddio llif cyson o ynni i'w chynnal. Mae'n ofynnol cyrchu llif di-dor o ynni i gadw celloedd yn bell o gydbwysedd gyda'u hamgylchedd, gan sicrhau bod eu cyfundrefn fewnol yn fanwl gyson ac yn addas i gynnal bywyd. Yn benodol, rhaid wrth y cysondeb mewnol hwn i alluogi darllen yn gywir y negeseuon mewn DNA (mewn ribosomau) ac i ddefnyddio'r wybodaeth i gynhyrchu prodinau – catalyddion pob gweithgaredd yn y gell. I gyflawni hyn rhaid i bob cell feddu ar allu 'homeostatig' manwl. Yn ôl damcaniaeth enwog Erwin Schrödinger (*What is Life?*, 1944), y gallu i greu ynys o drefn (sef y gell) mewn llif o ynni yw hanfod bywyd, er bod y tueddiad ffisegol yn arwain at anhrefn.

Yn ôl damcaniaeth Nick Lane (*The Vital Question: Why is Life the Way It Is?*, 2015), y ffynhonnell ynni a alluogodd y gell gyntaf i fodoli ac, yn ddiweddarach, i atgynhyrchu, oedd graddiant naturiol pH, sef graddiant o brotonau [H+]. Mae graddiannau naturiol tebyg yn bodoli heddiw mewn llif o ddŵr alcali. Mae'r dŵr alcali hwn yn tasgu o grombil y ddaear drwy dyrau thermol o greigiau rhydyllog sy'n codi o wely'r cefnforoedd. Yn ôl y ddamcaniaeth, esgorodd y peirianwaith naturiol hwn ar amgylchiadau addas wnaeth ganiatáu i'r proto-gelloedd cyntaf esblygu'r peirianwaith bywydegol gwreiddiol. Defnyddiwyd y graddiant o brotonau [H+] ac electronau [e-] i yrru nanobeiriant – yr ensim ATPase – i gynhyrchu ATP, sef yr 'arian ynni' sy'n egnïo ac yn rheoli gwaith y gell; hynny yw, troi'r ynni ffisegol (yn y graddiannau o brotonau) yn ynni cemegol. Hwn, yn ôl y ddamcaniaeth, oedd y cam tyngedfennol wnaeth alluogi bywyd i fodoli ar ein planed ac a gynhaliodd y gofod mewnol biocemegol sefydlog.

Mae'r dystiolaeth yn awgrymu i'r broses hon ddatblygu ar blaned Daear oddeutu 4 biliwn o flynyddoedd yn ôl. Yn rhyfeddol, goroesodd bron yn ddigyfnewid yn y celloedd sy'n rhan o adeiladwaith pob creadur heddiw, gan gynnwys dyn.

Yn y cyfnodau cynnar – yr Hadean a'r Archaean – ychydig o ffynonellau electronau egnïol oedd yn bodoli ac yr oedd y derbynyddion angenrheidiol i fewnsugno electronau llai egnïol hefyd yn brin. Yn ôl pob tebyg, yr oedd y potensial i greu amrywiaeth bywyd yn gyfyng. Eto, hwn oedd y cam tyngedfennol a egnïodd fywyd a dechrau'r broses o drawsnewid ynni i fod yn gymhlethdod materol dyrys a rhyfeddol, sef yn gelloedd byw. O ganlyniad, newidiwyd ein planed, nid yn unig yn fywydegol ond hefyd yn gemegol, megis yng nghyfansoddiad cemegol y môr, yr awyr a'r ddaear yn ei chreigiau gwaddod.

Am gyfnod o ddau biliwn o flynyddoedd, dim ond bywyd ungell syml, a elwir yn gelloedd procariotig, a fodolodd. Nid oeddynt, ac yn wir nid ydynt, yn meddu ar saernïaeth faterol fewnol amlwg. Er hynny, datblygodd dau uwch raniad o fywyd procariotig, sef bacteria, enw lled gyfarwydd i ni bellach, ac archaea, enw llai cyfarwydd ar fath o fywyd procariotig a ddarganfuwyd yn gymharol ddiweddar.

Yr ail chwyldro – Cynaeafu'r haul ac esblygiad y peirianwaith ffotosynthetig

Er bod tystiolaeth i rai bacteria ac archaea wneud defnydd cyfyngedig o ynni'r haul, yr ail chwyldro tyngedfennol oedd esblygiad ffotosynthesis llawn i egnïo'r graddiannau o brotonau ac electronau.

Yn ystod ffotosynthesis, defnyddir ynni'r haul i hollti dŵr gan gynhyrchu nwy ocsigen (sy'n dianc i'r awyr) a rhyddhau electronau [e-] a phrotonau [H+] egnïol. Defnyddir yr ynni hwn trwy'r un gyfundrefn ar gyfer troi'r ynni trydanol yn ynni cemegol a ddisgrifiwyd yn fras yn Chwyldro 1. Yna defnyddir yr ynni cemegol i fachu nwy cabron deuocsid o'r awyr i gynhyrchu siwgrau. Yr ynni cemegol yn y siwgrau hyn yw sylfaen y gadwyn fwyd sy'n cynnal bron holl fywyd ein planed.

Gelwir y bacteria a esblygodd y gallu i ffotosyntheseiddio yn seianobacteria [cyanobacteria] gan eu bod yn cynnwys cyflawnder o sylweddau gwyrdd eu lliw – sef y cloroffyl sy'n cipio ac yn adweithio â ffotonau egnïol o'r haul (gw. Tim Lenton, Andrew Watson, *Revolutions that Made the Earth*, 2011 a Park S Nobel, *Physicochemical and Environmental Plant Physiology*, 1991). Adeiladwyd Chwyldro 2 yn rhannol ar gefn Chwyldro 1, felly. Ceir trafodaeth fanwl ar y camau angenrheidiol a alluogodd ffotosynthesis i ddigwydd ac ar ei ddylanwad yn ocsigeneiddio ein Daear yn llyfr Lenton a Watson. Ac mae'n werth nodi wrth fynd heibio fod enghreifftiau o hen drefn fiolegol gyn-ffotosynthetaidd yn bodoli hyd heddiw, megis mewn ogofâu a mwyngloddiau ger Llanrwst.

Y trydydd chwyldro – Esblygiad celloedd ewcariotig egnïol

Er bod y trydydd chwyldro yn anghyfarwydd i lawer, hwn oedd y cam sylfaenol a arweiniodd at fywyd amlgellog cymhleth fel y gwelir mewn planhigion ac anifeiliaid. Mae gan gelloedd ewcariotig saernïaeth fewnol hynod gywrain. Ni cheir bodau byw amlgellog heb iddynt fod wedi eu hadeiladu o gelloedd ewcariotig. O safbwynt bioleg y gell, mae llawer yn gyffredin nid yn unig rhwng dyn ac epa, ond rhwng dyn a banana a malwoden a ffwng. Nid oes tystiolaeth i gelloedd ewcariotig fodoli yn y cofnod ffosil am y ddau biliwn o flynyddoedd cyntaf, sef tua hanner oes hanes ein planed. Yn y cyfnod maith hwn dim ond bywyd 'syml' ungellog oedd yn bodoli. Eto, yr oedd gan y celloedd 'syml' procariotig hyn alluoedd biocemegol neilltuol, rhai sydd ymhell y tu hwnt i alluoedd bodau amlgellog fel anifeiliaid.

Beth felly a esgorodd ar gelloedd ewcariotig a'r cymhlethdod materol a ddaeth yn eu sgil? Dengys tystiolaeth enetegol a biocemegol ddiamwys mai mewn uniad symbiotiadd rhwng dwy gell brocariotig – un yn facteria a'r llall yn archaea – y ffurfiwyd y gell ewcariotig.

Hyd y gwyddom, dim ond unwaith mewn hanes y digwyddodd yr uniad hwn, a hynny oddeutu 1.8 biliwn o flynyddoedd yn ôl. Oherwydd gofod cyfyngedig yr erthygl hon, rhaid hepgor trafod anferthedd a chymhlethdod yr uniad tyngedfennol hwn, ond gellir nodi rhai ffeithiau arwyddocaol. Yn dilyn traflyncu'r 'bacteriwm' gan yr 'archaea', esblygodd y bacteria mewnol dros amser i fod yn feitocondria. Rydym yn gyfarwydd â meitocondria fel y pwerdai mewn celloedd sydd hefyd yn meddu ar ychydig o DNA oherwydd y drafodaeth ddiweddar am blant tri rhiant. O'r fam yn unig yr etifeddir y meitocondria ac mae'r darganfyddiad hwn yn eithriadol bwysig i olrhain y camau yn esblygiad dynoliaeth a'r ddamcaniaeth ynghylch yr Efa wreiddiol. Yn dyngedfennol, o ganlyniad i'w draflyncu, collodd y bacteriwm bron y cyfan o'i DNA wrth esblygu i fod yn feitocondrion. Yn arferol, mewn celloedd procariotig, defnyddir tua 80% o holl ynni'r gell ganddi er mwyn atgynhyrchu ei hun ac i egnïo prosesau darllen negesau'r DNA a'u defnyddio i syntheseiddio'r prodinau angenrheidiol. Golyga hyn gyfyngiad egnïol enfawr ac ymddengys mai byw i epilio, a dim llawer mwy, y mae bywyd procariotig, heb yr 'ynni sbâr' i arbrofi gyda ffurfiau gwahanol ar fywyd.

Symbiosis oedd wrth wraidd esblygiad celloedd ewcariotig, felly, a chynhwysai'r rhain niwclews gwirioneddol yn ogystal â chromosomau a ddatblygodd ryw. Yn sgil rhyw, cafwyd y potensial i gyfnewid ac i gymysgu genynnau'r gwryw a'r fenyw drwy feiosis a meitosis. Hyn sydd, gyda chyfraniad mwtanau, yn creu'r deunydd crai ar gyfer detholiad naturiol a'i rym trawsnewidiol, fel y datgelwyd yn namcaniaeth Darwin a Wallace. Lynn Margulis a gynigiodd y ddamcaniaeth am esblygiad y gell ewcariotig yn wreiddiol, ond diweddarwyd y ddamcaniaeth gan Nick Lane drwy danlinellu pwysigrwydd ynni yn y broses hon. Mae'n honni bod gan bob genyn mewn cell ewcariotig hyd at 300,000 gwaith yn fwy o ynni i weithredu na'r genynnau cyfatebol mewn celloedd procariotig. Casgliad Lane, felly, yw mai'r chwyldro ynni hwn a alluogodd i fywyd ffynnu, i ledaenu ac i amlhau, gan arwain yn y diwedd at greu bodau amlgellog soffistigedig fel anifeiliaid a phlanhigion uwch.

Parhaodd y galw am homeostasis mewn celloedd ewcariotig unigol ac, wrth gwrs, mewn creaduriaid cymhleth amlgellog. Rhaid oedd datblygu systemau rheoli mwy soffistigedig i alluogi'r celloedd ewcariotig unigol, gyda'u hisraniadau mewnol megis meitocondria a gwagolion, i gydlynu ac i ffynnu. Rhaid hefyd gymathu a chydlynu gweithgareddau'r celloedd unigol mewn bodau amlgellog, yn ogystal â sicrhau bod gweithgareddau'r uned integredig yn ymateb i'w hanghenion mewnol ac i'w hamgylchedd. Trosglwyddir felly wahanol negeseuon cemegol a thrydanol (e.e. nerfau) mewn bodau amlgellog. Y negeseuon hyn sy'n caniatáu iddynt synhwyro ac ymateb ac, i raddau, i reoli eu hamgylchedd ar lefel y celloedd unigol yn ogystal ag ar lefel y cyfan integredig.

Mae'n bwysig tanlinellu'r goblygiadau canlynol. Yn y byd procariotig trosglwyddir DNA o un gell i'r llall yn lled hawdd; proses sy'n cyflymu addasiad microbau i wrthsefyll gwrthfiotigau, er nad yw'n tueddu at ddetholiad Darwinaidd. Nid arweiniodd chwaith at greu amlgellogrwydd dros gyfnod o biliynau o flynyddoedd. Yn y byd ewcariotig, ar y llaw arall, arweiniodd esblygiad Darwin/ Wallace at greu math newydd o gystadleuaeth, y gystadleuaeth ffyrnig sy'n hawlio'r dychymyg poblogaidd, 'nature red in tooth and claw'. Ond ni ddylid anghofio pwysigrwydd hanfodol cydweithio a symbiosis yn esblygiad a pharhad y drefn ewcariotig, a hefyd yn y cydweithio rhwng celloedd sy'n nodweddu amlgellogrwydd.

Er i'r celloedd ewcariotig cyntaf ymddangos oddeutu 1.8 biliwn o flynyddoedd yn ôl, araf iawn oedd ymlediad y newidiadau mawr a ddaeth yn eu sgil, gan gynnwys amlgellogrwydd. Bu sawl cyfnod o drai a llanw yn fiolegol a daearyddol dros y milenia, a wynebwyd rhai argyfyngau enfawr megis y Cyfnod Pelen Eira [Snowball Earth]. Yn yr Oes Gambriaidd, oddeutu 540 miliwn o flynyddoedd yn ôl, gwelir yn y cofnod ffosil olion rhai o ragflaenwyr cyntefig y rhywogaethau sy'n bodoli heddiw. Digwyddodd cam symbiotaidd arall eithriadol bwysig oddeutu 440 miliwn o flynyddoedd yn ôl. Ceir tystiolaeth i rai celloedd ewcariotig draflyncu celloedd seianobacteria i greu planhigion syml (algae) ac, yn ddiweddarach, planhigion amlgellog. Y mesurau hyn a alluogodd i blanhigion uwch gychwyn coloneiddio tiroedd sych y Ddaear.

Yn y cofnod ffosil ceir tystiolaeth bod o leiaf bum difodiant bywyd trychinebus wedi digwydd ar ein planed. Digwyddodd y mwyaf dinistriol ohonynt oll yn yr Oes Bermaidd, er i'r mwyaf adnabyddus ddigwydd ar ddiwedd yr Oes Gretasig gyda diflaniad y deinosoriaid. Rhwng y pum trychineb, diflannodd dros 99% o'r holl rywogaethau a fu ar y blaned erioed. Pendiliodd y Ddaear rhwng cyfnodau eithriadol boeth a rhai hynod oer, ac yn ystod y gwahanol gyfnodau daearegol bu newidiadau sylweddol yn lefelau'r ocsigen yn yr awyr cyn iddo gyrraedd y lefel bresennol. Ffurfiwyd creigiau gwaddodol amrywiol eu cyfansoddiad, o galch i siâl, ac erydwyd creigiau eraill mewn cynyrfiadau folcanig ffrwydrol. Yn araf erydwyd y pinaclau yn wastadeddau ac yn ddyffrynnoedd. Crwydrodd y platiau tectonig ar hyd wyneb y blaned gan ddifrodi rhai cyfandiroedd cyfan a chodi eraill yn eu lle. Cafwyd cyfresi o oesoedd rhewlifol ac, o ganlyniad, newidiodd lefel y môr ddegau o fetrau. Ond er gwaetha'r holl amrywiol rwystrau, llwyddodd bywyd i addasu ac i oresgyn, i amlhau ac i ddatblygu mathau mwy cymhleth o fywyd, a hyn oll yn ategu pŵer ynni'r haul a detholiad naturiol Wallace a Darwin. Drwy'r holl gyfnewidiadau, parhaodd y cylchrediadau geocemegol angenrheidiol o nitrogen, carbon, ocsigen a sylffwr. Yr oedd y rhain yn hanfodol er mwyn cynnal bywyd.

Ar ôl y trydydd chwyldro, a thros gyfnod o un biliwn a hanner o flynyddoedd, esblygodd bywyd ar ein Daear i ymdebygu i'r cyfundrefnau bywydegol ac ecolegol yr ydym yn lled gyfarwydd â hwy heddiw. Yn anffodus, yn ein byd trefol, drwy ffilmiau a theledu yn unig y mae canran sylweddol ohonom yn profi cyfoeth naturiol ein byd.

Toddi'r tir dan ein traed', Yr Ynys Las, 2014

Y pedwerydd chwyldro – Gallu, nid grym, a dyfodiad yr homo cyntefig

Oddeutu chwe deg miliwn o flynyddoedd yn ôl, ymddangosodd ffosilau a fyddai'n gynseiliau i deulu amrywiol y primatiaid, gan gynnwys y llinach a arweiniodd at y primat 'doeth', sef *Homo sapiens*. Ond oddeutu dwy filiwn o flynyddoedd yn ôl yn unig y daeth newid tyngedfennol – dyma'r Pedwerydd Chwyldro. Er i'r chwyldro hwn ymddangos yn dila ar y cychwyn, ymhen amser newidiodd gwrs a chydbwysedd ein byd mewn ffyrdd nas gwelwyd yn y pedwar biliwn o flynyddoedd blaenorol.

Yn ôl damcaniaeth Richard Wrangham (*Catching Fire: How Cooking Made Us Human*, 2009), y cam tyngedfennol ar y llwybr esblygiadol hwn oedd datblygiad coginio, neu, yn fwy manwl gywir, y gallu i ddefnyddio a rheoli ynni tân i goginio bwyd. Drwy wneud hynny ychwanegwyd yn sylweddol at dreuliant y bwyd a'r ynni a'r maeth a oedd ynddo. Buddsoddwyd yr ynni ychwanegol hwn mewn datblygu ymennydd mwy o ran ei faint, ei gymhlethdod a'i allu. Mae Wrangham yn cysylltu'r cam hwn ag ymddangosiad Homo erectus yn y cofnod ffosil oddeutu 1.8 miliwn o flynyddoedd yn ôl.

Mae Wrangham a gwyddonwyr eraill, megis Suzana Herculano-Houzel (*The Human Advantage*, 2016) a Jared Diamond (*The World Until Yesterday*, 2012), yn cyflwyno'r ffeithiau canlynol. Buddsoddir 25% o ynni dynol yn yr ymennydd, sy'n fuddsoddiad sylweddol fwy na'r 8–10% a fuddsoddir gan anifeiliaid eraill. I ddeall hyn yn nhermau ein bwydlen heddiw, mae'n cyfateb i fuddsoddiad o tua 500 cilocalori y person y dydd o'r 2,000 cCal sy'n angenrheidiol i fywyd dynol.

Mae tystiolaeth ddietegol gan anthropolegwyr yn awgrymu, oni choginnid bwyd, na allai'r drefn gyntefig o hela gyfrannu digon o galorïau i gynnal ymennydd dyn sylweddol ei faint, yn arbennig yn ystod cyfnodau llwm. Byddai bwyta cnawd amrwd o gymorth, ond heb ei goginio, anodd iawn fyddai ei dreulio. Adlewyrchir hyn yn ffurf gyntefig yr hominidau cynnar oedd â phenglogau enfawr, ymennydd bychan a dannedd a safnau cydnerth. Awgryma Herculano-Houzel y buasai bwydlen gyfyngedig yr hominidau cynharaf, o fyw drwy 'hela/hel' a chan roi blaenoriaeth i oroesi ac atgynhyrchu, ddim ond yn caniatáu i ymennydd o tua 30,000 biliwn niwron ddatblygu. (I'w cynnal, roedd ein cyndadau, am filoedd o flynyddoedd, yn gwbl ddibynnol ar allu'r llwyth bach a'r teulu i ddal a lladd anifeiliaid gwyllt, i hel planhigion ac, ar yr arfordir, pysgod cregyn; o ganlyniad, roedd eu bwyd yn dymhorol ac yn ansicr.) Mewn cymhariaeth, awgryma y byddai gan *Homo habilis* (cyn-ddyn cynnar a fodolai 2.5 i 2 filiwn o flynyddoedd yn ôl yn llinach *Homo erectus*) ymennydd o tua 40–50 biliwn niwron, tra byddai gan yr ymennydd dynol heddiw oddeutu 90 biliwn niwron. O ganlyniad i'r buddsoddiad ynni yng nghelloedd yr ymennydd ac yn y trosglwyddiadau trydanol sy'n egnïo ein 'meddyliau', mae'r ymennydd dynol yn caniatáu i tua 100 triliwn (sef miliwn miliwn) o gysylltiadau synaptig ddigwydd rhwng y niwronau. Mae maint a chymhlethdod y 'tynlapio' yn yr adeiladwaith yn nodwedd unigryw o'r hil ddynol. Golyga hyn fod yn rhaid wrth fuddsoddiad sylweddol o ynni i fwydo'r peirianwaith ymenyddol – buddsoddiad sydd wedi talu ar ei ganfed.

Yn ôl y ddamcaniaeth, o fuddsoddi ynni mewn ymennydd gyda

galluoedd amgenach, ychwanegwyd at allu'r hominidau cynnar i sicrhau mwy o fwyd, a thrwy hynny lansio rhywogaeth *Homo* ar lwybr esblygiadol effeithiol a chyffrous. Ar drywydd esblygiadol gwahanol, parhaodd yr epaod i fwyta bwydydd amrwd gan fuddsoddi'r ynni mewn nerth ac, o reidrwydd, mewn datblygu safnau a boliau mawr i gnoi a threulio'r bwyd. Awgryma data cymharol Herculano-Houzel mai'r epaod yw'r eithriadau ac iddynt ddilyn llwybr a oedd yn *cul de sac* esblygiadol. Yn ôl ei chanfyddiadau mae'r genws *Homo* wedi ymestyn ar hyd y llwybr a ragfynegir mewn ymennydd mwncïod llai.

Yn achos *Homo*, ei allu i reoli tân, sef troi ynni cemegol yn wres, a'i dysgodd i goginio ei fwyd ac, yn ddiarwybod iddo, i ryddhau hyd at ddwywaith yn fwy o ynni mewn caloriau nag y byddai wedi'i dderbyn o fwyta bwyd amrwd. Cyfrannodd Daniel Everett (*How Language Began: The Story of Humanity's Greatest Invention*, 2017) ychwanegiad hynod gyffrous at ddamcaniaeth y chwyldro hwn drwy faentumio bod y newid yn ymennydd *Homo erectus* wedi ei alluogi i ddatblygu iaith, gan roddi iddo alluoedd esblygiadol newydd i ymdopi â gofynion mwy cymhleth ei gymdeithas. Dadleua fod gofynion cymdeithas estynedig yn ddibynnol ar alluoedd i gydgynllunio, i gyd-hela a chasglu, i gyd-goginio a bwyta, pan fyddai cyfathrebu ar raddfa led syml yn sgîl anhepgor i lwyddiant a pharhad y gymdeithas. Dyma, yn ôl dehongliad Everett, pryd y daeth symbolau gweledol ac ieithyddol yn rhan o gymdeithas dyn am y tro cyntaf, i'w datblygu dros gyfnod o filiwn o flynyddoedd wrth i ddiwylliant *Homo erectus* ledaenu a phoblogi'n llwyddiannus gyfandiroedd Affrica ac Ewrasia.

Yn groes i syniadaeth Noam Chomsky, mae Everett yn maentumio i ddatblygiad iaith, yn sgil y newid mawr yn yr ymennydd, alluogi datblygu galluoedd newydd yn yr hominidiau cynnar hyn (cymharer Robert C Berwick, Noam Chomsky, *Why Only Us? Language and Evolution*, 2015). Dadleua Everett fod gan *Homo erectus*, gyda'i alluoedd newydd, yr hyn sy'n angenrheidiol, yn esblygiadol, i ddatblygu 'iaith', oherwydd pwysau a chymhlethdodau ei fywyd cymdeithasol. Fel llwyth neu deulu estynedig yn cyd-hela/hel, yn gyd-ddibynnol, yn cyd-goginio a chyd-fwyta, onid oedd cyfathrebu a mesur o gyd-gynllunio lled soffistigedig yn anhepgor? Felly cyfyd yr angen am symbolau, am eirfa ac am iaith gynyddol gymhleth. Dros filiwn o flynyddoedd, datblygodd *Homo erectus* declynnau Acheulean a phentrefi y gwelir olion ohonynt hyd heddiw, a theithiodd o'i g/chartref tebygol yn Affrica ar draws hemisffer y Dwyrain i Tsieina. Yn ôl pob golwg, llwyddodd i groesi moroedd! Gwêl Everett hyn oll fel tystiolaeth o'r gallu i ddefnyddio iaith i gyfathrebu.

Y buddsoddiad chwyldroadol o ynni, felly, a arweiniodd, gam wrth gam, a thros tua miliwn a hanner o flynyddoedd, at allu deallusol yr hil ddynol ac, yn y diwedd, at ei goruchafiaeth ar blaned Daear. Mae'r dystiolaeth am ymddygiad epaod a dyn cyntefig yn awgrymu cystadleuaeth frwd am fwyd a chymar. O anghenraid, yr oedd cydweithio o fewn y teulu a'r llwyth hefyd yn anhepgorol bwysig i sicrhau bwyd. Roedd pwysau yn ogystal ar i ddynion a merched gydweithio a rhannu dyletswyddau er mwyn sicrhau goroesiad y teulu.

Mewn cymhariaeth â'r chwyldroadau biolegol a daearyddol blaenorol, cyfnod byr o amser, 1.5 miliwn o flynyddoedd, sy'n gwahanu *Homo erectus* oddi wrth ein llinach ni, sef *Homo sapiens*. Ymddengys i'r llwybr esblygiadol, o'r hominidau cynnar i *Homo sapiens*, fod yn droellog a chymhleth, ac yn eithriadol gystadleuol ar brydiau. Ceir tystiolaeth gadarn i'r dyn modern ymddangos ar diroedd safana dwyrain Affrica oddeutu 200,000 o flynyddoedd yn ôl (Damcaniaeth Efa a'i meitocondria), er bod tystiolaeth ddiweddar o Foroco yn awgrymu bod llinach *Homo sapiens* yn ymestyn yn ôl i gyfnod cynharach rhwng 250,000 a 350,000 o flynyddoedd yn ôl. Yn rhyfeddol, ymddengys i'n cefndryd agos iawn, sef dynion Neanderthalaidd, Denisofan a Ffloresiensis, gyd-fyw ac, i raddau, baru gyda ni cyn iddynt ddarfod â bod tua 40,000 o flynyddoedd yn ôl. Ymddengys i'n hil gynnar lwyddo i oroesi mewn byd o rwystrau cyntefig yn cynnwys anifeiliaid rheibus a hominidau cystadleuol.

Ynni'r haul a ffotosynthesis oedd yn cyflenwi'r ynni a oedd mewn bwyd, a'r tanwydd i'w goginio, a thrwy hynny yn cynnal eu bywyd. Ond, yn y Pedwerydd Chwyldro, yn ogystal â throsi ynni'r haul yn gelloedd ac yn greaduriaid materol cymhleth a soffistigedig, gwelir cnewyllyn y broses o droi ynni yn waith i greu celfi ac addurniadau ac i ysgogi cymhlethdod cymdeithasol.

Y pumed chwydro – Amaethu i sicrhau mwy o fwyd

Ar ôl ei ymddangosiad, parhaodd *Homo sapiens* i fyw am o leiaf 180,000 o flynyddoedd yn y drefn 'hel/hela/coginio', a hynny mewn grwpiau bach gwasgaredig. Gyda diflaniad rhewlifau olaf Oes yr Iâ, oddeutu 12,000 CC, cychwynnodd y Chwyldro Amaethyddol mewn ardaloedd addas eu bywydeg, eu hinsawdd a'u priddoedd. Ar fryniau yn y Dwyrain Canol, oddeutu 9,000 o flynyddoedd yn ôl, ceir tystiolaeth i blanhigion gael eu dethol gan ddynoliaeth ar gyfer ailblannu a bridio, er mwyn sicrhau cnydau gwell. Digwyddodd cyfnewidiadau cyffelyb ond annibynnol oddeutu'r un amser mewn nifer o ranbarthau eraill yn y byd – yn Ne a Chanolbarth America, yn Affrica, ac yn Tsieina. Mewn ardaloedd tra gwahanol, meistrolwyd technegau tyfu a bridio planhigion a'u trosi o'u ffurfiau cyntefig yn gnydau ac yn fwydydd safonol megis gwenith, haidd, corn, tatws a reis, sy'n parhau i gynnal ein cymdeithas hyd heddiw. Ar gyfandir Ewrasia, llwyddwyd i ddofi anifeiliaid fel defaid, geifr, moch a gwartheg a'u hychwanegu at y fwydlen drwy eu cig a'u llaeth. Ychydig yn ddiweddarach, oddeutu 5,000 o flynyddoedd yn ôl, dofwyd cyfres o anifeiliaid pwn a gwaith megis ychen, camelod a cheffylau a'u defnyddio i lafurio dros eu meistri dynol.

Llwyddodd y Chwyldro Amaethyddol i grynhoi mwy o ynni ffotosynthetig i greu cyflenwadau mwy parhaol o fwyd at ddefnydd dyn. Ysgogodd hyn gymhlethdodau materol a diwylliannol newydd. Chwyddodd y boblogaeth yn sylweddol gan arwain at sefydlu cymdeithasau trefol sefydlog mewn rhai ardaloedd breintiedig, megis yn yr Aifft, Gorllewin Asia, Tsieina, a De a Chanolbarth America.

Bellach yr oedd angen gweinyddu ac amddiffyn yr adnoddau. Datblygodd sgiliau cyfrif ac ysgrifennu. Cyfoethogwyd yr economi drwy gynnal crefftau arbenigol a masnach. Datblygodd yr angen i weinyddu buddiannau'r gymdeithas drwy gyfraith a threfn a llywodraeth gan arwain at greu rhaniadau hierarchaidd yn y gymdeithas. Gwelwyd twf mewn credoau a oedd, ar yr un llaw, yn cynnig cysur i'r difreintiedig ac, ar y llall, yn cyfiawnhau awdurdod a golud y brenhinoedd a'u cynghreiriaid. Datblygodd grym milwrol i amddiffyn adnoddau'r cymdeithasau sefydlog ac i geisio cipio adnoddau cymdeithasau cyfagos. Datblygodd mathau ychwanegol o gystadleuaeth a'r rheini, i raddau helaeth, yn gystadleuaeth am ynni.

Rydym yn dystion i'r gwychder materol a ddeilliodd o waith corfforol diflino'r dynion a'r merched ac uchelgais yr arweinwyr a amlygir yn eu palasau, eu temlau a'u beddrodau ysblennydd. Tystiant i gystadleuaeth a gorthrwm yn ogystal ag i'w gallu i sefydlu cydymdrechion cymdeithasol. Profodd lleiafrif bychan olud ond nid oedd bwydlen a maeth y bobl gyffredin yn llawer gwell na'u cyndeidiau, yr helwyr, ac roeddynt, o bosib, yn waeth os rhywbeth (gw. Yuval N Harari, *Sapiens: A Brief History of Humankind*, 2014). Er yr ymddengys i'r gyfundrefn newydd ddibynnu ar waith didostur y mwyafrif, tybiaf i'r drefn drefol gynnig cyfleoedd am well bywyd i amryw, a phosibiliadau cymdeithasol llawnach. Wrth gwrs, parhaodd yr hen fywyd dros ran helaeth o'r byd, gan barhau hyd at ein hoes ni mewn ardaloedd diarffordd fel Papua Guinea Newydd a fforestydd yr Amason.

Ni ellir gwadu i lwyddiant y Chwyldro Amaethyddol yn rhwydo mwy o ynni ffotosynthetig newid ffawd dynoliaeth ac iddo ysgogi datblygiad dysg, llenyddiaeth a chelfyddyd. Ond yr oedd i'r chwyldro ei wendidau yn ogystal â'i ragoriaethau. Un ffaeledd amlwg oedd y distryw a achosodd amaeth i ecoleg y byd drwy ddymchwel fforestydd, troi'r peithiau'n borfeydd, sychu gwlypdiroedd a dinistrio cynefinoedd naturiol anifeiliaid rheibus.

Y chweched chwyldro – Ynni nerthol a phŵer i greu diwydiant

Ymledodd y drefn amaethyddol drwy Ewrop yn weddol gyflym gan gyrraedd Cymru oddeutu 5,000 o flynyddoedd yn ôl. Gwelwyd datblygiadau araf ac anghyson mewn technolegau cynhyrchu ynni, megis y gallu i ddefnyddio ynni'r gwynt a llif dŵr i yrru melinau a llongau i hwylio'r cefnforoedd. Defnyddiwyd peth glo i gynhesu tai ac, yn Tsieina, i gynhyrchu haearn. Drwyddi draw, ynni ffotosynthetig cyfoes oedd yn gyrru cymdeithas hyd nes cyn lleied â thri chan mlynedd yn ôl (gw. Hugh Thomas, *An Unfinished History of the World*, 1981 a Jared Diamond, *The World Until Yesterday*, 2013). Ond gyda chwilfrydedd, dyfeisgarwch ac ariangarwch yn cael eu hymgorffori mewn dysg, gwyddoniaeth, technoleg a chyfalafiaeth, agorwyd y drws i chwyldro newydd, sef y Chwyldro Diwydiannol.

Nid oes gofod i drafod y chwyldro yn fanwl yn yr erthygl hon ond gellir mesur ardrawiad y chwyldro yng ngeiriau'r masnachwr Matthew Boulton yn 1776. Boulton oedd noddwr James Watt, a ddatblygodd yr injan stêm led effeithlon gyntaf o'i bath, ac meddai Boulton wrth James Boswell, newyddiadurwr o'r cyfnod, 'I sell, Sir, what the world desires to have – power.' Mae ei eiriau yn adlewyrchu dyfeisgarwch ac uchelgais y cyfnod i weddnewid rheolaeth y byd drwy ddefnyddio ffynhonnell newydd o ynni; yr ynni hwnnw oedd i'w gynhyrchu drwy losgi tanwydd ffosil, sef gwaddol ffotosynthesis a storiwyd yng nghrombil y ddaear ers degau o filiynau o flynyddoedd. Hyn a newidiodd gwrs y byd ac a ryddhaodd ddynoliaeth o'i dibyniaeth ar ffotosynthesis cyfoes.

Carlamodd y chwyldro yn ei flaen gyda pheiriannau amgenach na dyfais flaengar Watt. Yn fuan dyfeisiwyd peiriannau trydan, olew, diesel a phetrol. Trwy losgi'r tanwydd cafwyd digonedd o ynni cyfleus at ddibenion diwydiant a thrafnidiaeth, gan newid effeithiolrwydd cynhyrchu a chrebachu pellteroedd y byd. Gweddnewidiwyd masnach a hamdden o ganlyniad. Addaswyd cynlluniau dinasoedd i dderbyn trenau, a cheir yn ddiweddarach. Mae'n werth nodi, yn y cyswllt

Cymreig, y disodlwyd glo gan olew a nwy fel y prif danwydd ffosil, ac o ganlyniad symudodd y pŵer a'r cyfoeth o Gaerdydd a'r Rhondda i Dhahran a Doha a chyfnewidiwyd grym Protestaniaeth gyfalafol am Islam Wahabaidd. Daeth cyfalafiaeth ac ariangarwch yn fodelau i'w hefelychu a chyfrannodd golud, ar raddfa nas gwelwyd o'r blaen, at gynnydd sylweddol yn y boblogaeth, er i'r mwyafrif ohonynt aros yn dlawd. Sicrhawyd 'tra-arglwyddiaeth dyn', gan ddefnyddio geiriau'r Beibl. Gwelwyd ymhelaethu sylweddol ar y broses o droi'r ynni newydd toreithiog yn waith ac yn bŵer, a arweiniodd at greu cymhlethdodau materol a chymdeithasol newydd i ddynoliaeth.

Ystyriwn un enghraifft o ddylanwad pellgyrhaeddol llosgi tanwydd, sef mewn gyriant ceir a lorïau. Rhaid oedd buddsoddi er mwyn adeiladu ffyrdd a systemau i reoli'r cerbydau hyn a chreu isadeiledd eang i gyflenwi tanwydd ar eu cyfer ac i ddiweddaru'r cerbydau yn gyson. Cafwyd damweiniau lu a datblygodd yr angen i yswirio'r gyrwyr ac i roi triniaeth i'r anffodusion. Yr oedd y systemau yn agored i dwyll, y teiars yn llygru a'r gyriant yn cyfrannu at wenwyno byd-eang. Newidiwyd ein dulliau o fasnachu ac o gymryd ein gwyliau – bychan fuasai'r galw am dai haf ym mryniau Cymru heb y car, a gwahanol iawn fyddai bywydau ffermwyr mynydd. Buasai'n hawdd ychwanegu at y gadwyn uchod ac amlinellu cadwyni tebyg ar gyfer llongau, trenau ac awyrennau ac i'r holl ddyfeisiadau a ddaeth yn sgil ynni'r Chwyldro Diwydiannol. Hynny yw, yn sgil yr ynni a'r dechnoleg, cododd yr angen i ddatblygu systemau rheoli er budd yr unigolyn a'r gymdeithas ac, i raddau, er budd yr amgylchedd. Gellir cyffelybu systemau rheoli'r chwyldro gyda'r systemau homeostatig a esblygodd yn fiolegol gyda dyfodiad bodau amlgellog.

Yn sgil yr holl gyfnewidiadau, tyfodd poblogaeth y byd o gwta biliwn yn oes Boulton i'r ffigwr presennol sy'n nesu at wyth biliwn. Drwy'r ail chwyldro amaethyddol, a gynhelir yn rhannol gan danwydd ffosil a'r gwyddorau biolegol newydd, llwyddwyd, yn groes i ddarogan Malthus, i fwydo canrannau helaeth o'r boblogaeth ychwanegol, ond ysywaeth nid pawb. Roedd yr ail chwyldro amaethyddol hefyd yn dibynnu ar y grym i feddiannu tiroedd estron yn yr Amerig ac ardaloedd eraill.

I grynhoi fy nadl, mae ynni ffosil nid yn unig wedi creu twf materol ond hefyd dwf mewn cymhlethdod cymdeithasol. Mae hwn yn ganlyniad anochel i'r gallu i gyflawni mwy o waith materol ac i'r pŵer a'r grym a gyfyd ohono. O ganlyniad, cododd yr angen i greu trefn, rheolaeth a disgyblaeth (homeostasis). Galluogodd hyn berchnogion y pŵer, sef gwladwriaethau a chwmnïau'r Gorllewin, i wladychu ac i ddominyddu gweddill y byd am ddegawdau. Nodir, wrth fynd heibio, i arfau milwrol ddatblygu yn ddull o gyfeirio ynni a phŵer at ddibenion rheoli a gorfodi.

Effeithiwyd yn ddwys ar ecoleg y blaned gan achosi, mewn perthynas ag ynni, ganlyniadau niweidiol allyriadau nwy carbon deuocsid a ryddheir o losgi'r holl danwydd ffosil. O ganlyniad, newidiwyd cydbwysedd mewnlif ac all-lif ynni'r haul, sy'n golygu bod mwy a mwy o'i wres yn cronni yn y moroedd a'r awyr. Y ffenomen hon sy'n arwain at gynhesu byd-eang ac at y cyfnewidiadau dwys yn hinsawdd y blaned. Hyn sydd wrth wraidd y Seithfed Chwyldro a fydd yn ganolbwynt fy nhrafodaeth yn ail ran yr erthygl hon, i'w chyhoeddi yn rhifyn Haf 2018 *O'r Pedwar Gwynt*.

Ynni, gwaith a chymhlethdod II

Yn rhan gyntaf yr erthygl hon, a gyhoeddwyd yn rhifyn Gwanwyn 2018 *O'r Pedwar Gwynt*, amlinellais chwe chwyldro ynni yn hanes y Ddaear (gw. y tabl yn Rhan I). Dadleuais i'r chwyldroadau ynni hyn fod yn greiddiol i ddatblygiad ac esblygiad bywyd ar ein planed. Buont yn gyfrifol am drawsffurfio cyfansoddiad yr atmosffer a'r cefnforoedd ac, i raddau, y lithosffer, dros gyfnod o 4 biliwn o flynyddoedd.

Wrth wraidd fy nadansoddiad mae'r berthynas ffisegol rhwng llif ynni a'r gallu a ddaw yn ei sgil i wneud gwaith ac, o ganlyniad, i greu pŵer – sef y gwaith a gyflawnir mewn uned o amser. O wneud gwaith a defnyddio pŵer, cynhyrchir cymhlethdod materol ac, yn fwy diweddar, gymhlethdod cymdeithasol ac economaidd. Dadleuaf yn y rhan hon o'r erthygl bod fframwaith ddeallusol y chwe chwyldro ynni yn ein cynorthwyo nid yn unig i ddehongli datblygiad bywyd ar y Ddaear ond i ddeall yn well yr argyfwng presennol, hynny yw, her yr hyn a alwaf y seithfed chwyldro, sydd eisoes ar ein gwarthaf.

Cyflenwodd y chweched chwyldro – y Chwyldro Diwydiannol – ynni rhad ar sail llosgi tanwydd ffosil. Arweiniodd hynny at olud anghredadwy i ganran sylweddol o boblogaeth y Ddaear. Ond mae'n glir ers tro, bellach, bod ein defnydd o danwydd ffosil nid yn unig yn halogi ein Daear ond yn peryglu gwareiddiad. Rhyddheir tunelli di-ri o nwyon tŷ gwydr anthropogenaidd i'r atmosffer; gwelir simsanu difrifol yn y cydbwysedd rhwng yr ynni a dderbynnir gan ein planed oddi wrth y haul a'r ynni a adlewychir yn ôl gan y Ddaear i'r gofod. O ganlyniad mae'r byd yn araf gynhesu a'r hinsawdd yn newid ac yn mynd yn llawer mwy anwadal. Byr iawn o amser sydd gennym i ddiddyfnu dynoliaeth o'i dibyniaeth ar ynni hydrocarbonau ffosil.

Yn hanfodol i'n dealltwriaeth o'r berthynas rhwng ynni a chymhlethdod mae'r cysyniad o 'homeostasis'. Hon yw'r broses sy'n sefydlogi'r stwythurau cymhleth a gyfyd o'r gwaith a ddaw o'r llif ynni. Yn hanesyddol datblygodd ein dealltwriaeth o homeostasis ar sail ymchwil fywydegol. Ond gan ddilyn gweledigaeth y niwrofiolegydd blaenllaw o Bortiwgal, Antonio Damasio, dadleuaf y dylid dehongli llawer o strwythurau cymdeithasol a chyfreithiol dyn fel ymestyniad o'r un cysyniad hwn.

Ar lefel fiolegol, gwyddom fod pob cell fyw yn dibynnu ar homeostasis i sefydlogi a chynnal ei hun fel uned o drefn gywrain, er bod yr amgylchedd o'i chwmpas yn tueddu at anhrefn. Rhaid wrth gyfres o adweithiau soffistigedig i sicrhau cydbwysedd mewnol er mwyn i'r gell oroesi a chenhedlu. Dadlennol yw cymharu dylanwad homeostasis ar hirhoedledd cell unigol gyda hynt trobwll mewn afon neu gorwynt trofannol. Adeiladir trobwll lled sefydlog a deinamig mewn afon o grych mewn llif cyflym; hynny yw, mae'r 'cymlethdod materol' lled sefydlog sy'n cael ei greu yn y dŵr yn ddibynnol ar ynni cinetig yr afon. Ond diflanna'r trobwll yn syth os newidir grym y llif. Yn yr un modd, perthyn adeiladwaith rhyfeddol o gywrain – a pheryglus – i gorwyntoedd. Dibynna eu strwythur ar sugno ynni o gefnforoedd cynnes. Eto, os yw'r storm yn croesi dyfroedd oerach neu'r tir mawr ac yn colli ei chyflenwad o ynni, yna diflanna'r adeiladwaith a gostegir y gwyntoedd. Syml a byrhoedlog yw'r strwythurau ffisegol hyn. Dibynna strwythurau ffisegol a bywydegol fel ei gilydd ar lif o ynni ond, yn dyngedfennol, bodau byw yn unig sy'n meddu ar beirianwaith homeostatig – peirianwaith sy'n eu sefydlogi a'u cynnal ac yn eu

Llun: Ştefan Ungureanu

galluogi i ymateb i newidiadau yn yr amgylchedd allanol.

Mewn celloedd 'syml' procariotig, megis bacteria ac archaea, ni welir isadeiledd amlwg o fewn y gell a amgylchynir gan bilen allanol. O ganlyniad, cyfyngir yr angen am beirianwaith homeostatig i'r berthynas rhwng gofod mewnol trefnus y gell a'r amgylchedd allanol. Mae'r sefyllfa yn fwy cymhleth mewn celloedd ewcariotig, fel a drafodwyd yn rhan gyntaf yr erthygl hon. Mewn celloedd ewcariotig gwelir israniadau o fewn y gell megis meitocondria, cloroplastau, niwclei a'r gwagolion; a phob organel wedi ei hamgylchynu gan bilen. Cyfyd yr angen, felly, am gyfundrefnau homeostatig i sicrhau cydlyniad a chydweithrediad boddhaol rhwng yr organelau a'i gilydd a gyda'r gell yn ei chyfanrwydd. Wrth gwrs, parhau mae'r angen hefyd am reolaeth homeostatig rhwng y gell a'i hamgylchedd.

Hyd heddiw, ceir llaweroedd o greaduriaid ungell ewcariotig (y protistiaid), ond dros sawl can miliwn o flynyddoedd esblygodd creaduriaid aml-gell yn ogystal. Cyfyd her homeostatig newydd yn sgil amlgelledd. Esblygwyd y prosesau homeostatig er mwyn cydlynu nid yn unig weithgaredd y celloedd unigol a'u horganelau, ond swyddogaethau gwahanol fathau o gelloedd ac organau hefyd. Enghraifft led gyfarwydd yw'r corff dynol (a mamaliaid eraill o waed cynnes). Gŵyr pawb am yr angen i reoli cyfansoddiad, tymheredd a phwysau ein gwaed o fewn terfynau cyfyngedig ac am y negeseuon a drosglwyddir trwy gyfrwng yr hormonau yn y gwaed a thrwy ein nerfau i gydlynu gweithgareddau ein cyrff – cyrff sy'n cynnwys cannoedd o wahanol fathau o gelloedd gyda'u swyddogaethau penodol eu hunain a thriliynau o gelloedd unigol. Gwelwn fod yr egwyddor homeostatig yn weithredol ar lefelau uwch, felly, yn ogystal. Syniaf am y camau homeostatig hyn fel cyfres o risiau 'Grisiau Hierarchiaeth Homeostatig' – am fod pob gris yn dibynnu ac yn adeiladu ar y gyfundrefn lai cymhleth islaw.

Eisoes yn y bedwaredd ganrif ar bymtheg ysgrifennodd y ffisiolegydd enwog o Ffrancwr, Claude Bernard, bod bywyd rhydd ac annibynnol yn

Claude Bernard yn y Collège de France, 1888. Darlun gan Léon Augustin L'hermitte

dibynnu ar sefydlogrwydd yr amgylchfyd mewnol ('La fixité du milieu intérieur est la condition d'une vie libre et indépendante.'). Mewn geiriau eraill, ni all anifail fodoli heb homeostasis. Ychwanegwyd dimensiwn newydd i'r cysyniad hwn gan Antonio Damasio. Yn ôl Damasio, rhaid ystyried camau uwch, sef y galwadau homeostatig sy'n codi o'r ffyrdd mae anifeiliaid, gyda'u galluoedd ymenyddol amgenach, yn ymateb ac yn adweithio i'w gilydd ac i'w hamgylchedd. Mae'r prosesau hyn hefyd yn hanfodol i sefydlogrwydd, iechyd a 'boddhad' y creadur pan fydd yn cydfodoli mewn grwpiau neu o fewn 'cymdeithas'. Gan ddilyn teithi meddwl yr athronydd Spinoza, ystyria Damasio bod emosiynau a theimladau mewn anifeiliaid yn deillio o'r awydd creiddiol i osgoi poen ac i goleddu pleser, sef gris arall ar y grisiau homeostatig. Maent yn rhan o'r drefn ffisiolegol/seicolegol sy'n sicrhau parhad a boddhad y creadur. Yr enghraifft glasurol yw'r reddf mewn dyn neu anifail i ddianc neu i ymladd pan fo peryg. Diddorol cymharu hyn gyda gallu cell brocariotig i nofio oddi wrth fygythiad neu tuag at fwyd.

Dengys gwaith Damasio bod modd gwahaniaethu yn ffisiolegol

rhwng ein hemosiynau a'n teimladau a bod ein hymatebion emosiynol yn digwydd yn lled 'ddifeddwl'. Dadleua fod emosiwn yn deillio – yn niwrolegol a ffisiolegol – o integreiddio gweithgaredd ein hymennydd a'n cyrff (yn groes i ddeuoliaeth Descartes). I raddau helaeth mae emosiynau yn codi yn 'awtomatig' yn yr ystyr eu bod yn digwydd heb i'r anifail 'ystyried' llawer, os o gwbl.

Mae casgliadau niwrofiolegol Damasio yn ategu ac yn cryfhau gwaith seicolegwyr ymddygiadol megis Daniel Kahneman, Amos Tversky a Jonathan Haidt. Dangosodd Kahneman (enillydd Gwobr Nobel Economeg 2002 ac awdur y gyfrol boblogaidd *Thinking, Fast and Slow*, 2011) fod ymatebion dynol i heriau allanol yn ffurfio dwy ffrwd a elwir ganddo yn Systemau 1 a 2. Mae Kahneman yn galw ein hymatebion sydyn, awtomatig yn ymatebion System 1, tra bod ymatebion System 2 yn dibynnu ar ymbwyllo, myfyrio a rhesymu.

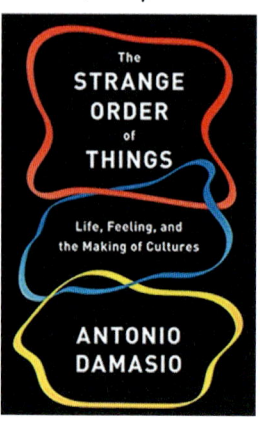

Rydym yn hynod amharod, mae'n debyg, i ymwneud â thybiaethau System 2. Tybiaethau System 1, gyda'i hoptimistiaeth, ei rhagdybiaethau a'i rhagfarnau, piau ein penderfyniadau bron yn ddieithriad. O bryd i'w gilydd, trown at dybiaethau System 2 ond gan amlaf, gwneir hynny i geisio cyfiawnhad am benderfyniadau a wnaethpwyd eisoes gan ein tybiaethau System 1. O ystyried y prosesau emosiynol ac ymddygiadol hyn, ynghyd â thystiolaeth Damasio, mae'n rhesymol awgrymu iddynt ddatblygu er mwyn cynnal bywyd mewn grwpiau bychan cyntefig mewn amgylchiadau heriol a pheryg yn ystod y ddwy filiwn o flynyddoedd ers y Pedwerydd Chwyldro.

Fel y gwelsom yn Rhan I, lleihaodd yr amser rhwng y chwyldroadau ynni trawsnewidiol yn rhyfeddol. Mae cannoedd o filiynau o flynyddoedd yn gwahanu'r pedwar chwyldro cyntaf; dim ond tua dwy filiwn o flynyddoedd sydd rhwng chwyldroadau pedwar a phump, a chwta 12,000 o flynyddoedd rhwng y pumed a'r chweched chwyldro. Wynebwn y seithfed chwyldro o fewn 250 mlynedd i'r chweched. Cymerodd y chwyldroadau cynnar amser sylweddol i aeddfedu ac i ymestyn dros y Ddaear. Ond erbyn y chweched chwyldro, gweddnewidwyd ein planed, er gwell neu er gwaeth, mewn ychydig dros ganrif. Mae'r berthynas rhwng ynni, pŵer, cyflymder y datblygiadau a throsiant y cymhlethdodau yn ystyriaeth bwysig. Mae mwy o ynni a phŵer, o gofio'r ffiseg, yn anochel yn cyflymu'r newidiadau yn y byd biolegol a dynol; hynny yw, cyflymir y trosiant.

Dros y blynyddoedd gwelwyd hefyd newid mawr yn neinameg y gystadleuaeth oesol i oroesi ac i genhedlu. Yn y byd procariotig ni cheir rhyw; rhennir genynnau a DNA yn lled rwydd rhwng rhywogaethau ymhlith y bacteria a'r archaea. Yn y byd ewcariotig, ar y llaw arall, cyfyd cyfathrach rywiol a detholiad naturiol Darwinaidd/ Wallacaidd clasurol sydd hefyd yn sicrhau goroesiad y genynnau mwyaf defnyddiol.

Ar ôl y Pedwerydd Chwyldro, roedd y mân rywogaethau *Homo* hefyd yn wynebu cystadleuaeth Ddarwinaidd. Llwyddodd rhai i oroesi yn rhannol oherwydd eu gallu i resymu ac i gydweithio a, thrwy hynny, gadarnhau'r buddsoddiad ynni mewn esblygiad ymenyddol. Trosglwyddwyd y wybodaeth a'r sgiliau yn y tylwyth o un genhedlaeth i'r llall. Ond fe ddaeth datblygiad i ddibynnu ar etifeddiaeth ddiwylliannol hefyd. Datblygodd y gallu i ddethol a throsglwyddo syniadau a sgiliau, neu 'memes' i ddefnyddio terminoleg Richard Dawkins, yn gynyddol arwyddocaol. Gyda datblygiad cymdeithasau trefol yn dilyn y Chwyldro Amaethyddol dwysaodd y gystadleuaeth gymdeithasol rhwng y dinasoedd a rhwng eu harweinwyr. Gwelwyd ymerodraethau milwriaethus yn cystadlu am adnoddau, am bŵer a rhwysg, gan feddiannu'r tiroedd gorau (hynny yw, ennill adnoddau ychwanegol o ynni ffotosynthetig) a gormesu poblogaethau (sicrhau gwaith dynol i'w dibenion eu hunain). Yn sgil y Chwyldro Diwydiannol, breiniwyd cystadleuaeth mewn gwleidyddiaeth, busnes a diwydiant yn enw Darwiniaeth Gymdeithasol. Parhaodd rheolaeth dros dir a chynnyrch ffotosynthetig i fod yn bwysig ond tyfodd cystadleuaeth gyfalafol, ddiwydiannol a diwylliannol yn gynyddol arwyddocaol.

Er y pwyslais ar 'gystadleuaeth', ni ddylid ychwaith anghofio'r elfennau o gydweithio a chydlynu fu'n rhan annatod o'r holl chwyldroadau. Rhain sy'n nodweddu'r uno symbiotaidd sy'n caniatáu meitocondria a chloroplastau o fewn celloedd ewcariotig; y cydweithio soffistigedig rhwng celloedd mewn bodau amlgellog; y cyd-fyw sy'n nodweddu cen a chwrel a llu o enghreifftiau eraill; a'r gallu dynol i resymu a chydweithio. Yn baradocsaidd, felly, mae cydweithio wedi bod, ac yn parhau i fod, nid yn unig yn rhinwedd ynddo'i hun ond yn arf cystadleuol.

Explosion 3, 2011

Newidiodd natur 'cystadleuaeth' o un chwyldro i'r llall ond nid oes tystiolaeth i'r gyneddf i ymgiprys am bŵer leihau. Os rhywbeth, i'r gwrthwyneb: yn ein byd presennol gorseddir cystadlaethau fel nod rhinweddol. Yn ddadlennol, pan gollir rheolaeth homeostatig ar dwf a chydweithrediad celloedd yn ein cyrff, gelwir hynny yn ddadfeiliad, yn 'gancr'. Ond pan gollir rheolaeth yn y byd masnach neu ym myd gwleidyddiaeth, er i hyn arwain at ormes a rhyfeloedd, yn lled aml adnabyddir yr unigolion sy'n gyfrifol fel 'arwyr'.

Mae tair agwedd arall yn berthnasol i'r patrymau a ddatgelir gan y chwe chwyldro. Yn gyntaf, er i ddynoliaeth ffynnu yn y Chwyldro Diwydiannol ar gefn ynni ffosil, ni leihaodd ein dibyniaeth ar ynni ffotosynthetig. Dengys gwaith Marc Imhoff a Lahouari Bounoua (2006) bod dyn wedi hawlio, ar gyfartaledd, tuag 20 y cant o gynnyrch ffotosynthetig y blaned ar gyfer ei ddibenion personol erbyn diwedd yr ugeinfed ganrif. Yn Ewrop, roedd y ffigwr yn uchel, yn 70 y cant; yn Ne Asia roedd yn uwch fyth, yn 80 y cant, er ei fod yn llawer is mewn ardaloedd llai poblog heb ddiwydiannau trwm. Gyda'r twf mewn cyfoeth ac ym mhoblogaeth y byd (o tua 6 biliwn pan gynhaliwyd yr ymchwil gyntaf, gryn ugain mlynedd yn ôl, i dros 7.5 biliwn erbyn heddiw), ofnaf i'n dibyniaeth ar ynni ffotosynthetig chwyddo i'r fath raddau nes peryglu adnoddau byw pob creadur.

Yn ail, mae cymhariaeth o botensial datblygiadol y tri chwyldro cyntaf yn awgrymu bod nenfwd i'r cymhlethdod materol a ganiateir ar bob gris. Analluog oedd bywyd procariotig (cyn ac ar ôl datblygiad ffotosynthesis), yn yr un modd â bywyd wedi esblygiad celloedd ewcariotig, i ddatblygu y tu hwnt i gyrhaeddiad arbennig. Yr oedd cyfyngiad hefyd ar ddatblygiad materol a chymdeithasol yr helwyr-gasglwyr cyn iddynt ddysgu sut i reoli tân, coginio ac ymelwa o'i fanteision. Awgrymwyd gan Ian Morris fod cyfyngiadau tebyg ynghlwm wrth y Chwyldro Amaethyddol. Yn ôl ei ddadansoddiad ef, nid oedd modd i'r gyfundrefn amaethyddol ddatblygu i gynnal mwy na'r llwyddiant economaidd a chymdeithasol a gyrhaeddwyd gan Ymerodraeth Rhufain yn y gorllewin a Brenhinlin Song yn Tsieina. O gyplysu'r dadansoddiadau hyn, ceir awgrym cryf bod i bob un o'r chwe chwyldro ynni ei nenfwd ei hun.

Yn drydydd, rhaid tanlinellu pwysigrwydd llif gwybodaeth. Perthyn trosglwyddiad gwybodaeth i bob un o'r chwyldroadau. Fe'i gwelir yn nhrosglwyddiad y cod genynnol a'r gallu i gynaeafu ynni'r haul a'r fframwaith ewcariotig. Yn dilyn esblygiad ymenyddol *Homo* gwelir

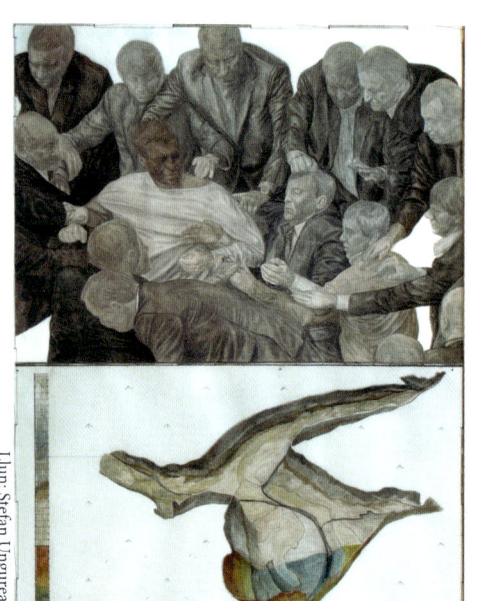

Llun: Ștefan Ungureanu, 2011

Battle for Anthropocene

trosglwyddiad gwybodaeth bersonol. Yn ôl Daniel Everett, sbardunodd y pedwerydd chwyldro angen dybryd am ffyrdd amgenach i gyfathrebu. Yn ei dyb ef, i ateb y galw hwn y datblygwyd yr ieithoedd cynharaf, cyntefig. Roeddynt yn fodd i alluogi cydweithio a chyd-drefnu mwy effeithiol. Am bron y ddwy filiwn o flynyddoedd hyn cynhelid y gadwyn esblygiadol nid yn unig trwy gyfrwng rhyw ond hefyd gan y genhedlaeth hŷn yn trosglwyddo gwybodaeth i'r ieuanc. Yn dilyn y Chwyldro Amaethyddol atgyfnerthwyd y trosglwyddiad cymdeithasol trwy ddatblygiad recordio, ysgrifennu a chyfrif. O ganlyniad i ddatblygu'r gallu i ysgrifennu daeth yn bosib, i raddau, i ddiogelu holl sgiliau, hanesion, chwedlau a chredoau y llwythau a'r cenhedloedd. Ar ben hyn, daeth cyfleon i drethu, i reoli'n fwy llym ac i hybu busnes a masnach. Ac erbyn ein cyfnod ni, a hynny'n syfrdanol o gyflym, daeth byd yr algorithmau, y cyfrifiaduron, y ffonau bach a throsglwyddiadau electronig i dra-arglwyddiaethu ar brosesau cyfnewid gwybodaeth, gan glymu'r Ddinas yn Llundain i'r un gyfundrefn â'r pentref mwyaf diarffordd yn y Trydydd Byd.

Rhaid pwysleisio maint y newid yn ein defnydd o ynni a ddaeth yn sgil y Chwyldro Diwydiannol. I gynnal y corff dynol defnyddir isafswm o 2,000 cilocalori y dydd, gydag oddeutu eu chwarter yn cael eu neilltuo ar gyfer yr ymennydd. I gynhyrchu a chyflenwi'r mewnbwn hwn o gilocalorïau defnyddir tua deg gwaith mwy o ynni o'r haul ac o ynni ffosil. Ond i gynnal safon byw presennol Cymro neu Brydeiniwr mae galw am yn agos at 240,000 o gilocalorïau y dydd. Canlyniad y gofyn anferthol am ynni yw cynhyrchu allyriadau enfawr a pheryglus o garbon deu

Rydym i gyd heddiw yn ymwybodol o'r dystiolaeth ddiymwad ynghylch newid hinsawdd anthropogenaidd a'i berthynas â llosgi tanwydd ffosil. Mae gwneud y newidiadau angenrheidiol yn ein ffordd o fyw yn fater o frys ond hefyd yn gymhleth. Yn y seithfed chwyldro, wynebwn her gwbl newydd. Yng Nghytundeb Paris Rhagfyr 2015 cytunodd 195 o wledydd y byd y dylid atal cynnydd cyfartalog o 2°C yn nhymheredd ein byd ac ymdrechu i gadw'r cynnydd at ddim uwch nag 1.5°C. Yn ystod yr El Niño diweddaraf yn 2016, roedd y tymheredd eisoes ~1.2°C yn uwch nag yn Oes Fictoria, sef ar gychwyn cyfnod ymlediad y Chwyldro Diwydiannol drwy'r byd. Wrth reswm, felly, hyd yn oed heb y llusgo traed a di-frawder yr Arlywydd Trump a'i debyg, mae'r amser i weithredu yn frawychus o brin. Cwta ugain mlynedd sydd gennym i gyflawni uchelgais Paris a gwireddu'r seithfed chwyldro; ugain mlynedd i fabwysiadu ffyrdd di-garbon o gynhyrchu ynni, i reoli'r cynnydd yn nhymheredd y byd ac osgoi trychinebau cyfan gwbl apocalyptaidd. Mae cymharu'r amserlen hon gyda'r cyfnod o 250 o flynyddoedd ers dechrau'r Chwyldro Diwydiannol a'r 12,000 o flynyddoedd ers y Chwyldro Amaethyddol yn sobri dyn.

Rhaid tynnu sylw at rai ffeithiau ysgytwol eraill. Mae'r carbon deuocsid cynyddol a gynhyrchir gan ein ffordd o fyw yn dosbarthu'n gyflym trwy'r atmosffer ac yn parhau yn yr atmosffer am gyfnod cymharol hir (›100 mlynedd). Mae'r allyriadau mewn un rhan o'r byd yn cael effaith ar y Ddaear yn ei chyfanrwydd ac yn parhau yn yr atmosffer am genedlaethau. Cyfanswm yr allyriadau ers y chweched chwyldro yw'r ystadegyn tyngedfennol. Goblygiadau hynny yw po fwyaf poblogaeth y byd, lleiaf y ddogn a ganiateir i bob unigolyn. Hefyd, po fwyaf ein hallyriadau presennol, lleiaf yr allyriadau y bydd yn ddiogel i'n plant ac i blant ein plant eu cynhyrchu. Nid yw'r allyriadau rhyngwladol a gynhyrchir y pen yn agos at fod yn gyfartal. Yn gyffredinol, po gyfoethocaf y wlad (a'r unigolyn), mwyaf yw ei hôl-troed carbon (gwerth nodi bod Gwledydd Llychlyn, i raddau, yn eithriad). Felly, i gyrchu ateb i her yr hinsawdd ac i geisio rheoli a rhannu ein hadnoddau cyffredin yn rhesymol deg, rhaid wynebu'r anghyfartaledd dwys sy'n bodoli rhwng pobl, gwledydd a chenedlaethau.

Soniais eisoes am waith Damasio a Kahneman a'n hymateb greddfol, homeostatig i symbyliadau neu gyffyrddiadau allanol. Nodais fod y rhain, i raddau helaeth, yn awtomatig ac yn dilyn tybiaethau System 1 Kahneman. Y tueddiad yw i ni fod yn llwythol-blwyfol, yn or-optimistaidd, yn rhagfarnllyd, i fyw i'r funud ac yn gyndyn neu'n analluog i drin gwybodaeth ystadegol a risg hir-dymor. Rydym yn blaenoriaethu cadw yr hyn sydd gennym lawer mwy na chwenychu enillion posib yn y dyfodol. Bathodd Kahneman y term WYSIATI i grynhoi ein hymddygiad: 'What you see is all there is!' Mae ymgodymu â phroblemau newid hinsawdd yn herio ein rhagdybiaethau, yn gofyn am ddadansoddiad manwl ystadegol, asesiad o risg anweledig, hir-dymor, a chydymdeimlad trawsffiniol a thrawsgenedlaethol. Y tebygrwydd yw y bydd rhaid i fywydau y mwyaf breintiedig hefyd newid. Nid syndod felly bod sicrhau consenws gweithredol rhyngwladol yn anodd – mae'r drwg yn ein genynnau.

Yn ail, er i gyfalafiaeth a chwmnïau ragddyddio'r Chwyldro Diwydiannol, heb os bu'r ynni a'r pŵer a ddatblygodd yn sgil y chwyldro hwn yn gatalydd grymus i'r gyfundrefn gyfalafol, i'r twf anhygoel yn ei chyraeddiadau ac, yn llawn mor rhyfeddol, yn y boblogaeth ddynol. Datblygodd cyfundrefn economaidd a gwleidyddol sydd, fel 'top' plentyn, yn ansefydlogi wrth arafu. Yn waeth, o fewn ein trefn economaidd rhaid i'r economi dyfu a chyflymu'n *flynyddol* i sefydlogi. O ganlyniad, yn ystod yr ugeinfed ganrif, traflyncwyd mwy a mwy o adnoddau ac ynni. Mae'n hynod anodd, felly, dianc oddi ar yr olwyn droed hon heb ddatgymalu'r holl gyfundrefn. Ymateb y rhai sy'n credu'n gryf yn rhagoriaethau'r drefn gyfalafol, megis yr Arglwydd Lawson a'r Global Warming Policy Foundation, yw gwadu realiti newid hinsawdd a chynghreirio gyda'r rhai sydd yn elwa o ddiwydiannau llosgi hydrocarbon i hau amheuon.

Yn drydydd, mae'n werth ystyried beth fyddai'r canlyniadau pe datblygid digonedd o ynni rhad, di-garbon. Oni fyddai hyn, yn ôl tystiolaeth y chwe chwyldro blaenorol, yn cyflymu'r holl brosesau sy'n ychwanegu at gymhlethdodau materol a chymdeithasol ein byd a'n bywydau unigol? Un canlyniad anochel fyddai diflaniad llawer o'r hyn sy'n weddill o fyd natur. Soniais eisoes ein bod, fel dynoliaeth, er ein dibyniaeth ar danwydd ffosil, yn manteisio ar fwy a mwy o ynni ffotosynthetig y byd. Yn ôl pob tebyg, byddai ffynhonnell ynni ddi-garbon a diwaelod yn cyflymu trosiant yr economi a'n defnydd o adnoddau eraill y blaned ac yn caniatáu poblogaeth ddynol ymhell dros ddeg, o bosib hyd at ugain biliwn. Byddai hyn, yn anochel, ar draul gweddill y biosffer.

Beth fyddai effaith naid arall mewn cyflymdra a chystadleuaeth ar foddhad a lles y ddynoliaeth? Pregethir cenadwri cystadleuaeth gan ein gwleidyddion ond prin y cydnabyddir bod i bob cystadleuaeth ei chollwyr yn ogystal â'i henillwyr. Pregethir cenadwri twf diddiwedd heb gydnabod ei oblygiadau. Rhaid cofio bod gan gyfundrefnau cymhleth eu nodweddion arbennig eu hunain a ymgorfforir yn namcaniaeth Caos. Ystyr hyn yw y gallasai cryniadau bychan lleol ysgogi cyfnewidiadau anferthol, annisgwyl ym mhen arall y byd. Gallasai peryglon cymdeithasol, gwleidyddol ac economaidd anrhagweladwy ddeillio o'r cymhlethdod ychwanegol a'i ansadrwydd. Yn ogystal, rhaid codi amheuon am yr effeithiau ar iechyd meddwl *Homo sapiens* – 'y dyn doeth'. Heb os, mae Homo sapiens yn greadur hynod hyblyg gyda galluoedd annisgwyl ac amrywiol ond mae'r dystiolaeth o draul y byd aflonydd presennol ar foddhad ac iechyd meddwl oedolion a phlant yn cynyddu.

Dengys ymchwil anthropolegol bod grwpiau cyntefig yn cydweithio'n glòs er mwyn goroesi ond, o berchnogi adnoddau ychwanegol, maent yn amddiffyn eu mantais hyd yr eithaf (gw. Jared Diamond, *The World until Yesterday*). Gwelir meddylfryd debyg yng nghestyll y Canol Oesoedd neu mewn cymunedau heddiw y'u gwarchodir gan giatiau caeedig, er enghraifft yn Ne Affrica a'r Unol Daleithiau, ac yn rhethreg rhai o'n gwleidyddion – 'taking back control' neu 'America first' – sloganau System 1 perffaith. Defnyddiwyd Darwiniaeth Gymdeithasol i gyfiawnhau'r cysyniad fod rhai pobloedd yn fwy teilwng na'i gilydd ac yn haeddu adnoddau gwell na'u cydddyn llai ffodus. O goleddu'r athroniaeth hon yng nghyd-destun newid hinsawdd, wynebir galanastra. Mae Cytundeb Paris, er gwaethaf ei ffaeleddau, yn ymwrthod â'r feddylfryd hon ac yn derbyn bod cydweithio a rheolaeth ryngwladol yn hanfodol.

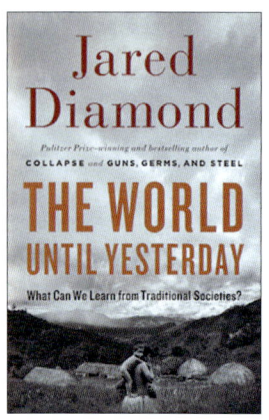

Er gwaethaf fy nadansoddiad tywyll, mae gan y chwe chwyldro ynni, a'r 'Grisiau Hierarchiaeth Homeostatig', eu negeseuon gobeithiol. Drwy'r milenia gwelir bod llinyn aur cydweithredu a chyd-dynnu hefyd yn rhedeg drwy hanes. Sylfaen hyn oll, mi dybiaf, yw gofynion homeostatig celloedd a bodau byw, gan gynnwys *Homo sapiens* a'i gymdeithasau cymhleth. Dyfynna Domasio o waith Spinoza: 'Mae pob un peth, i'r graddau y mae hynny o fewn ei allu, yn ymdrechu i warchod ei fodolaeth ei hun' ac 'Yn greiddiol i'r cysyniad o rinwedd mae'r ymdrech i warchod yr hunan ac mae hapusrwydd yn gyfystyr â gallu dyn i warchod yr hunan.'

Fe ellir dehongli hyn fel sylfaen hunanoldeb a thrachwant, ond camddealltwriaeth ddybryd fyddai hynny o syniadaeth Spinoza a

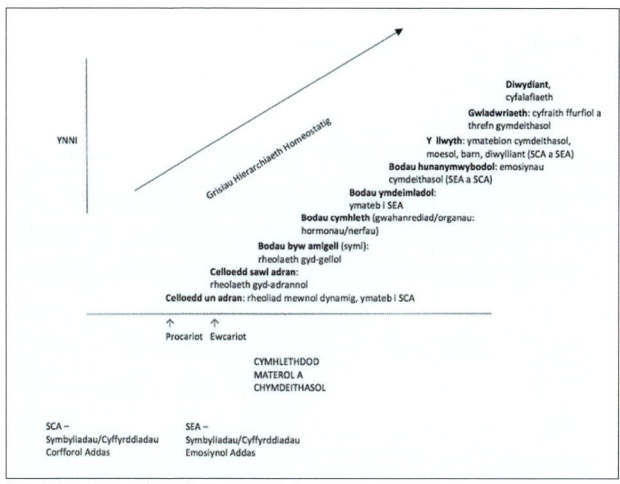

Y Grisiau Hierarchiaeth Homeostatig

Domasio. Fel y nodais, ystyria Domasio fod ein hymatebion emosiynol a theimladol yn binacl pyramid rheolaeth homeostatig fywydegol. Eu prif amcan yw diogelu ein dynoliaeth, ein hunaniaeth a'n bodlonrwydd. Y gamp yw byw bywyd iach, dymunol a theilwng ac i wneud hynny o fewn cydbwysedd homeostatig sy'n ymestyn o lefel y gell unigol i'r unigolyn, ac i'r gymdeithas a'i hamgylchfyd.

Ymddengys nad yw'r cyfundrefnau homeostatig a esblygodd i ateb galwadau bywyd bregus llwythau bach ein hynafiaid yn addas i'n byd integreiddiedig, egnïol, tra-chystadleuol, digidol a thechnolegol. Ymddengys hefyd nad yw ein cyfundrefn gyfalafol, brynwriaethol yn gydnaws ag iechyd ein planed nac, o bosib, iechyd dynion a merched, er i filiynau freuddwydio y byddai'r drefn hon yn eu rhyddhau o gadwyni tlodi.

Athrylith dynoliaeth, ar y llaw arall, yw ei gallu i ddadansoddi (System 2), i ailfeddwl ac i ddilyn llwybrau amgenach. Mae gennym y gallu i ddatblygu cyfundrefnau yn ogystal â thechnolegau newydd. Yr her greiddiol yw argyhoeddi pobl bod hyn nid yn unig yn bosib ond yn gwbl angenrheidiol.

Ofnaf nad oes llawer o obaith osgoi cynhesu byd-eang i o leiaf 2°C, ar gyfartaledd. Bydd yr ymateb i'r seithfed chwyldro yn rhy araf ac, o ganlyniad, bydd rhaid addasu i newidiadau niweidiol. Ond erys rhai cwestiynau mawr. A gyraeddasom benllanw'r gyfres o chwyldroadau ynni – chwyldroadau sydd wedi ymestyn dros 4 biliwn o flynyddoedd, pob un ohonynt yn trosi mwy fyth o ynni yn waith a phŵer, yn gymhlethdod materol neu gymdeithasol, ac yn cyflymu curiad bywyd? Pe llwyddid i gynhyrchu digonedd o ynni a phŵer di-garbon, oni fyddai hynny'n arwain at greu cymhlethdodau materol a chymdeithasol pellach a fyddai mor enbyd nes lleihau, yn hytrach na chynyddu, ein buddiannau a'n lles?

Oni fyddai pwyllo a byw yn effeithlon ar lawer llai o ynni yn gam i'r cyfeiriad iawn ac yn cynnig gwell dyfodol i'r ddynoliaeth? Mae wynebu'r seithfed chwyldro yn gyfrwng i ailystyried taith dynoliaeth a'n blaenoriaethau. A all y 'primat doeth' ddarganfod ffyrdd amgenach o drefnu cymdeithasau'r dyfodol?

Diolch i John Llew Williams am ei gymorth anhepgor gyda'r erthygl hon.

Y saith chwyldro - y cyfweliad llawn

Cynog Dafis yn holi R Gareth Wyn Jones

Rydych chi'n trafod hanes y blaned a hanes y ddynoliaeth yn eich darlith *Ynni, Gwaith a Chymhlethdod*. **Mae'r cyfnod amser sydd dan sylw yn rhychwantu pedwar biliwn o flynyddoedd, o gyfnod ffurfio'r blaned hyd at heddiw. Mae gyda chi ddamcaniaeth hynod o uchelgeisiol, os ca' i fentro dweud hynny. Rydych wedi rhoi perspectif cwbl newydd i mi ar bethau. Dwi'n credu bod cyhoeddi'r ddarlith hon yn ddigwyddiad o bwys mawr. A dwi wrth fy modd bod y Coleg Cymraeg Cenedlaethol ac** *O'r Pedwar Gwynt* **wedi penderfynu rhoi sylw i'r gwaith. Eich thesis chi yw bod ynni wedi chwarae rhan hollol allweddol mewn chwe chwyldro yn hanes y ddynoliaeth. Rydych yn sôn am seithfed chwyldro ond canolbwyntiwn ni i ddechrau ar y chwe chwyldro cyntaf. Beth sy'n gyffredin rhwng y chwe chwyldro hyn; ym mha ystyr maen nhw'n debyg?**

Y prif reswm maen nhw'n debyg ydi'r ddibyniaeth ar ffynhonnell newydd o ynni, neu newid sylweddol yn yr ynni sydd ar gael i'w ddefnyddio. Y ddamcaniaeth, yn syml iawn, ydi hyn: o gael ffynhonnell o ynni, galluogir gwneud gwaith, ac mae'r gwaith hwnnw'n creu cymhlethdod; mae hynny'n wir o safbwynt y ffiseg wreiddiol hefyd. O safbwynt bywyd ar y ddaear, er mwyn creu'r gell, sef y gyfundrefn sy'n galluogi bywyd i fodoli, mae'n rhaid i chi gael ffynhonnell o ynni. A dyna i chi sail y chwyldro cyntaf: ffynhonnell o ynni sydd yn galluogi creu cyfundrefn barhaol, sef cell sydd yn medru cynnal ac atgynhyrchu ei hun ac ymateb i newidiadau allanol mewn ffyrdd cadarnhaol, er mwyn goroesi; hynny trwy fanteisio, er enghraifft, ar ffynonellau bwyd neu trwy ddianc rhag peryglon. Dros gyfnod maith y chwe chwyldro cafwyd newid sylfaenol yn y ffynonellau ynni hyn ac yn effeithlonrwydd y defnydd o'r ffynonellau, hyd at y dydd heddiw. Gyda llaw, ar ôl chwilota, dwi'n darganfod rŵan bod eraill wedi dweud pethau nid annhebyg!

Llun: Ştefan Ungureanu

Rydych chi'n sôn bod y chwyldro ynni ym mhob achos yn galluogi cyflawni gwaith ychwanegol a hynny, yn ei dro, yn creu cymhlethdod. Yn chwyldroadau'r cyfnod cynharaf, cymhlethdod biolegol sydd dan sylw, ac yn y chwyldroadau mwyaf diweddar, cymhlethdod cymdeithasol.

Mae hynny'n hollol gywir, ond mae yna elfen o gymhlethdod materol, o greu mater, yn perthyn i bob un. Pan ystyrir y bywydegol, amlgelledd sydd dan sylw, o lewod i goed derw, ond yn achos datblygiad dynol mae'r cymhlethdod yn un cymdeithasol yn ogystal â materol.

Fedrwch chi esbonio pob un o'r chwyldroadau yma yn eu tro i ni? Mae'n werth nodi'r hyn rydych chi'n ei bwysleisio yn eich darlith [gw. 'Ynni, gwaith a chymhlethdod', Rhifyn Gwanwyn 2018], sef bod y cyfnod amser rhwng y chwyldroadau hyn, drwy'r oesoedd, yn byrhau'n ddifrifol. Rydyn ni'n dechrau gyda chyfnodau o biliynau o flynyddoedd rhwng chwyldroadau, wedyn miliynau o flynyddoedd, ac yna miloedd – nes, yn y diwedd, mae'r gwahaniaeth amser rhyngddyn nhw yn gannoedd o flynyddoedd yn unig.

Yn achos y seithfed chwyldro, rydan ni'n sôn am ddegau o flynyddoedd, sy'n frawychus! Ond dowch i ni ddechrau efo'r chwyldro cyntaf: er mwyn cael cell fyw, rhaid cael ynni parhaus a pharhaol. Damcaniaeth ddiweddar, gan Nick Lane o Brifysgol Llundain, ydi bod y ffynhonnell honno o ynni i'w chael mewn tyrau o greigiau tyllog sy'n codi ar waelod y môr. Ynddynt, ceir hylif alcali sy'n pistyllio i'r môr. Gan fod pH y môr yn agos at 7, yn niwtral, a pH hylif alcali o gwmpas 10, mae hyn yn creu graddiant o brotonau H+. Rydan ni wedi arfer efo'r syniad o raddiant o *electronau*, sef llif o drydan (mae yna olau dros eich ysgwydd rŵan yn adlewyrchu hynny, er enghraifft); mae graddiant o brotonau, sef gronynnau positif, nid negatif, yn debyg iawn ac, fel trydan, yn cario ynni a'r gallu i wneud gwaith. Y ddamcaniaeth ydi bod y graddiant hwn o brotonau wedi galluogi datblygiad cell mewn ffordd sy'n gyson â deddfau thermodeinameg. Ond roedd yr ynni oedd i'w gael o hynny yn gyfyngedig i ychydig o lefydd ar y ddaear. Ac yn y llefydd hyn y datblygodd bywyd.

Rydyn ni'n sôn am bedwar biliwn o flynyddoedd yn ôl, felly?
Mae'r dystiolaeth orau'n dangos bod y blaned wedi ffurfio rhyw 4.5 biliwn o flynyddoedd yn ôl a bywyd wedi dechrau ar y ddaear yma tua pedwar biliwn o flynyddoedd yn ôl – yn lled fuan wedi ffurfio'r blaned. Felly roedd y byd yn cael ei ffurfio dros gyfnod o 500,000 miliwn o flynyddoedd. Yr ail chwyldro oedd datblygiad y gallu i greu graddiant y protonau hyn trwy ddefnyddio ynni'r haul i hollti dŵr. Mi ddigwyddodd hynny o gwmpas biliwn o flynyddoedd yn ddiweddarach. Datblygwyd, yn ara' deg ac yn fiocemegol, y gallu i fanteisio ar ynni'r haul. Roedd hyn yn dibynnu ar gael cloroffyl – y cemegolion gwyrdd rydan ni'n eu gweld mewn bacteria ac mewn planhigion – i alluogi bachu ynni'r haul, ond hefyd i greu yr un graddiant o brotonau dwi newydd ei drafod. Dyma'r graddiant a ddefnyddir i syntheseiddio ATP, sef y cemegolyn mewn celloedd sy'n galluogi symud yr ynni o gwmpas y gell – nid yn annhebyg i bres mewn cell.

A'r trydydd chwyldro?

Y trydydd chwyldro ydi un o'r gwyrthiau mwyaf sydd wedi digwydd. Yn wreiddiol, dim ond celloedd syml a elwir yn rhai 'procariotig' a ddatblygodd. Roedd yna ddau fath o gelloedd procariotig: rhai a elwir yn facteria, sy'n enw cyfarwydd i ni, a rhai a elwir yn archea, sy'n ddarganfyddiad cymharol ddiweddar. Doedd neb yn gwybod am yr archaea pan oeddwn i'n dechrau astudio gwyddoniaeth. Ond ddiwedd y saithdegau a dechrau'r wythdegau, darganfuwyd procariot oedd yn tyfu mewn llefydd arbennig o galed, fel y Môr Marw.

Llefydd heriol, felly?

Dyna chi. Archaea ydi llawer o'r rheiny ac mae eu defnydd metabolaidd nhw'n wahanol i facteria. Y ddamcaniaeth ydi hyn: rhyw 1.7 biliwn o flynyddoedd yn ôl, hanner ffordd trwy oes y byd, unwyd un gell facteriol ac un gell archaeol i greu cell ewcariotig, ac mae'r celloedd ewcariotig hyn yn cynnwys niwclews a meitocondria. Esblygodd y meitocondria o facteria a draflyncwyd gan yr archaea ac a gollodd y rhan fwyaf o'i enynnau. Mae'r meitocondria hwn yn gweithredu fel pwerdy bach yn y gell. O astudio celloedd eich iau, er enghraifft, mi ellir gweld bod o gwmpas dwy fil ohonyn nhw mewn un gell. Felly, ers yr uno hwn, mae cell ewcariotig yn medru cynhyrchu o gwmpas 300,000 gwaith yn fwy o ynni i bob genyn yn y gell. Mae hynny wedi newid ecoleg ynni celloedd yn y byd yn syfrdanol. O'r newid mawr hwn y datblygodd pob creadur aml-gell. Mae'r bacteria yn dal yma, yr archaea yn dal hefo ni, sef parhad y byd procariotig; maen nhw'n dal yn holl bwysig, yn hanfodol i'r byd hwn. Ond o greu'r naid i'r byd ewcariotig, crëwyd potensial ar gyfer bywyd llawer mwy cymhleth, bywyd amlgellog. Ar ôl oedi hir yn yr Oes Gambriaidd, rhyw 450 miliwn o flynyddoedd yn ôl, mi gafwyd ffrwydrad o greaduriaid a 'adeiladwyd' o gelloedd ewcariotig. Ac o hynny y mae pob un ohonom yn tarddu.

Rydyn ni'n sôn fan hyn am ddechrau rhyw a dechrau cystadleuaeth?

Llawer mwy o gystadleuaeth. Y dystiolaeth ydi bod celloedd bacteria a'r archea yn medru trosglwyddo a rhannu genynnau, ac, wrth gwrs, mae hynny'n creu problem i ni heddiw. Er enghraifft, mewn ysbytai, os ydi un math o facteria yn datblygu y gallu i wrthsefyll gwrthfiotigau, maen nhw'n medru rhannu'r 'wybodaeth' yna efo'i gilydd heb gyfathrach rywiol, yn sydyn ofnadwy. Mae'n cynnig potensial i'r celloedd procariotig, ond mae'n bosib fod hyn wedi eu dal nhw'n ôl hefyd, oherwydd doedd yr elfen gystadleuol ddim yno i roi hwb iddynt.

Mae'r chwyldro nesaf, y pedwerydd chwyldro, yn agosach atom ni o lawer.

Yn ôl y dystiolaeth, er bod mwncïod wedi dechrau datblygu ymennydd mwy galluog ymhell yn ôl, mi oedd fforch esblygiadol yn eu hwynebu. Roedd hi'n bosib buddsoddi mewn cryfder, nerth bôn braich felly, mewn bol mawr mi fedrwn ni ddweud, neu mewn ymennydd, mewn gallu. Cyn rhyw ddwy filiwn o flynyddoedd yn ôl, doedd yna ddim digon o galorïau yn y bwydydd i ganiatáu buddsoddiad mewn ymennydd drudfawr, galluog. A'r ddamcaniaeth, yn ôl Richard Wrangham o Brifysgol Harvard, ydi bod ein cyndeidiau, *Homo erectus* yn ôl pob tebyg, wedi darganfod sut i ddefnyddio tân, a'i ddefnyddio i goginio, a bod y calorïau ychwanegol sy'n dod o goginio bwyd yn cael eu buddsoddi yn yr ymennydd. Mae hyn wedyn yn ein gwneud yn fwy galluog i ddal mwy o fwyd a'i goginio'n fwy effeithlon, gan sefydlu cylch cadarnhaol. Ystyrir i hyn ddigwydd o gwmpas rhyw 2 filiwn o flynyddoedd yn ôl. O ganlyniad, mae canran sylweddol iawn o'n calorïau yn cael ei defnyddio gan yr ymennydd; dyna lle mae tua chwarter y calorïau yn mynd. Yn ara' deg, rhwng 2 filiwn o flynyddoedd yn ôl ac o gwmpas 300,000 o flynyddoedd yn ôl, mi ddatblygodd yr amrywiol homonidau, sef Homo erectus, y soniais amdano eisoes, *Homo neanderthalensis* a *Homo heidelbergensis* ac eraill, ac yn y pen draw,

Homo sapiens, sef ni. Y naid hanfodol, felly, oedd rheoli tân, coginio bwyd a buddsoddi ynni yn yr ymennydd. A dyma ni gymlethdod yn dechrau lledu o'r bywydegol i'r cymdeithasegol.

Wedyn rydyn ni'n dod at amaethyddiaeth.
Am gannoedd o filoedd o flynyddoedd, mae'n debyg bod poblogaeth *Homo erectus*, ac yna *Homo sapiens*, yn fychan iawn. O gwmpas 70,000 o flynyddoedd yn ôl, ar ôl ffrwydrad mawr mewn llosgfynydd yn yr ardal a elwir yn Indonesia heddiw, roedd y boblogaeth ddynol i lawr i fil, yn ôl rhai. Ychydig iawn oedd wedi goroesi. Grwpiau bach, bach oedd y rhain, yn gorfod ymladd am eu bodolaeth. Cofiwch, yn y cyfnod hwn, mi oedd yna o leiaf bedair oes iâ, a'r iâ yn cynyddu ac yn cilio, a'r cnewyllyn bach yma o *Homo erectus*, wedyn Homo sapiens, wedi goroesi trwy hynny i gyd. Yn ara' deg, roedd eu galluoedd nhw'n cynyddu, wrth i ynni gael ei fuddsoddi yn yr ymennydd ac, felly, mewn gallu.

Rydyn ni'n symud fan hyn nawr o'r miliynau i'r miloedd o flynyddoedd yn ôl. Rhyw bum i ddeng mil o flynyddoedd yn ôl y digwyddodd y pumed chwyldro, y chwyldro amaethyddol?
Dyna pryd y dechreuwyd dofi anifeiliaid ac y gwelwyd cynnydd yng nghnwd rhai mathau o laswellt, er enghraifft gwenith a haidd yn y Dwyrain Canol, reis yn Tsieina ac india-corn yn ardal Mecsico heddiw. Dechreuwyd gwella'r planhigion hyn a'u cynhyrchu mewn ffyrdd amaethyddol ar ddiwedd yr oes iâ ddiweddaraf. Ac wedyn, datblygwyd pentrefi a chymunedau a oedd yn rhannol amaethyddol ond hefyd yn parhau i ddibynnu ar hela. Doedd dim naid sydyn, felly, o hela i amaethu. Dechreuwyd defnyddio tir a oedd yn cael ei ddyfrio, er enghraifft i lawr yn Mesopotamia, a'r un peth yn y dyffrynnoedd yn Tsieina. Wrth i fwy o fwyd gael ei gynhyrchu yn y ffordd hon, roedd y boblogaeth yn cynyddu a gwelwyd datblygiad o fath arall: pentrefi a threfi sefydlog, a threfn a hierarchiaeth yn y trefi hyn. Roedd yna lawer mwy i'w amddiffyn o ganlyniad. Ar y cyfan, does yna ddim tystiolaeth

bod y bobl gyffredin ar eu hennill rhyw lawer; doedd eu hiechyd ddim cystal ag iechyd y rhai a oedd yn dal i hela, ond roedd y rhifau'n cynyddu a'r ynni ffotosynthetig yn cael ei ddefnyddio at waith y gallai pobl ei gyflawni. Roedd y gwaith hwnnw wedyn yn cael ei ddefnyddio i gynyddu grym, yn bŵer i'r uchelwyr yn y gymdeithas, nes creu ymerodraethau bach.

Dyna lle mae ynni'n chwarae ei ran, felly? Trwy amaethu, roedd y ddynoliaeth yn medru tynnu ar ffynhonnell ynni ychwanegol a dyna oedd yn gyrru popeth arall?

Dwi wedi bod yn meddwl am hyn gryn dipyn yn ddiweddar, yng nghyddestun ystadau mawr Ewrop, ac yng Nghymru hefyd – beth oedd ystad ar un adeg ond ffordd o reoli ynni, ynni ffotosynthetig? Os oedd gynnoch chi ystad o dir da, roeddech yn bwydo llawer o bobl, a oedd wedyn yn dibynnu arnoch, ac yn gweithio i chi; roedd hon yn ffordd o roi grym i chi'ch hun. Dychmygwch y dylanwad fyddai gan Gymru petai tir da i'r gorllewin, petai Cantre'r Gwaelod heb gael ei foddi ac ystadau ffrwythlon gynnon ni i'r gorllewin. Byddai hanes Cymru yn wahanol.

Byddai hanes Prydain yn wahanol. Dyna pam rydyn ni'n galaru cymaint am Gantre'r Gwaelod!

Mi oedd o'n newid sylweddol a gymerodd filoedd o flynyddoedd i ddatblygu drwy'r byd. Mi oedd pobl yn dal i hela tra oedd hyn i gyd yn mynd ymlaen. Ac maen nhw'n dal i wneud mewn rhannau o'r byd heddiw, wrth gwrs. Cafwyd datblygiadau eraill yn y cyfnod hwn: dechreuwyd defnyddio ynni adnewyddadwy, trwy gyfrwng melinau gwynt a dŵr ac ati. Mae yna lyfr diddorol gan Ian Morris, *Why the West Rules – For Now* (2010), sy'n cymharu datblygiad yn y gorllewin, sef yn Rhufain ac yn y Dwyrain Canol, efo datblygiad yn Tsieina. Mae'n dod i'r casgliad bod yna benllanw i ddatblygiad cymhlethdod cymdeithasol sydd yn bosib o ddibynnu ar ynni ffotosynthetig amaethyddol yn unig; bod y penllanw a welwyd yn Rhufain yn y gorllewin, a'r diwylliant Song yn Tsieina, wedi digwydd oherwydd bod y gymdeithas honno wedi dibynnu ar amaethyddiaeth ac ynni ffotosynthetig yn unig.

Mae hyn yn dod â ni at y chweched chwyldro. Cannoedd o flynyddoedd sydd dan sylw nawr, a'r chwyldro diwydiannol wedi digwydd rhyw ddau gan mlynedd a hanner yn ôl.
Dyma'r darganfyddiad bod ynni yn cuddio yng nghrombil y ddaear, wedi bod yn cuddio yno am gannoedd o filoedd o flynyddoedd, a bod modd defnyddio'r adnoddau hyn – glo yn wreiddiol, wedyn olew a nwy, fel ffynhonnell ynni. Yn y pen draw, roedd hyn i gyd yn dibynnu ar yr haul. Mi fedrwch ddeall pam roedd pobl yn addoli'r haul yn eu hamser! Ond trwy ddefnyddio gwaddol yr oesoedd, gweddnewidiwyd y byd.

Mae'n deg dweud bod yna newidiadau pwysig iawn wedi digwydd cyn y chwyldro diwydiannol, wrth gwrs, yn enwedig o safbwynt ein hymwneud â thechnoleg a gwyddoniaeth a hefyd cyfalafiaeth. Ond y cyfuniad o gyfalafiaeth, technoleg a'r gallu i ddefnyddio tanwydd ffosil i gynhyrchu pŵer sydd wedi gweddnewid y byd, nid dim ond ein defnydd o danwydd ffosil.

Cyn i ni droi at y seithfed chwyldro: yn y stori hon, mae dyfodiad yr homonid, datblygiad y ddynoliaeth, yn ffactor trawsnewidiol ynddo'i hun. Fyddech chi'n hoffi dweud rhywbeth ynghylch sut rydych yn gweld y ddynoliaeth, am nodweddion dyn fel creadur, am ei hynt drwy'r oesoedd fel petai, a'i effaith ar y byd, ar yr amgylchfyd naturiol, ar rywogaethau eraill? A ydych chi'n gweld dyn yn greadur unigryw?
Mae'n gwestiwn llawer mwy anodd nag mae'n ymddangos. Mae'r berthynas rhwng y defnydd o ynni a bywydeg dros gyfnod o biliynau o flynyddoedd wedi trawsnewid y byd. Yr ail chwyldro, sef ffotosynthesis, sydd gen i dan sylw fan hyn. O ocsigeneiddio'r awyr, mi newidiodd hynny'r byd yn gemegol; mae yna newidiadau syfrdanol wedi digwydd yn ddaearegol ac yn ddaearyddol oherwydd hynny. Ond erbyn hyn,

mae'r sefyllfa wedi newid eto. Rydan ni'n sôn am ddylanwad un rhywogaeth yn unig ar y ddaear gyfan. Beth sy'n syfrdanol i mi ydi, er nad ydan ni'n dibynnu'n uniongyrchol ar ffotosynthesis heddiw, rydan ni'n dal i fachu mwy fyth o'r defnydd ffotosynthetig yn y byd.

O'r tanwyddau ffosil felly?
Yn ogystal â'r tanwyddau ffosil, mae'r ganran o diroedd y byd rydan ni'n eu ffarmio, o'r coedwigoedd a dorrir i'r dŵr a ddefnyddir i ddyfrio'n cnydau, yn dal i gynyddu. Ar ben y defnydd o danwydd, rydan ni'n defnyddio adnoddau eraill, ac felly, rhwng popeth, yn cystadlu efo bron pob anifail yn y byd, ar wahân i'r rhai sydd wedi bod yn ddigon doeth i ddod i gytundeb symbeiotig efo ni, fel defaid a geifr. Mi rydan ni'n ddigon parod i edrych ar eu holau nhw! Ond wrth gwrs, maen nhw'n cael eu lladd yn ifanc, yn talu pris go fawr am y berthynas efo'r ddynoliaeth. Felly erbyn heddiw, mae'n dylanwad ni'n sylweddol iawn a does dim gofod mwyach i'r hen anifeiliaid mawr oedd yn crwydro'r ardal hon ddim ond ychydig filoedd o flynyddoedd yn ôl.

Mae'n anodd iawn gweld sut mae modd newid hynny. Un o'r elfennau eraill sydd yn llawn mor bwysig ag ynni yw'r berthynas rhwng ynni a chymhlethdod. Mae yna bont yn eu cysylltu, sef homeostasis. Rhaid i unrhyw fath o gell reoli ei *milieu*, ei chynefin mewnol, er mwyn cynnal cysondeb. Rhaid i gell hefyd ymateb yn barhaus i newidiadau allanol. Homeostasis ydi enw'r broses hon; mae'n ganolog i'r cysylltiad rhwng ynni, gwaith a chymhlethdod. Dros y blynyddoedd, datblygodd un gell syml yn gell gymhleth, yna'n anifail neu greadur amlgellog nes creu'r ddynoliaeth. Mae'n bwysig iawn gosod y datblygiad hwn yng nghyd-destun homeostasis, sef y broses sy'n gwarchod 'bodlonrwydd' y gell.

Fel ei bod yn gydlynol, felly?
Fel ei bod yn medru dal ei thir, yn fewnol; heb hynny, mi fyddai ar chwâl.

Mae homeostasis yn gysyniad allweddol, felly, ac mae hynny'n

dod â ni at drothwy'r seithfed chwyldro. Rydyn ni'n byw mewn oes anthropogenaidd. Mae tra-arglwyddiaeth ein rhywogaeth ni ar y blaned yn llethol. Goblygiadau hynny yw ein bod ni'n llethu popeth arall. Rhan o effaith hyn yw'r argyfwng amgylcheddol, sy'n llawer mwy amlweddog na dim ond newid hinsawdd, e.e. colli bioamrywiaeth. Rydych chi'n trafod hyn yn fanwl yn eich darlith ac yn rhagweld ein bod ar drothwy seithfed chwyldro. Ai'r hyn ry'ch chi'n ei olygu gan hynny yw ein bod yn ymwrthod â defnydd o danwydd ffosil, er bod digon ar ôl, yn harneisio ynni adnewyddadwy ar raddfa helaeth iawn, iawn? Dydych chi ddim yn sôn yn eich darlith o gwbl am ynni niwclear. Ydw i'n iawn i gymryd mai dyna sylwedd y seithfed chwyldro?

Mae'n fwy cymhleth na hynny. Rydan ni wedi creu dibyniaeth anhygoel ar danwydd ffosil a hynny'n syfrdanol o sydyn. Mae'n gwbl amlwg bod rhaid symud ar fyrder oddi wrth yr arfer o losgi tanwydd ffosil – ugain mlynedd, dyna'r cyfan, sydd gynnon ni i wneud hynny. Y cwestiwn ydi sut. Mae'n gwestiwn technegol. Mae ynni adnewyddadwy yn rhan o'r ateb, wrth gwrs. Pwy a ŵyr. Dwi ddim yn ffafrio ynni niwclear, nid oherwydd fy mod yn meddwl y byddai'n beryg rhoi ynni niwclear yn yr Wylfa fel y cyfryw ond oherwydd nad ydw i'n gweld sut, ar lefel fyd-eang, mae'n bosib datblygu ynni niwclear mewn ffordd ddibynadwy o fewn ugain mlynedd. Dwi wedi gweithio mewn llefydd fel Aleppo yn Syria ac yng ngogledd Nigeria, lle mae Boko Haram rŵan, ac mae'r syniad o gael atomfeydd yn y llefydd ansefydlog hynny yn erchyll i mi. Dwi'n ofni bod newid hinsawdd am greu byd mwy ansefydlog fyth yn yr ugain neu'r deng mlynedd ar hugain nesaf yma; mae'r syniad ein bod ni'n ceisio ateb y broblem trwy ddefnyddio technoleg sydd â'i phroblemau ei hun yn beryg.

Ond mae hyn yr un mor bwysig: os edrychwch chi ar y patrwm dwi'n ei olrhain yn fy narlith, mae pob chwyldro wedi cyflymu prosesau bywyd; mae'n arbennig o wir am y chwyldro diwydiannol. Felly, os darganfyddir ffynonellau sy'n cynnig digonedd o ynni di-garbon yn y dyfodol, mi fydd hynny'n arwain at fwy o ynni, mwy o waith a mwy o rym. Os ydi'r ddamcaniaeth yn gywir, mi fydd hynny'n golygu mwy o ganoli grym ac mi fydd bywyd yn mynd yn fwyfwy anodd i'r rhan fwyaf o bobl. Felly, os llwyddwn trwy ddatrysiadau technegol, dwi'n ofni mai canlyniad hynny fydd creu pwn mawr i ni'n hunain. Dyma'r cwestiwn mawr: a ydi hi'n bosib defnyddio'r seithfed chwyldro i ailystyried ein perthynas efo'r ynni rydan ni'n ei ddefnyddio a'r byd rydan ni'n byw ynddo?

Mae hynny'n golygu mai dim ond rhan o'r ateb ydi'r ffics technolegol?
Rhaid wrth ddatrysiad technolegol, er enghraifft i leihau'r methan a gynhyrchir gan anifeiliaid. Ond tu hwnt i hynny, dwi'n amheus y bydd yn ddigonol. Dwi'n meddwl bod yr argyfwng yn llawer mwy dwys. Dwi'n besimistaidd yn yr ystyr yna. Dyma sialens fwyaf y ddynoliaeth. Rydan ni wedi llwyddo i wella'n byd am gannoedd os nad am filoedd o flynyddoedd, trwy ddefnyddio ynni yn y ffordd hon. A oes yna ben draw i hynny?

Applied magnetism, 2013, acrylic ar ganfas

Llun: Ştefan Ungureanu

Ac mae twf ym mhob ystyr yn digwydd o'n cwmpas, a twf yn sacrosanct i bolisïau economaidd. Sut ydyn ni'n symud o'r dryswch presennol? Rydych chi'n dweud, os ydyn ni'n parhau i dyfu, bod i hynny ganlyniadau go ddifrifol.

Mae yna effeithiau clir ar yr amgylchfyd, ond beth am y goblygiadau arnom ni fel pobl? Rydan ni'n gweld eisoes y pwysau y mae cymaint o'r genhedlaeth iau yn ei ysgwyddo, a'r hyn sy'n debyg o ddigwydd ydi y bydd grŵp bach yn elwa ar y sefyllfa. Dwi'n ofni mai dyna ddigwyddith. Bydd y ddynoliaeth yn goroesi trwy leihau carbon, ond bydd llawer iawn o'r byd yn mynd yn ddiffeithwch a grŵp bach yn rheoli.

Rydych chi wedi ysgrifennu am y model economaidd cyfredol ac wedi sôn mewn llefydd eraill am fargen Galbraith. Ar sail y fargen honno, ry'n ni'n derbyn bod twf yn digwydd, oherwydd bod rhyfaint o'r twf hwnnw yn gallu cael ei ddefnyddio i wella lles y lliaws. Ond rydych chi o'r farn bod hynny'n rysáit amheus, oherwydd bod twf economaidd diddiwedd yn creu problemau economaidd difrifol yn y pen draw.

Mae Galbraith a Keynes yn dweud ein bod wedi taro bargen Ffawstaidd efo traha. Rhaid i chi gymryd gofal mawr wrth weithredu yn y ffordd hon, meddai Galbraith, neu fel arall ceir problemau dybryd. A dyna sydd wedi digwydd. Mae'n amlwg bod rhaid rheoli'r broses hon ond mae yna gwestiwn mwy dwys na hynny, sef ydi'r model yn galluogi rhannu'n deg pan mae'r adnoddau'n brin. Mae newid hinsawdd yn enghraifft glasurol o'r hyn a elwir 'the tragedy of the commons'. Rydan ni'n anorfod yn rhannu effeithiau'r carbon deuocsid ychwanegol sy'n cael ei greu gan y ddynoliaeth – mae'n cael ei drosglwyddo i bob rhan o'r byd. Mae hyn yn effeithio ar bawb yn y pen draw ond tydi'r amgylchiadau cyfredol ddim yn caniatáu i ni ddod i delerau ag effeithiau hyn i gyd.

Beth am y syniad yma o'r 'tir comin byd-eang'? Pan o'n i'n Aelod Seneddol, buais yn gweithio gyda dyn o'r enw Aubrey Meyer

oedd yn rhedeg sefydliad bychan ond hynod ddiddorol o'r enw y Global Commons Institute. O safbwynt Meyer, y dasg yw sicrhau ein bod ni, fel dynoliaeth, yn defnyddio'r amgylchedd – gan gynnwys yr holl elfennau hynny sy'n perthyn i ni gyd: hinsawdd, bioamrywiaeth, cyfoeth natur – fel 'tir comin'. Y peryg yw bod lleiafrif pwerus yn cipio'r 'tir comin' hwn ac yn ei ddefnyddio at eu pwrpas eu hunain, fel a ddigwyddodd yng nghefn gwlad Cymru ddechrau'r bedwaredd ganrif ar bymtheg.

Tydi'r rhagolygon ddim yn dda. Tan yn ddiweddar iawn, roeddan ni'n byw mewn llwythau bychain, mewn grwpiau bach teuluol, a'n hymlyniad at ein llwyth bach ni, ein cenedl, yn bopeth; hwyrach nad oeddem yn edrych tu hwnt i hynny, a hwyrach heddiw, am y tro cyntaf yn ein hanes, ein bod ni'n gorfod meddwl amdanom ein hunain fel dynoliaeth. Synio am ein hunain fel Cymry, Ffrancwyr ac ati ydi'r arfer wedi bod, ac mi rydan ni'r Cymry, wrth gwrs, yn hoff o sôn am ein milltir sgwâr ...

Ond y cwestiwn nawr yw pa un ai a yw'r ddynoliaeth yn mynd i allu gweithredu fel cymuned gydwladol, fyd-eang gan rannu adnoddau a chyfocth yn deg er lles pawb, nid dim ond er lles y lleiafrif, a chydnabod, wrth gwrs, yr hyn nad yw wedi ei gydnabod ddigon tan yn gymharol ddiweddar, sef bod adnoddau'n gyfyngedig. Oes yna rai arwyddion gobeithiol?

Beth am gytundeb Paris?

Mae cytundeb Paris wedi bod yn wirioneddol wych o safbwynt y byd diplomataidd, o safbwynt gwleidyddol, ond o safbwynt y wyddoniaeth does neb wedi ymrwymo i wneud y newidiadau angenrheidiol ac mae pethau wedi gwaethygu oddi ar hynny. Mae Trump wedi wfftio'r peth yn gyfan gwbl, ac felly rydan ni mewn sefyllfa lle mae un o wledydd mwyaf pwerus y byd yn tynnu i'r cyfeiriad arall. Mae'n anodd bod yn optimistaidd. Fy marn i ydi bod newid hinsawdd eithaf damniol am ddigwydd, a bod hynny'n anochel. Dwi ddim yn ein gweld ni'n osgoi codiad tymheredd llai na dwy radd Celsius.

Clywais glerigwr unwaith, Bruce Kent, yn dweud: un peth yw optimistiaeth, ond mae bod yn ddiobaith yn waharddedig. Fedrwn ni ddim caniatáu rhoi'r gorau i obeithio. Rydyn ni'n gorfod byw mewn gobaith am fyd lle mae dynoliaeth yn cyd-dynnu, yn derbyn egwyddor cydweithrediad yn hytrach na chystadleuaeth noeth, ddilyffethair.

Ond rhaid derbyn bod chwyldroadau dymchweliadol wedi digwydd. Ystyriwch, er enghraifft, y Pla Du, pan gollodd Ewrop draean ei phoblogaeth, neu 70,000 o flynyddoedd yn ôl, pan grebachodd y ddynoliaeth yn fil o bobl. Dwi'n ffyddiog y bydd y ddynoliaeth yn parhau ond dwi ddim yn sicr y bydd y drefn bresennol, sy'n gyfleus i ni heddiw, yn goroesi. Tydi'r dystiolaeth ddim yn ddigon cadarn i awgrymu fel arall.

Ond rydych chi hefyd yn credu bod angen model economaidd newydd, i ddisodli neoryddfrydiaeth?

Oes, yn sicr, dyna pam dwi'n sgwennu'r pethau hyn, achos mae'n rhaid creu'r drafodaeth ar sail y wybodaeth orau sydd ar gael wrth edrych i'r dyfodol.

A bydd rhaid i hynny olygu adnewyddiad moesegol? Dwi'n gwybod bod gyda chi ddiddordeb mewn crefydd. Rydych chi'n wyddonydd i waelod eich traed ac i fêr eich esgyrn ond mae gyda chi ddiddordeb mewn crefydd hefyd.

Tydi bod yn rhesymegol yn unig ddim yn ddigonol. Tydan ni ddim yn medru dibynnu ar ein hymateb rhesymegol; rydan ni angen ystyried llawer mwy ar ein hymateb emosiynol a'r gwerthoedd emosiynol sy'n llywio bywydau. Mae'r byd rydan ni'n byw ynddo fo heddiw yn pwysleisio ac yn gwobrwyo materoliaeth ac mae angen ailfeddwl hynny'n llwyr. Heb ailfeddwl hynny, mae'n anodd iawn ailfeddwl y dimensiwn economaidd.

Mae angen newid diwylliannol, felly?
Mae'n bwysig iawn ystyried yn ofalus sut i symud gam ymlaen, gan gofio ein bod yn paratoi er lles y dyfodol – er y bydd amgylchiadau anodd dros ben yn siŵr o ddatblygu. I roi enghraifft i chi: pan oeddwn i yn Aleppo, Syria rhyw ugain mlynedd yn ôl – yn cynrychioli llywodraeth Prydain yno, ar ryw olwg – roedd newid hinsawdd eisoes ar y bwrdd, a ninnau'n trafod beth fyddai'n digwydd pan fyddai pobl yn trio dianc i Ewrop. Ond ychydig ddigwyddodd yn sgil y trafodaethau hynny i baratoi at y newidiadau hyn, er bod y llywodraethau yn llawn sylweddoli'r datblygiadau tebygol. Cyfrifoldeb pobl fel fi ydi canu'n groch am y problemau, gan obeithio y bydd y bobl mewn grym yn gwrando.

Rydych yn llais proffwydol oherwydd rydych yn herio doethineb a moeseg eich cyfnod ac yn dweud: os ewch chi ymlaen fel hyn, mi fydd hi'n annibendod mawr arnom ni.
Dwi ddim yn ddyn digalon ar y cyfan, ond mae'n rhaid bod yn ymarferol.

Mae gobaith yn well cysyniad nag optimistiaeth.
Ond mae gobaith yn debyg o godi ar ôl tipyn o gwymp. Dwi ddim yn meddwl bod modd osgoi rhyw fath o drychineb.

'Dyw'r ffics technolegol ddim yn mynd i fod yn ddigon. 'Dyw'r model economaidd presennol ddim yn mynd i fod yn gynaladwy. Ond cyn bod trawsnewidiad economaidd yn bosib, mae angen adnewyddiad moesol. Ac mae hynny'n troi Marcsiaeth wyneb i waered: mae Marx yn dweud mai o realiti materol mae popeth yn tarddu a bod diwylliant yn perthyn i ail lefel, fel petai, ond efallai ein bod ni wedi cyrraedd man nawr lle mae'n rhaid i'r diwylliannol a'r moesegol orfodi newid yn y

patrwm economaidd, a bod hynny'n mynd i orfod canlyn rhyw fath o adnewyddiad.

Fuaswn i ddim yn eu gwahanu nhw. Dwi'n ffan mawr o'r niwrolegydd Antonio Damasio. Tuedd pob creadur ydi amddiffyn ei hun a chreu'r gofod iddo fedru byw yn fodlon. Mae modd dehongli hynny mewn mwy nag un ystyr, er enghraifft yn yr ystyr ymosodol, sef amddiffyn ein hunain a naw wfft i bawb arall. Ond mae Damasio yn dadlau, o safbwynt niwrolegol, ffisiolegol a bywyd y gell, bod angen i ni fyw mewn ffordd sydd yn gytûn efo'n cyd-ddyn. Ac o safbwynt y cysyniad o homeostasis, mae'n angenrheidiol i bob un ohonom fod yn fodlon ein byd. Hynny ydi, os nad ydan ni'n rhan o gymdeithas fodlon, does dim modd cyflawni ein potensial fel bodau dynol. Dyna fy man cychwyn i, ac mae hynny'n deillio o'r fywydeg llawn cymaint ag o'r gymdeithaseg. Yr hyn sydd wedi digwydd ydi ein bod ni wedi gorbwysleisio'r dimensiwn cystadleuol.

Yn sgil Darwin?

Mae Darwin wedi rhoi cyfiawnhad deallusol i'r hyn oedd yn digwydd yn Oes Fictoria. Roedd yr amseru'n anffodus ar un olwg oherwydd dyma pryd yr oedd yr Ymerodraeth Brydeinig ar ei hanterth ac mi roedd damcaniaeth Darwin yn rhoi cyfiawnhad i'r safbwynt mai ni, Prydain, yw'r goruchaf, mai ni sy'n haeddu cymryd drosodd diroedd a phobl yr India. Mi ges i hyn yn yr ysgol, yn Lerpwl. Ystyrid mai ein swyddogaeth ni oedd mynd allan i redeg yr Ymerodraeth.

Mae hynny i'w weld yng ngwaith Herbert Spencer ac yn ein harwain yn syth at Natsïaeth.

Mae'n bwysig ofnadwy cau pen mwdwl y ddadl honno, mae'n gamddealltwriaeth ac mae'n cyfiawnhau Trump a'i fyd. Rhaid i ni greu deialog newydd sydd yn symud tu hwnt i hynny ond sy'n dal yn seiliedig ar y wyddoniaeth. Mae yna lyfr diddorol iawn gan Robert Skidelsky, *How Much is Enough* (2012). Yn y tridegau, mi ragwelodd Keynes, arwr i Skidelsky, y buasem wedi cyrraedd cyflwr o ddigonedd erbyn rŵan, yn gweithio tridiau'r wythnos, yn canolbwyntio ar y celfyddydau a ddim yn gweld angen y dimensiwn cystadleuol i'r fath raddau, ond mi wnaeth o gamddeall natur y gwryw alffa. I hwnnw, does yna byth ddigon, yn yr un modd nad oes yna byth ddigon i'r bobl sy'n rhedeg y cwmnïau mawr yma.

Twf parhaol felly ...

Nid jyst twf parhaol ond y gynneddf i ddangos eich bod chi'n fwy pwerus. Rydan ni wedi creu cymdeithas sy'n mawrygu hynny, a dyna sydd raid ei newid. Mae'n anodd iawn gwneud. Ond dwi'n gweld newid hinsawdd fel prawf arnom ni: a ydan ni'n medru newid pethau? Petawn i'n efengylaidd, mi fyddwn i hyd yn oed yn dweud bod Duw wedi gyrru prawf i ni. Dyna pam na fedr Trump dderbyn realiti newid hinsawdd: mae'n herio ei holl werthoedd sylfaenol o oherwydd does dim modd datrys problemau newid hinsawdd o gwbl o fewn y model economaidd hunanol presennol. Yr hyn sy'n fy mhoeni i efo Brexit ydi ein bod ni'n cael ein gorfodi i fynd ar hyd y llwybr economaidd hwn. Mae hynny'n codi arswyd arna i.

Rydych nid yn unig yn feddyliwr theoretig ac yn wyddonydd ond yn weithredol hefyd yn gymdeithasol. Rydych chi wedi bod yn ymwneud â phrosiectau datblygu ynni cymunedol yn eich ardal chithau, yn do?

Dwi wedi bod yn cynorthwyo'r bobl sy'n gwneud y gwaith ymarferol. Mae gynnon ni brosiect datblygu adnoddau Cymru ar fynd efo'r Sefydliad Materion Cymreig. Rydan ni'n ceisio defnyddio datblygu adnoddau ynni adnewyddadwy fel hwb i greu economi fwy gwydn, cymdeithasol a lleol. Rydan ni wedi bod yn gwneud y gwaith theoretig fel sail i ddatblygu'r polisïau ac er mwyn ystyried sut mae asio'r ffynhonnell ynni leol efo'r grid trydan. Mewn ffordd, ceisio gwneud gwaith y llywodraeth drostyn nhw yr ydan ni. Maen nhw'n trafod y syniad ond tydyn nhw ddim yn gweithredu hanner digon. Mae yna baradocs mawr yn hynny. Os edrychwch chi ar yr holl gyfundrefn ynni, mae'n amlwg bod rhaid trydaneiddio ceir, bysys, tai, a symud oddi wrth losgi olew a nwy ar fyrder. Mae'n sefyll i reswm felly bod y galw am drydan am gynyddu. Dylai fod modd cynhyrchu canran o'r trydan yna yn lleol. Ond, o safbwynt y llywodraeth a chwmnïau mawr, mae'n haws dibynnu ar gyflenwadau canolog. Yn anffodus, mae ffiseg a pheirianneg cynhyrchu trydan yn dueddol o fod yn groes i'r hyn sy'n llesol yn gymdeithasol a'r peryg efo ynni niwclear ydi bod grym yn cael ei roi mewn ychydig o ddwylo. Ond o gyplysu cynhyrchu trydan yn lleol efo storio'n lleol, mae modd creu cyfundrefn lle mae cymdeithas rywfaint yn fwy annibynnol ar y gyfundrefn ryngwladol. Mae bod yn hunangynhaliol yn dod â'i beryglon ei hun, wrth gwrs, ac mae ganddo'r potensial hefyd i fagu grwpiau hunanol. Mae angen nythaid o bethau, gweithredol yn lleol, gan gydnabod ein bod fel dynoliaeth hefyd yn gyd-ddibynnol.

Ydych chi'n gweld y gall Cymru, yn ei ffordd ei hun, wneud cyfraniad o bwys?
Petai gynnon ni fôr-gloddiau llanw, byddai modd i ni fod yn gwbl annibynnol o ran ein cyflenwad ynni. Mae'n biti garw nad ydi'r llywodraeth yn ymrwymo i ariannu'r cynllun hwnnw; mae'n anodd iddyn nhw wneud, wrth gwrs, gan mai Llundain sy'n rheoli'r grid a'r broses drwyddedu. Mae gan Gymru gyfraniad i'w wneud: mae Ynni Ogwen yn gynllun campus; mae ieuenctid Dyffryn Ogwen yn credu bod yna ddyfodol iddyn nhw, a'r un tueddiad i'w weld yn Llanberis. Denmarc sydd yn enghraifft dda o hyn, lle rhoddwyd ynni gwynt mewn dwylo lleol, yn groes i'r patrwm Prydeinig. Mae yna lawer y gall Cymru ei wneud.

Mae **Cynog Dafis** yn fab i weinidog Methodist, ac yn gyn-athro a chyn-wleidydd.

Profiles

R Gareth Wyn Jones

Short History

After graduating from the then University College of North Wales in 1961, I studied at the universities of British Columbia, Oxford, UCLA and UC Berkeley before returning to Bangor with my wife and young family in 1968 as a lecturer in my old department. I worked for many years on soil-plant relations and plant eco-physiology, especially the mechanisms of plant adaptation to drought and soil salinity and alkalinity. The work led to new understanding of the cellular mechanisms in plants, and indeed in all eukaryotic organisms, of tolerance to related stresses (salt, drought and, to an extent, heat). In this period I was also supported by Fellowships at the Universities of Tel Aviv, Würzburg and Sydney. In 1983 I was awarded a DSc by Oxford University and a personal chair by the University of Wales. I have published well over 200 peer-reviewed papers and book chapters and edited or written three books.

In the course of this work I developed a growing and continuing interest in the sustainable welfare of communities seeking to live in and adapt to stressful environments as well as the edaphic and climatic issues themselves. This was reinforced by a large ODA/ British Council project with the University of Agriculture in Faisalabad in Pakistan, focussed on 'Bio-saline Agriculture'. In 1983, with the late Dr Tony Chamberlain, I established the Centre for Arid Zone Studies in Bangor and, in 1989, became manager and later chair of the Strategic Plant Science Research Programme of UK Department for International Development. This resulted in my working in many countries in West Asia and Africa, both north and south of the Sahara.

On its establishment in 1991 I was appointed Chief Scientist and Deputy Chief Executive for the Countryside Council of Wales. In this post I was responsible for establishing the first major agri-environmental scheme based on payments for specific environmental goods (a pilot project, Tir Cymen, covered 10% of Wales) as well as other sustainable development projects. I later returned to my University post following deep cuts imposed on the Council by the then Secretary of State for Wales, John Redwood.

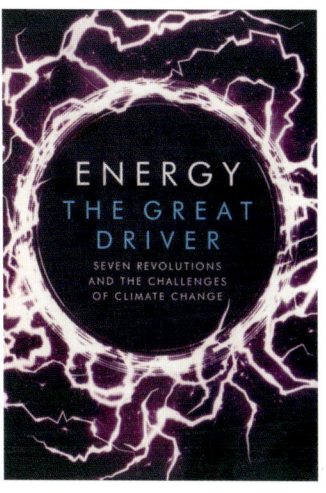

I have been a member of the Board of the International Centre for Agricultural Research in Dry Areas (ICARDA) in Aleppo, Syria, chairing the Programme Committee and served for many years as Secretary General of the International Dryland Development Commission based in Alexandria.

Jointly with Dr Einir Young, I co-ordinated an EU interdisciplinary, multinational projects on the sustainability of communal rangelands in southern Africa (GLOBALS/MAPOSDA).

Nearer home I chaired the Amlwch Industrial Heritage Trust, the Wales Rural Forum and Coed Cymru, was President of Bardsey Island Trust and was a member of the Board of the National Museum of Wales; chairing the Natural History sub-committee.

Jointly with Dr Havard Prosser, I was commissioned to produce a report on 'Land Use and Climate Change' to Welsh Government detailing ways to mitigate greenhouse gas emissions from land use and the food chain.

I have published a number of popular articles in Welsh and English on aspect of anthropogenic global warming and initiated the IWA project "Re-energising Wales". My book "*Energy – The Great Driver: Seven Revolutions and the Challenges of Climate Change*" was published by the University of Wales Press in 2019.

On retirement in 2006, I was made an Emeritus Professor and more recently an Honorary Fellow of my old University. In 2011 I was elected a Fellow of the Learner Society of Wales.

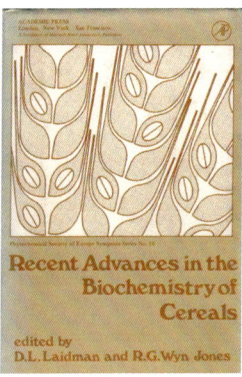

Robin Grove-White

From an Anglo Welsh background, in the 60s Robin Grove-White was a freelance scriptwriter including for the famous 'That was the week that was'. He later graduated from Oxford University in Politics, Economics and Philosophy before committing to environmental issues and becoming Director of the Campaign for the Protection of Rural England and Chairman of Greenpeace. In 1991 he established the Centre for the Study of Environmental Change (CSEC) at Lancaster University where he is now an Emeritus Professor. He is also an Honorary Fellow of Bangor University and very active in many local initiatives including Anglesey Antiquarian Society and Llanfechell community cafe and shop.

Acknowledgements

I am deeply grateful to Ali Anwar and the H'mm Foundation for their support in making the dream of a presentation in the Senedd and this volume a reality. Similarly I am greatly indebted to Jon Gower whose idea it was and whose efforts and support made it possible.

I would like to thank a number of colleagues for their advice, criticisms and encouragement and, importantly, drawing specific papers and books to my attention in a range of academic disciplines – I wish, especially, to thank Robin Grove-White and his Lancastrian colleagues, Brian Wynne, Bronislaw Szerszynski and David Tyfield; Ron Pethig; James Intriligator and his colleagues at Tufts University; Timm Hoffman; the late John Raven; Bob Ayres; Isabelle Winder; John Llywelyn Williams; Tony Rippin and Roger Leigh.

My heartfelt thanks to my wife, Ella who has supported me in this endeavour when, by rights, I should have been concentrating on the garden. I've also received invaluable help from my son, Huw, and grandson, Aled, with the preparation of the slides for the Senedd lecture and navigating and picking-up material from the web. Similarly Meg Elis, John and Lowri Llywelyn Williams have done much to improve and correct the Welsh. Our thanks also to Ant Evans for his Welsh translations of the Foreword and Preface.

Errors, be they of commission or omission, are of course mine as are any mistakes in either language.

Diolchiadau

Rwy'n hynod ddiolchgar i Ai Anwar a'r Sefydliad H'mm am eu cefnogaeth. Oherwydd eu haelioni a'i gweledigaeth daeth cyfle i gyflwyno fy neges yn y Senedd ac i gyhoeddi llyfr swmpus hwn sy'n amlinellu fy namcaniaethau. Yn yr un modd mae fy nyled i Jon Gower yn enfawr. Yfo bia'r uchelgais wreiddiol i drefnu'r ddarlith ac i gyhoeddi'r llyfr hwn.

Diolch hefyd i nifer o gymdogion am eu cyngor a'u hanogaeth a fy arwain at ddeunydd newydd pwysig. Rhaid enwi yn arbennig Robin Grove-White a'i gyfeillion o Prifysgol Lancaster, Brian Wynne, Bronislaw Szerszynski a David Tyfield; Ron Pethig; James Intriligator; Timm Hofmann; y diweddar John Raven; Bob Ayres; Isabell Winder; John Llywelyn Williams, Tony Rippin a Roger Leigh.

Fy niolch didwyll hefyd i fy ngwraig Ella am ei chefnogaeth yn fy antur pan, yn rhesymegol, dylwn ganolbwyntio ar y chwynnu. Cefais gymorth hanfodol oddi wrth fy mab, Huw, a fy ŵyr, Aled, i baratoi'r sleidiau ac i dathlwytho gwybodaeth o'r we. Diolch i'r ddau.

Mawr hefyd yw fy nyled i Meg Elis am gyfieithu rhai darnau a John a Lowri Llywelyn Williams am loywi gweddill fy Nghymraeg. Diolch hefyd i Ant Evans am ei gyfieithiadau o'r Rhagair a'r Rhagarweiniad.

Derbyniaf y cyfrifoldeb am y gwallau a erys yn y testun naill ai o esgeulustod neu o ddiofalwch neu o gamsynio ac am gamgymeriadau yn y ddwy iaith.

Other publications from the H'mm Foundation

 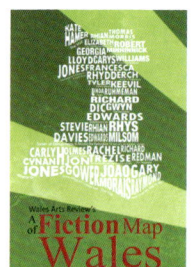  

Encounters with Nigel Jenkins
Edited by Jon Gower

CONTRIBUTORS:
Edwina Hart AM, John Barnie, Stevie Davies, Steve Griffiths, Angharad Jenkins and Branwen Jenkins, Noel Witts, Deborah Llewelyn, Robert Minhinnick, Peter Finch, John Davies, Menna Elfyn, Delyth Jenkins, Daniel G. Williams, Dave Hughes, Margo Morgan, Janet Dube, Jane James, Steve Griffiths, Ivor McGregor, Benjamin Palmer, Dave Oprava, Iwan Bala and Twm Morys, Janice Moore Fuller, Mike Parker and Ceri Wyn Jones, Ifor Thomas, Jane Fraser, Martyn Jenkins, Carey Knox, Steve Dube, Humberto Gatica, Tom Jenkins, D.J. Britton, Fflur Dafydd, Anne Lauppe-Dunbar, M. Wynn Thomas, Jon Gower and Peter Gruffydd.

ISBN 978-0-9927560-4-8

A Fiction Map of Wales
Edited by John Lavin

CONTRIBUTORS:
Rachel Trezise, Thomas Morris, Stevie Davies, Cynan Jones, Francesca Rhydderch, Joao Morais, Jon Gower, Rhian Elizabeth, Carly Holmes, Lloyd Jones, Gary Raymond, Tyler Keevil, Richard Redman, Georgia Carys Williams, Rhian Edwards, Rhys Milsom, Dic Edwards, Linda Ruhmeman, Richard Gwyn, Kate Hamer and Robert Minhinnick.

ISBN 978-0-9927560-6-2

Encounters with R.S. Thomas
Edited by John Barnie

CONTRIBUTORS:
Gillian Clarke, Fflur Dafydd, Grahame Davies, Gwyneth Lewis, Peter Finch, Jon Gower, Menna Elfyn, Osi Rhys Osmond, Jeff Towns, Archbishop of Wales Barry Morgan, M. Wynn Thomas and First Minister of Scotland Alex Salmond.

ISBN 978-0-9927560-0-0

Encounters with Dylan Thomas
Edited by Jon Gower

CONTRIBUTORS:
Rachel Trezise, Michael Bogdanov, Kaite O'Reilly, D.J. Britton, Dafydd Elis-Thomas AM, Dai George, Sarah Gridley, Sarah King, Jeff Towns, George Tremlett, Steve Groves, Gary Raymond, Guy Masterson, Jon Gower, Horatio Clare and Andrew Lycett.

ISBN 978-0-9927560-2-4

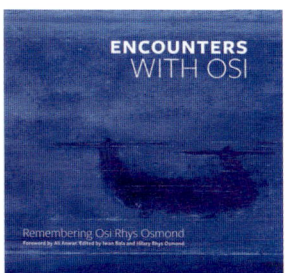
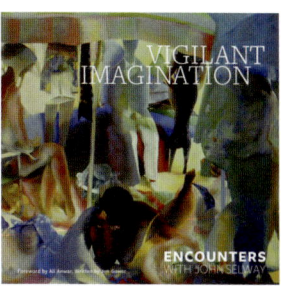

It is as if... fragments
Essays about recent work by Iwan Bala, with images and DVD of the PROsiect hAIcw performances in collaboration with musician Angharad Jenkins, based on the poetry of her father, the late Nigel Jenkins.

CONTRIBUTORS:
Iwan Bala, Dr Anne Price-Owen, Osi Rhys Osmond, Twm Morys, Aneirin Karadog and Angharad Jenkins.

ISBN 978-0-9927560-8-6

Encounters with Osi
Essays about Osi Rhys Osmond, edited by Iwan Bala and Hilary Rhys Osmond.

CONTRIBUTORS:
Iwan Bala, Hilary Rhys Osmond, Ivor Davies, David Alston, M. Wynn Thomas, John Osmond, Christine Kinsey, Dai Smith, Karl Francis, Wyn Morris, David Parfitt, Mick and Thea Arnold, Hedley Jones, Noelle Francis, Susanne Schüeli, Teilo Trimble, Bella Kerr, Steve Wilson, Sam Vicary, Tina Carr, Siân Lewis, Nathan Osmond, Sara Rhys-Martin, Luke Osmond, Simon Thirsk, Lynne Crompton, Gwenan Rhys Price, Linda Sonntag, Rolf Jucker, Ché Osmond, Macsen Osmond, Colin Brewster, Ben Dressel, Megan Crofton, Lesley Davies, Beverley Oosthuizen-Jones, John Barnie, Menna Elfyn, Richard Pawelko and Mary Simmonds, Bethan John, Mererid Hopwood, Ann Oosthuizen,

ISBN 978-0-9927560-9-3

Encounters with John Selway
By Jon Gower

CONTRIBUTORS:
Paul Bowen, Derek Butler, Ivor Davies, Ken Elias, Richard Frame, Karl Francis, Brian Gardiner, Jonathan Glasbrook-Griffiths, Robert Alwyn Hughes, Alison Howard, David Hurn, Julian Meek, Phil Muirden, Osi Rhys Osmond, Brian Rice, Dai Smith, Marion Sprackling, Norman Toynton, Keith Underwood, Peter Wakelin, Phil Watkins, Roger and Den Wolfe.

ISBN 978-1-9999522-0-4

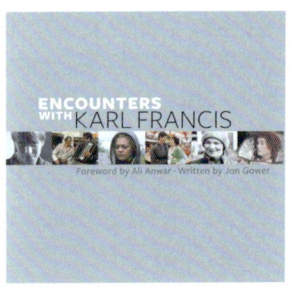

Encounters with Karl Francis

CONTRIBUTORS:
Foreword by Ali Anwar, written by Jon Gower

ISBN 978-1-9999522-1-1

Nigel Jenkins
Damned for Dreaming and other essays

CONTRIBUTORS:
Foreword by Ali Anwar and Jon Gower

ISBN 978-1-9999522-8-0

Osi Rhys Osmond
Cultural Alzheimers and other essays

CONTRIBUTORS:
Foreword by M. Wynn Thomas and Ali Anwar

ISBN 978-1-9999522-9-7

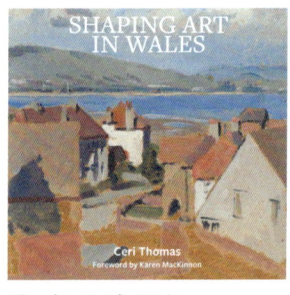

IMPACT*Ardrawiad*
Angharad Pearce Jones

Foreword by Ffion Rhys
Afterword by Ali Anwar

CONTRIBUTORS:
Dylan Huw, Beca Brown

ISBN 978-1-9999522-2-8

Shaping Art in Wales
Ceri Thomas

Foreword by Karen Mackinnon

ISBN 978-1-9999522-3-5

Penbanc
Gaynor Funnell

ISBN 978-1-9999522-4-2

 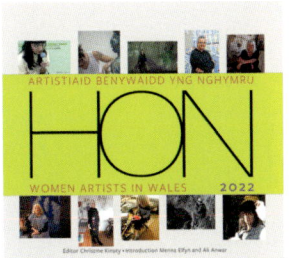 

Wicked Words
£2.95
In partnership with The Wales Millenium Centre in 2015
ISBN 978-0-9927560-7-9

HON 2022
Artistiaid Benywaidd
Yng Nghymru
Women Artists In Wales

ARTISTIAID/ARTISTS
Marian Delyth
Sadia Pineda Hameed
Angharad Pearce Jones
Julia Griffiths Jones
Christine Kinsey
Sian Parri
Sarah Rhys
Catrin Webster
Sarah Williams
Sarah Younan

Editor Christine Kinsey
Introduction by Menna Elfyn
and Ali Anwar

ISBN 978-1-9999522-6-6

Truth, Lies & Alibis
Encounters with
Christine Kinsey

Foreword by Ali Anwar and M. Wynn Thomas.
Introduction by Menna Elfyn

ISBN 978-1-9999522-5-9

The H'mm Foundation is:
c/o Bevan Buckland, Langdon House,
Langdon Road, Swansea SA1 8QY.
info@thehmmfoundation.co.uk
www.thehmmfoundation.co.uk